Notes on Diffy Qs

Differential Equations for Engineers

by Jiří Lebl

March 21, 2017
(version 5.1)

2

Typeset in LaTeX.

Copyright ©2008–2017 Jiří Lebl
ISBN-13: 978-1541329058

During the writing of these notes, the author was in part supported by NSF grant DMS-0900885 and DMS-1362337.

The date is the main identifier of version. The major version / edition number is raised only if there have been substantial changes. Edition number started at 5, that is, version 5.0, as it was not kept track of before.

See http://www.jirka.org/diffyqs/ for more information (including contact information).

Contents

Introduction

0.1 Notes about these notes

This book originated from my class notes for teaching Math 286 at the University of Illinois at Urbana-Champaign in Fall 2008 and Spring 2009. It is a first course on differential equations for engineers. I also taught Math 285 at UIUC and Math 20D at UCSD using this book. The standard book at UIUC was Edwards and Penney, *Differential Equations and Boundary Value Problems: Computing and Modeling* [EP], fourth edition. The standard book at UCSD is Boyce and DiPrima's *Elementary Differential Equations and Boundary Value Problems* [BD]. As the syllabus at UIUC was based on [EP], the early chapters in the book have some resemblance to [EP] in choice of material and its sequence, and examples used. Among other books I used as sources of information and inspiration are E.L. Ince's classic (and inexpensive) *Ordinary Differential Equations* [I], Stanley Farlow's *Differential Equations and Their Applications* [F], now available from Dover, Berg and McGregor's *Elementary Partial Differential Equations* [BM], and William Trench's free book *Elementary Differential Equations with Boundary Value Problems* [T]. See the Further Reading chapter at the end of the book.

I taught the UIUC courses using the IODE software (http://www.math.uiuc.edu/iode/). IODE is a free software package that works with Matlab (proprietary) or Octave (free software). Unfortunately IODE is not kept up to date at this point, and may have trouble running on newer versions of Matlab. The graphs in the book were made with the Genius software (see http://www.jirka.org/genius.html). I used Genius in class to show these (and other) graphs.

This book is available from http://www.jirka.org/diffyqs/. Check there for any possible updates or errata. The LATEX source is also available from the same site for possible modification and customization.

Firstly, I would like to acknowledge Rick Laugesen. I used his handwritten class notes the first time I taught Math 286. My organization of this book through chapter 5, and the choice of material covered, is heavily influenced by his notes. Many examples and computations are taken from his notes. I am also heavily indebted to Rick for all the advice he has given me, not just on teaching Math 286. For spotting errors and other suggestions, I would also like to acknowledge (in no particular order): John P. D'Angelo, Sean Raleigh, Jessica Robinson, Michael Angelini, Leonardo Gomes, Jeff Winegar, Ian Simon, Thomas Wicklund, Eliot Brenner, Sean Robinson, Jannett Susberry, Dana Al-Quadi, Cesar Alvarez, Cem Bagdatlioglu, Nathan Wong, Alison Shive, Shawn White,

Wing Yip Ho, Joanne Shin, Gladys Cruz, Jonathan Gomez, Janelle Louie, Navid Froutan, Grace Victorine, Paul Pearson, Jared Teague, Ziad Adwan, Martin Weilandt, Sönmez Şahutoğlu, Pete Peterson, Thomas Gresham, Prentiss Hyde, Jai Welch, Simon Tse, Andrew Browning, James Choi, Dusty Grundmeier, John Marriott, Jim Kruidenier, Barry Conrad, Wesley Snider, Colton Koop, Sarah Morse, Erik Boczko, Asif Shakeel, Chris Peterson, Nicholas Hu, and probably others I have forgotten. Finally I would like to acknowledge NSF grants DMS-0900885 and DMS-1362337.

The organization of this book to some degree requires chapters be done in order. Later chapters can be dropped. The dependence of the material covered is roughly:

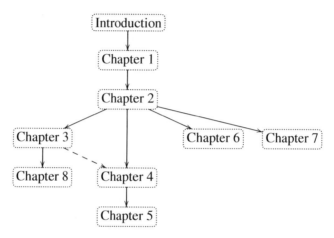

There are some references in chapters 4 and 5 to material from chapter 3 (some linear algebra), but these references are not absolutely essential and can be skimmed over, so chapter 3 can safely be dropped, while still covering chapters 4 and 5. The textbook was originally done for two types of courses. Either at 4 hours a week for a semester (Math 286 at UIUC):

Introduction, chapter 1, chapter 2, chapter 3, chapter 4 (w/o § 4.10), chapter 5 (or 6 or 7 or 8).

Or a shorter version (Math 285 at UIUC) of the course at 3 hours a week for a semester:

Introduction, chapter 1, chapter 2, chapter 4 (w/o § 4.10), (and maybe chapter 5, 6, or 7).

The complete book can be covered in approximately 72 lectures, depending on the lecturers speed and not accounting for exams, review, or time spent in computer lab. Therefore, a two quarter course can easily be run with the material, and if one goes a bit slower than I do, then even a two semester course.

The chapter on Laplace transform (chapter 6), the chapter on Sturm-Liouville (chapter 5), the chapter on power series (chapter 7), and the chapter on nonlinear systems (chapter 8), are more or less interchangeable time-wise. If chapter 8 is covered it may be best to place it right after chapter 3. If time is short, the first two sections of chapter 7 make a reasonable self-contained unit.

0.2 Introduction to differential equations

Note: more than 1 lecture, §1.1 in [EP], chapter 1 in [BD]

0.2.1 Differential equations

The laws of physics are generally written down as differential equations. Therefore, all of science and engineering use differential equations to some degree. Understanding differential equations is essential to understanding almost anything you will study in your science and engineering classes. You can think of mathematics as the language of science, and differential equations are one of the most important parts of this language as far as science and engineering are concerned. As an analogy, suppose all your classes from now on were given in Swahili. It would be important to first learn Swahili, or you would have a very tough time getting a good grade in your classes.

You already saw many differential equations without perhaps knowing about it. And you even solved simple differential equations when you took calculus. Let us see an example you may not have seen:

$$\frac{dx}{dt} + x = 2\cos t. \tag{1}$$

Here x is the *dependent variable* and t is the *independent variable*. Equation (1) is a basic example of a *differential equation*. In fact, it is an example of a *first order differential equation*, since it involves only the first derivative of the dependent variable. This equation arises from Newton's law of cooling where the ambient temperature oscillates with time.

0.2.2 Solutions of differential equations

Solving the differential equation means finding x in terms of t. That is, we want to find a function of t, which we will call x, such that when we plug x, t, and $\frac{dx}{dt}$ into (1), the equation holds. It is the same idea as it would be for a normal (algebraic) equation of just x and t. We claim that

$$x = x(t) = \cos t + \sin t$$

is a *solution*. How do we check? We simply plug x into equation (1)! First we need to compute $\frac{dx}{dt}$. We find that $\frac{dx}{dt} = -\sin t + \cos t$. Now let us compute the left hand side of (1).

$$\frac{dx}{dt} + x = (-\sin t + \cos t) + (\cos t + \sin t) = 2\cos t.$$

Yay! We got precisely the right hand side. But there is more! We claim $x = \cos t + \sin t + e^{-t}$ is also a solution. Let us try,

$$\frac{dx}{dt} = -\sin t + \cos t - e^{-t}.$$

Again plugging into the left hand side of (1)

$$\frac{dx}{dt} + x = (-\sin t + \cos t - e^{-t}) + (\cos t + \sin t + e^{-t}) = 2\cos t.$$

And it works yet again!

So there can be many different solutions. In fact, for this equation all solutions can be written in the form

$$x = \cos t + \sin t + Ce^{-t}$$

for some constant C. See Figure 1 for the graph of a few of these solutions. We will see how we find these solutions a few lectures from now.

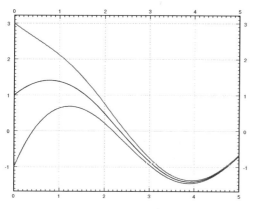

Figure 1: Few solutions of $\frac{dx}{dt} + x = 2\cos t$.

It turns out that solving differential equations can be quite hard. There is no general method that solves every differential equation. We will generally focus on how to get exact formulas for solutions of certain differential equations, but we will also spend a little bit of time on getting approximate solutions.

For most of the course we will look at *ordinary differential equations* or ODEs, by which we mean that there is only one independent variable and derivatives are only with respect to this one variable. If there are several independent variables, we will get *partial differential equations* or PDEs. We will briefly see these near the end of the course.

Even for ODEs, which are very well understood, it is not a simple question of turning a crank to get answers. It is important to know when it is easy to find solutions and how to do so. Although in real applications you will leave much of the actual calculations to computers, you need to understand what they are doing. It is often necessary to simplify or transform your equations into something that a computer can understand and solve. You may need to make certain assumptions and changes in your model to achieve this.

To be a successful engineer or scientist, you will be required to solve problems in your job that you never saw before. It is important to learn problem solving techniques, so that you may apply those techniques to new problems. A common mistake is to expect to learn some prescription for solving all the problems you will encounter in your later career. This course is no exception.

0.2.3 Differential equations in practice

So how do we use differential equations in science and engineering? First, we have some *real world problem* we wish to understand. We make some simplifying assumptions and create a *mathematical model.* That is, we translate the real world situation into a set of differential equations. Then we apply mathematics to get some sort of a *mathematical solution.* There is still something left to do. We have to interpret the results. We have to figure out what the mathematical solution says about the real world problem we started with.

Learning how to formulate the mathematical model and how to interpret the results is what your physics and engineering classes do. In this course we will focus mostly on the mathematical analysis. Sometimes we will work with simple real world examples, so that we have some intuition and motivation about what we are doing.

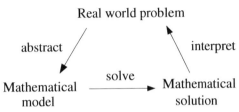

Let us look at an example of this process. One of the most basic differential equations is the standard *exponential growth model.* Let P denote the population of some bacteria on a Petri dish. We assume that there is enough food and enough space. Then the rate of growth of bacteria is proportional to the population—a large population grows quicker. Let t denote time (say in seconds) and P the population. Our model is

$$\frac{dP}{dt} = kP,$$

for some positive constant $k > 0$.

Example 0.2.1: Suppose there are 100 bacteria at time 0 and 200 bacteria 10 seconds later. How many bacteria will there be 1 minute from time 0 (in 60 seconds)?

First we have to solve the equation. We claim that a solution is given by

$$P(t) = Ce^{kt},$$

where C is a constant. Let us try:

$$\frac{dP}{dt} = Cke^{kt} = kP.$$

And it really is a solution.

OK, so what now? We do not know C and we do not know k. But we know something. We know $P(0) = 100$, and we also know $P(10) = 200$. Let us plug these conditions in and see what happens.

$$100 = P(0) = Ce^{k0} = C,$$
$$200 = P(10) = 100\, e^{k10}.$$

Therefore, $2 = e^{10k}$ or $\frac{\ln 2}{10} = k \approx 0.069$. So we know that

$$P(t) = 100\, e^{(\ln 2)t/10} \approx 100\, e^{0.069t}.$$

At one minute, $t = 60$, the population is $P(60) = 6400$. See Figure 2 on the next page.

Let us talk about the interpretation of the results. Does our solution mean that there must be exactly 6400 bacteria on the plate at 60s? No! We made assumptions that might not be true exactly, just approximately. If our assumptions are reasonable, then there will be approximately 6400 bacteria. Also, in real life P is a discrete quantity, not a real number. However, our model has no problem saying that for example at 61 seconds, $P(61) \approx 6859.35$.

Normally, the k in $P' = kP$ is known, and we want to solve the equation for different *initial conditions*. What does that mean? Take $k = 1$ for simplicity. Now suppose we want to solve the equation $\frac{dP}{dt} = P$ subject to $P(0) = 1000$ (the initial condition). Then the solution turns out to be (exercise)

$$P(t) = 1000\, e^t.$$

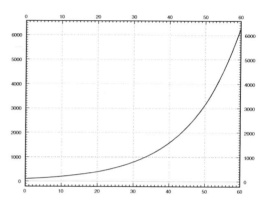

Figure 2: *Bacteria growth in the first 60 seconds.*

We call $P(t) = Ce^t$ *the general solution*, as every solution of the equation can be written in this form for some constant C. You will need an initial condition to find out what C is, in order to find the *particular solution* we are looking for. Generally, when we say "particular solution," we just mean some solution.

Let us get to what we will call the four fundamental equations. These equations appear very often and it is useful to just memorize what their solutions are. These solutions are reasonably easy to guess by recalling properties of exponentials, sines, and cosines. They are also simple to check, which is something that you should always do. There is no need to wonder if you remembered the solution correctly.

First such equation is,

$$\frac{dy}{dx} = ky,$$

for some constant $k > 0$. Here y is the dependent and x the independent variable. The general solution for this equation is

$$y(x) = Ce^{kx}.$$

We saw above that this function is a solution, although we used different variable names.

Next,

$$\frac{dy}{dx} = -ky,$$

for some constant $k > 0$. The general solution for this equation is

$$y(x) = Ce^{-kx}.$$

Exercise 0.2.1: *Check that the y given is really a solution to the equation.*

Next, take the *second order differential equation*

$$\frac{d^2y}{dx^2} = -k^2y,$$

for some constant $k > 0$. The general solution for this equation is

$$y(x) = C_1 \cos(kx) + C_2 \sin(kx).$$

Note that because we have a second order differential equation, we have two constants in our general solution.

Exercise 0.2.2: *Check that the y given is really a solution to the equation.*

And finally, take the second order differential equation

$$\frac{d^2y}{dx^2} = k^2y,$$

for some constant $k > 0$. The general solution for this equation is

$$y(x) = C_1 e^{kx} + C_2 e^{-kx},$$

or

$$y(x) = D_1 \cosh(kx) + D_2 \sinh(kx).$$

For those that do not know, cosh and sinh are defined by

$$\cosh x = \frac{e^x + e^{-x}}{2},$$

$$\sinh x = \frac{e^x - e^{-x}}{2}.$$

These functions are sometimes easier to work with than exponentials. They have some nice familiar properties such as $\cosh 0 = 1$, $\sinh 0 = 0$, and $\frac{d}{dx} \cosh x = \sinh x$ (no that is not a typo) and $\frac{d}{dx} \sinh x = \cosh x$.

Exercise 0.2.3: *Check that both forms of the y given are really solutions to the equation.*

An interesting note about cosh: The graph of cosh is the exact shape a hanging chain will make. This shape is called a *catenary*. Contrary to popular belief this is not a parabola. If you invert the graph of cosh it is also the ideal arch for supporting its own weight. For example, the gateway arch in Saint Louis is an inverted graph of cosh—if it were just a parabola it might fall down. The formula used in the design is inscribed inside the arch:

$$y = -127.7 \text{ ft} \cdot \cosh(x/127.7 \text{ ft}) + 757.7 \text{ ft}.$$

0.2.4   Exercises

Exercise 0.2.4: *Show that $x = e^{4t}$ is a solution to $x''' - 12x'' + 48x' - 64x = 0$.*

Exercise 0.2.5: *Show that $x = e^t$ is not a solution to $x''' - 12x'' + 48x' - 64x = 0$.*

Exercise 0.2.6: *Is $y = \sin t$ a solution to $\left(\frac{dy}{dt}\right)^2 = 1 - y^2$? Justify.*

Exercise 0.2.7: *Let $y'' + 2y' - 8y = 0$. Now try a solution of the form $y = e^{rx}$ for some (unknown) constant r. Is this a solution for some r? If so, find all such r.*

Exercise 0.2.8: *Verify that $x = Ce^{-2t}$ is a solution to $x' = -2x$. Find C to solve for the initial condition $x(0) = 100$.*

Exercise 0.2.9: *Verify that $x = C_1 e^{-t} + C_2 e^{2t}$ is a solution to $x'' - x' - 2x = 0$. Find C_1 and C_2 to solve for the initial conditions $x(0) = 10$ and $x'(0) = 0$.*

Exercise 0.2.10: *Find a solution to $(x')^2 + x^2 = 4$ using your knowledge of derivatives of functions that you know from basic calculus.*

Exercise 0.2.11: *Solve:*

a) $\dfrac{dA}{dt} = -10A, \;\; A(0) = 5$
b) $\dfrac{dH}{dx} = 3H, \;\; H(0) = 1$

c) $\dfrac{d^2 y}{dx^2} = 4y, \;\; y(0) = 0, \;\; y'(0) = 1$
d) $\dfrac{d^2 x}{dy^2} = -9x, \;\; x(0) = 1, \;\; x'(0) = 0$

Exercise 0.2.12: *Is there a solution to $y' = y$, such that $y(0) = y(1)$?*

Note: Exercises with numbers 101 and higher have solutions in the back of the book.

Exercise 0.2.101: *Show that $x = e^{-2t}$ is a solution to $x'' + 4x' + 4x = 0$.*

Exercise 0.2.102: *Is $y = x^2$ a solution to $x^2 y'' - 2y = 0$? Justify.*

Exercise 0.2.103: *Let $xy'' - y' = 0$. Try a solution of the form $y = x^r$. Is this a solution for some r? If so, find all such r.*

Exercise 0.2.104: *Verify that $x = C_1 e^t + C_2$ is a solution to $x'' - x' = 0$. Find C_1 and C_2 so that x satisfies $x(0) = 10$ and $x'(0) = 100$.*

Exercise 0.2.105: *Solve $\frac{d\varphi}{ds} = 8\varphi$ and $\varphi(0) = -9$.*

0.2, 1.1, 1.2, 1.3

4	3	1	1
↓	↓	↓	↓
11	7	6	7
	8		10
	10		12

0.3 Classification of differential equations

Note: less than 1 lecture or left as reading, §1.3 in [BD]

There are many types of differential equations and we classify them into different categories based on their properties. Let us quickly go over the most basic classification. We already saw the distinction between ordinary and partial differential equations:

- *Ordinary differential equations* or (ODE) are equations where the derivatives are taken with respect to only one variable. That is, there is only one independent variable.

- *Partial differential equations* or (PDE) are equations that depend on partial derivatives of several variables. That is, there are several independent variables.

Let us see some examples of ordinary differential equations:

$$\frac{dy}{dt} = ky, \qquad\qquad \text{(Newton's law of cooling)}$$

$$m\frac{d^2x}{dt^2} + c\frac{dx}{dt} + kx = f(t). \qquad\qquad \text{(Mechanical vibrations)}$$

And of partial differential equations:

$$\frac{\partial y}{\partial t} + c\frac{\partial y}{\partial x} = 0, \qquad\qquad \text{(Transport equation)}$$

$$\frac{\partial u}{\partial t} = \frac{\partial^2 u}{\partial x^2}, \qquad\qquad \text{(Heat equation)}$$

$$\frac{\partial^2 u}{\partial t^2} = \frac{\partial^2 u}{\partial x^2} + \frac{\partial^2 u}{\partial y^2}. \qquad\qquad \text{(Wave equation in 2 dimensions)}$$

If there are several equations working together we have a so-called *system of differential equations*. For example,

$$y' = x, \qquad x' = y$$

is a simple system of ordinary differential equations. Maxwell's equations governing electromagnetics,

$$\nabla \cdot \vec{D} = \rho, \qquad\qquad\qquad \nabla \cdot \vec{B} = 0,$$

$$\nabla \times \vec{E} = -\frac{\partial \vec{B}}{\partial t}, \qquad\qquad\qquad \nabla \times \vec{H} = \vec{J} + \frac{\partial \vec{D}}{\partial t},$$

are a system of partial differential equations. The divergence operator $\nabla\cdot$ and the curl operator $\nabla\times$ can be written out in partial derivatives of the functions involved in the x, y, and z variables.

The next bit of information is the *order* of the equation (or system). The order is simply the order of the largest derivative that appears. If the highest derivative that appears is the first derivative, the equation is of first order. If the highest derivative that appears is the second derivative, then the equation is of second order. For example, Newton's law of cooling above is a first order equation, while the Mechanical vibrations equation is a second order equation. The equation governing transversal vibrations in a beam,

$$a^4 \frac{\partial^4 y}{\partial x^4} + \frac{\partial^2 y}{\partial t^2} = 0,$$

is a fourth order partial differential equation. It is fourth order since at least one derivative is the fourth derivative. It does not matter that derivatives with respect to t are only second order.

In the first chapter we will start attacking first order ordinary differential equations, that is, equations of the form $\frac{dy}{dx} = f(x, y)$. In general, lower order equations are easier to work with and have simpler behavior, which is why we start with them.

We also distinguish how the dependent variables appear in the equation (or system). In particular, we say an equation is *linear* if the dependent variable (or variables) and their derivatives appear linearly, that is only as first powers, they are not multiplied together, and no other functions of the dependent variables appear. In other words, the equation is a sum of terms, where each term is some function of the independent variables or some function of the independent variables multiplied by a dependent variable or its derivative. Otherwise the equation is called *nonlinear*. For example, an ordinary differential equation is linear if it can be put into the form

$$a_n(x)\frac{d^n y}{dx^n} + a_{n-1}(x)\frac{d^{n-1} y}{dx^{n-1}} + \cdots + a_1(x)\frac{dy}{dx} + a_0(x)y = b(x). \tag{2}$$

The functions $a_0, a_1, \ldots, a_n$ are called the *coefficients*. The equation is allowed to depend arbitrarily on the independent variables. So

$$e^x \frac{d^2 y}{dx^2} + \sin(x)\frac{dy}{dx} + x^2 y = \frac{1}{x} \tag{3}$$

is still a linear equation as y and its derivatives only appear linearly.

All the equations and systems given above as examples are linear. It may not be immediately obvious for Maxwell's equations unless you write out the divergence and curl in terms of partial derivatives. Let us see some nonlinear equations. For example Burger's equation,

$$\frac{\partial y}{\partial t} + y\frac{\partial y}{\partial x} = \nu \frac{\partial^2 y}{\partial x^2},$$

is a nonlinear second order partial differential equation. It is nonlinear because y and $\frac{\partial y}{\partial x}$ are multiplied together. The equation

$$\frac{dx}{dt} = x^2 \tag{4}$$

is a nonlinear first order differential equation as there is a power of the dependent variable x.

A linear equation may further be called *homogeneous*, if all terms depend on the dependent variable. That is, if there is no term that is a function of the independent variables alone. Otherwise the equation is called *nonhomogeneous* or *inhomogeneous*. For example, Newton's law of cooling, Transport equation, Wave equation, above are homogeneous, while Mechanical vibrations equation above is nonhomogeneous. A homogeneous linear ODE can be put into the form

$$a_n(x)\frac{d^n y}{dx^n} + a_{n-1}(x)\frac{d^{n-1} y}{dx^{n-1}} + \cdots + a_1(x)\frac{dy}{dx} + a_0(x)y = 0.$$

Compare to (2) and notice there is no function $b(x)$.

If the coefficients of a linear equation are actually constant functions, then the equation is said to have *constant coefficients*. The coefficients are the functions multiplying the dependent variable(s) or one of its derivatives, not the function standing alone. That is, a constant coefficient ODE is

$$a_n\frac{d^n y}{dx^n} + a_{n-1}\frac{d^{n-1} y}{dx^{n-1}} + \cdots + a_1\frac{dy}{dx} + a_0 y = b(x),$$

where $a_0, a_1, \ldots, a_n$ are all constants, but b may depend on the independent variable x. The Mechanical vibrations equation above is constant coefficient nonhomogeneous second order ODE. Same nomenclature applies to PDEs, so the Transport equation, Heat equation and Wave equation are all examples of constant coefficient linear PDEs.

Finally, an equation (or system) is called *autonomous* if the equation does not depend on the independent variable. Usually here we only consider ordinary differential equations and the independent variable is then thought of as time. Autonomous equation means an equation that does not change with time. For example, Newton's law of cooling is autonomous, so is equation (4). On the other hand, Mechanical vibrations or (3) are not autonomous.

0.3.1 Exercises

***Exercise* 0.3.1:** *Classify the following equations. Are they ODE or PDE? Is it an equation or a system? What is the order? Is it linear or nonlinear, and if it is linear, is it homogeneous, constant coefficient? If it is an ODE, is it autonomous?*

a) $\sin(t)\dfrac{d^2 x}{dt^2} + \cos(t)x = t^2$

b) $\dfrac{\partial u}{\partial x} + 3\dfrac{\partial u}{\partial y} = xy$

c) $y'' + 3y + 5x = 0, \quad x'' + x - y = 0$

d) $\dfrac{\partial^2 u}{\partial t^2} + u\dfrac{\partial^2 u}{\partial s^2} = 0$

e) $x'' + tx^2 = t$

f) $\dfrac{d^4 x}{dt^4} = 0$

Exercise 0.3.2: *If $\vec{u} = (u_1, u_2, u_3)$ is a vector, we have the divergence $\nabla \cdot \vec{u} = \frac{\partial u_1}{\partial x} + \frac{\partial u_2}{\partial y} + \frac{\partial u_3}{\partial z}$ and curl $\nabla \times \vec{u} = \left(\frac{\partial u_3}{\partial y} - \frac{\partial u_2}{\partial z}, \frac{\partial u_1}{\partial z} - \frac{\partial u_3}{\partial x}, \frac{\partial u_2}{\partial x} - \frac{\partial u_1}{\partial y} \right)$. Notice that curl of a vector is still a vector. Write out Maxwell's equations in terms of partial derivatives and classify the system.*

Exercise 0.3.3: *Suppose F is a linear function, that is, $F(x, y) = ax + by$ for constants a and b. What is the classification of equations of the form $F(y', y) = 0$.*

Exercise 0.3.4: *Write down an explicit example of a third order, linear, nonconstant coefficient, nonautonomous, nonhomogeneous system of two ODE such that every derivative that could appear, does appear.*

Exercise 0.3.101: *Classify the following equations. Are they ODE or PDE? Is it an equation or a system? What is the order? Is it linear or nonlinear, and if it is linear, is it homogeneous, constant coefficient? If it is an ODE, is it autonomous?*

a) $\dfrac{\partial^2 v}{\partial x^2} + 3\dfrac{\partial^2 v}{\partial y^2} = \sin(x)$

b) $\dfrac{dx}{dt} + \cos(t)x = t^2 + t + 1$

c) $\dfrac{d^7 F}{dx^7} = 3F(x)$

d) $y'' + 8y' = 1$

e) $x'' + tyx' = 0, \quad y'' + txy = 0$

f) $\dfrac{\partial u}{\partial t} = \dfrac{\partial^2 u}{\partial s^2} + u^2$

Exercise 0.3.102: *Write down the general zeroth order linear ordinary differential equation. Write down the general solution.*

Chapter 1

First order ODEs

1.1 Integrals as solutions

Note: 1 lecture (or less), §1.2 in [EP], covered in §1.2 and §2.1 in [BD]

A first order ODE is an equation of the form

$$\frac{dy}{dx} = f(x, y),$$

or just

$$y' = f(x, y).$$

In general, there is no simple formula or procedure one can follow to find solutions. In the next few lectures we will look at special cases where solutions are not difficult to obtain. In this section, let us assume that f is a function of x alone, that is, the equation is

$$y' = f(x). \tag{1.1}$$

We could just integrate (antidifferentiate) both sides with respect to x.

$$\int y'(x)\, dx = \int f(x)\, dx + C,$$

that is

$$y(x) = \int f(x)\, dx + C.$$

This $y(x)$ is actually the general solution. So to solve (1.1), we find some antiderivative of $f(x)$ and then we add an arbitrary constant to get the general solution.

Now is a good time to discuss a point about calculus notation and terminology. Calculus textbooks muddy the waters by talking about the integral as primarily the so-called indefinite integral. The indefinite integral is really the *antiderivative* (in fact the whole one-parameter family of antiderivatives). There really exists only one integral and that is the definite integral. The only reason for the indefinite integral notation is that we can always write an antiderivative as a (definite) integral. That is, by the fundamental theorem of calculus we can always write $\int f(x)\,dx + C$ as

$$\int_{x_0}^{x} f(t)\,dt + C.$$

Hence the terminology *to integrate* when we may really mean *to antidifferentiate*. Integration is just one way to compute the antiderivative (and it is a way that always works, see the following examples). Integration is defined as the area under the graph, it only happens to also compute antiderivatives. For sake of consistency, we will keep using the indefinite integral notation when we want an antiderivative, and you should *always* think of the definite integral as a way to write it.

Example 1.1.1: Find the general solution of $y' = 3x^2$.

Elementary calculus tells us that the general solution must be $y = x^3 + C$. Let us check by differentiating: $y' = 3x^2$. We got *precisely* our equation back.

Normally, we also have an initial condition such as $y(x_0) = y_0$ for some two numbers x_0 and y_0 (x_0 is usually 0, but not always). We can then write the solution as a definite integral in a nice way. Suppose our problem is $y' = f(x)$, $y(x_0) = y_0$. Then the solution is

$$y(x) = \int_{x_0}^{x} f(s)\,ds + y_0. \tag{1.2}$$

Let us check! We compute $y' = f(x)$, via the fundamental theorem of calculus, and by Jupiter, y is a solution. Is it the one satisfying the initial condition? Well, $y(x_0) = \int_{x_0}^{x_0} f(x)\,dx + y_0 = y_0$. It is!

Do note that the definite integral and the indefinite integral (antidifferentiation) are completely different beasts. The definite integral always evaluates to a number. Therefore, (1.2) is a formula we can plug into the calculator or a computer, and it will be happy to calculate specific values for us. We will easily be able to plot the solution and work with it just like with any other function. It is not so crucial to always find a closed form for the antiderivative.

Example 1.1.2: Solve

$$y' = e^{-x^2}, \qquad y(0) = 1.$$

By the preceding discussion, the solution must be

$$y(x) = \int_{0}^{x} e^{-s^2}\,ds + 1.$$

Here is a good way to make fun of your friends taking second semester calculus. Tell them to find the closed form solution. Ha ha ha (bad math joke). It is not possible (in closed form). There is absolutely nothing wrong with writing the solution as a definite integral. This particular integral is in fact very important in statistics.

Using this method, we can also solve equations of the form

$$y' = f(y).$$

Let us write the equation in Leibniz notation.

$$\frac{dy}{dx} = f(y).$$

Now we use the inverse function theorem from calculus to switch the roles of x and y to obtain

$$\frac{dx}{dy} = \frac{1}{f(y)}.$$

What we are doing seems like algebra with dx and dy. It is tempting to just do algebra with dx and dy as if they were numbers. And in this case it does work. Be careful, however, as this sort of hand-waving calculation can lead to trouble, especially when more than one independent variable is involved. At this point we can simply integrate,

$$x(y) = \int \frac{1}{f(y)} \, dy + C.$$

Finally, we try to solve for y.

Example 1.1.3: Previously, we guessed $y' = ky$ (for some $k > 0$) has the solution $y = Ce^{kx}$. We can now find the solution without guessing. First we note that $y = 0$ is a solution. Henceforth, we assume $y \neq 0$. We write

$$\frac{dx}{dy} = \frac{1}{ky}.$$

We integrate to obtain

$$x(y) = x = \frac{1}{k} \ln |y| + D,$$

where D is an arbitrary constant. Now we solve for y (actually for $|y|$).

$$|y| = e^{kx-kD} = e^{-kD} e^{kx}.$$

If we replace e^{-kD} with an arbitrary constant C we can get rid of the absolute value bars (which we can do as D was arbitrary). In this way, we also incorporate the solution $y = 0$. We get the same general solution as we guessed before, $y = Ce^{kx}$.

Example 1.1.4: Find the general solution of $y' = y^2$.

First we note that $y = 0$ is a solution. We can now assume that $y \neq 0$. Write

$$\frac{dx}{dy} = \frac{1}{y^2}.$$

We integrate to get

$$x = \frac{-1}{y} + C.$$

We solve for $y = \frac{1}{C-x}$. So the general solution is

$$y = \frac{1}{C - x} \qquad \text{or} \qquad y = 0.$$

Note the singularities of the solution. If for example $C = 1$, then the solution "blows up" as we approach $x = 1$. Generally, it is hard to tell from just looking at the equation itself how the solution is going to behave. The equation $y' = y^2$ is very nice and defined everywhere, but the solution is only defined on some interval $(-\infty, C)$ or (C, ∞).

Classical problems leading to differential equations solvable by integration are problems dealing with velocity, acceleration and distance. You have surely seen these problems before in your calculus class.

Example 1.1.5: Suppose a car drives at a speed $e^{t/2}$ meters per second, where t is time in seconds. How far did the car get in 2 seconds (starting at $t = 0$)? How far in 10 seconds?

Let x denote the distance the car traveled. The equation is

$$x' = e^{t/2}.$$

We just integrate this equation to get that

$$x(t) = 2e^{t/2} + C.$$

We still need to figure out C. We know that when $t = 0$, then $x = 0$. That is, $x(0) = 0$. So

$$0 = x(0) = 2e^{0/2} + C = 2 + C.$$

Thus $C = -2$ and

$$x(t) = 2e^{t/2} - 2.$$

Now we just plug in to get where the car is at 2 and at 10 seconds. We obtain

$$x(2) = 2e^{2/2} - 2 \approx 3.44 \text{ meters}, \qquad x(10) = 2e^{10/2} - 2 \approx 294 \text{ meters}.$$

Example 1.1.6: Suppose that the car accelerates at a rate of $t^2 \text{ m/s}^2$. At time $t = 0$ the car is at the 1 meter mark and is traveling at 10 m/s. Where is the car at time $t = 10$.

Well this is actually a second order problem. If x is the distance traveled, then x' is the velocity, and x'' is the acceleration. The equation with initial conditions is

$$x'' = t^2, \qquad x(0) = 1, \qquad x'(0) = 10.$$

What if we say $x' = v$. Then we have the problem

$$v' = t^2, \qquad v(0) = 10.$$

Once we solve for v, we can integrate and find x.

Exercise 1.1.1: Solve for v, and then solve for x. Find x(10) to answer the question.

1.1.1 Exercises

***Exercise* 1.1.2:** *Solve $\frac{dy}{dx} = x^2 + x$ for $y(1) = 3$.*

***Exercise* 1.1.3:** *Solve $\frac{dy}{dx} = \sin(5x)$ for $y(0) = 2$.*

***Exercise* 1.1.4:** *Solve $\frac{dy}{dx} = \frac{1}{x^2-1}$ for $y(0) = 0$.*

***Exercise* 1.1.5:** *Solve $y' = y^3$ for $y(0) = 1$.*

***Exercise* 1.1.6** (little harder)*: Solve $y' = (y-1)(y+1)$ for $y(0) = 3$.*

***Exercise* 1.1.7:** *Solve $\frac{dy}{dx} = \frac{1}{y+1}$ for $y(0) = 0$.*

***Exercise* 1.1.8** (harder)*: Solve $y'' = \sin x$ for $y(0) = 0$, $y'(0) = 2$.*

***Exercise* 1.1.9:** *A spaceship is traveling at the speed $2t^2 + 1$ km/s (t is time in seconds). It is pointing directly away from earth and at time $t = 0$ it is 1000 kilometers from earth. How far from earth is it at one minute from time $t = 0$?*

***Exercise* 1.1.10:** *Solve $\frac{dx}{dt} = \sin(t^2) + t$, $x(0) = 20$. It is OK to leave your answer as a definite integral.*

***Exercise* 1.1.11:** *A dropped ball accelerates downwards at a constant rate 9.8 meters per second squared. Set up the differential equation for the height above ground h in meters. Then supposing $h(0) = 100$ meters, how long does it take for the ball to hit the ground.*

***Exercise* 1.1.12:** *Find the general solution of $y' = e^x$, and then $y' = e^y$.*

***Exercise* 1.1.101:** *Solve $\frac{dy}{dx} = e^x + x$ and $y(0) = 10$.*

***Exercise* 1.1.102:** *Solve $x' = \frac{1}{x^2}$, $x(1) = 1$.*

***Exercise* 1.1.103:** *Solve $x' = \frac{1}{\cos(x)}$, $x(0) = \frac{\pi}{2}$.*

***Exercise* 1.1.104:** *Sid is in a car traveling at speed $10t + 70$ miles per hour away from Las Vegas, where t is in hours. At $t = 0$, Sid is 10 miles away from Vegas. How far from Vegas is Sid 2 hours later?*

***Exercise* 1.1.105:** *Solve $y' = y^n$, $y(0) = 1$, where n is a positive integer. Hint: You have to consider different cases.*

***Exercise* 1.1.106:** *The rate of change of the volume of a snowball that is melting is proportional to the surface area of the snowball. Suppose the snowball is perfectly spherical. Then the volume (in centimeters cubed) of a ball of radius r centimeters is $\frac{4}{3}\pi r^3$. The surface area is $4\pi r^2$. Set up the differential equation for how r is changing. Then, suppose that at time $t = 0$ minutes, the radius is 10 centimeters. After 5 minutes, the radius is 8 centimeters. At what time t will the snowball be completely melted.*

***Exercise* 1.1.107:** *Find the general solution to $y'''' = 0$. How many distinct constants do you need?*

1.2   Slope fields

Note: 1 lecture, §1.3 in [EP], §1.1 in [BD]

As we said, the general first order equation we are studying looks like

$$y' = f(x, y).$$

In general, we cannot simply solve these kinds of equations explicitly. It would be nice if we could at least figure out the shape and behavior of the solutions, or if we could find approximate solutions.

1.2.1   Slope fields

The equation $y' = f(x, y)$ gives you a slope at each point in the (x, y)-plane. And this is the slope a solution $y(x)$ would have at the point (x, y). At a point (x, y), we plot a short line with the slope $f(x, y)$. For example, if $f(x, y) = xy$, then at point $(2, 1.5)$ we draw a short line of slope $2 \times 1.5 = 3$. That is, if $y(x)$ is a solution and $y(2) = 1.5$, then the equation mandates that $y'(2) = 3$. See Figure 1.1.

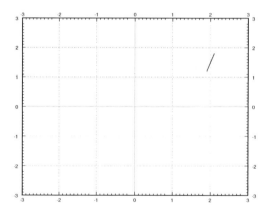

Figure 1.1: The slope $y' = xy$ at $(2, 1.5)$.

To get an idea of how the solutions behave, we draw such lines at lots of points in the plane, not just the point $(2, 1.5)$. Usually we pick a grid of such points fine enough so that it shows the behavior, but not too fine so that we can still recognize the individual lines. See Figure 1.2 on the next page. We call this picture the *slope field* of the equation. Usually in practice, one does not do this by hand, but has a computer do the drawing.

Suppose we are given a specific initial condition $y(x_0) = y_0$. A solution, that is, the graph of the solution, would be a curve that follows the slopes. For a few sample solutions, see Figure 1.3 on the facing page. It is easy to roughly sketch (or at least imagine) possible solutions in the slope field, just from looking at the slope field itself.

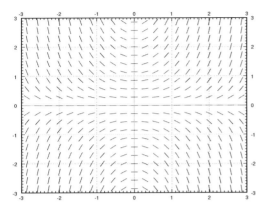

Figure 1.2: Slope field of $y' = xy$.

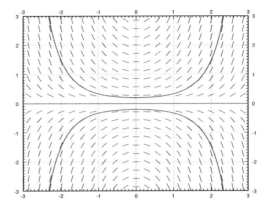

Figure 1.3: Slope field of $y' = xy$ with a graph of solutions satisfying $y(0) = 0.2$, $y(0) = 0$, and $y(0) = -0.2$.

By looking at the slope field we can get a lot of information about the behavior of solutions. For example, in Figure 1.3 we can see what the solutions do when the initial conditions are $y(0) > 0$, $y(0) = 0$ and $y(0) < 0$. Note that a small change in the initial condition causes quite different behavior. We can see this behavior just from the slope field imagining what solutions ought to do. On the other hand, plotting a few solutions of the equation $y' = -y$, we see that no matter what $y(0)$ is, all solutions tend to zero as x tends to infinity. See Figure 1.4. Again that behavior should be clear from simply from looking at the slope field itself.

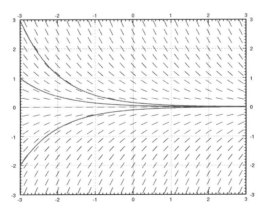

Figure 1.4: Slope field of $y' = -y$ with a graph of a few solutions.

1.2.2   Existence and uniqueness

We wish to ask two fundamental questions about the problem

$$y' = f(x, y), \qquad y(x_0) = y_0.$$

(i) Does a solution *exist*?

(ii) Is the solution *unique* (if it exists)?

What do you think is the answer? The answer seems to be yes to both does it not? Well, pretty much. But there are cases when the answer to either question can be no.

Since generally the equations we encounter in applications come from real life situations, it seems logical that a solution always exists. It also has to be unique if we believe our universe is deterministic. If the solution does not exist, or if it is not unique, we have probably not devised the correct model. Hence, it is good to know when things go wrong and why.

Example 1.2.1: Attempt to solve:

$$y' = \frac{1}{x}, \qquad y(0) = 0.$$

Integrate to find the general solution $y = \ln|x| + C$. The solution does not exist at $x - 0$. See Figure 1.5. The equation may have been written as the seemingly harmless $xy' = 1$.

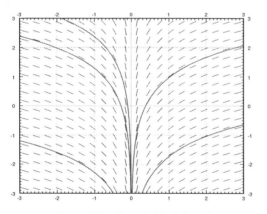

Figure 1.5: Slope field of $y' = 1/x$.

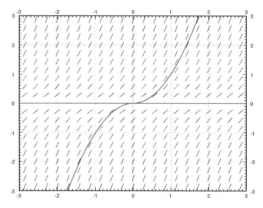

Figure 1.6: Slope field of $y' = 2\sqrt{|y|}$ with two solutions satisfying $y(0) = 0$.

Example 1.2.2: Solve:

$$y' = 2\sqrt{|y|}, \qquad y(0) = 0.$$

See Figure 1.6 on the preceding page. Note that $y = 0$ is a solution. But another solution is the function

$$y(x) = \begin{cases} x^2 & \text{if } x \geq 0, \\ -x^2 & \text{if } x < 0. \end{cases}$$

It is hard to tell by staring at the slope field that the solution is not unique. Is there any hope? Of course there is. We have the following theorem, known as Picard's theorem[*].

Theorem 1.2.1 (Picard's theorem on existence and uniqueness). *If $f(x, y)$ is continuous (as a function of two variables) and $\frac{\partial f}{\partial y}$ exists and is continuous near some (x_0, y_0), then a solution to*

$$y' = f(x, y), \qquad y(x_0) = y_0,$$

exists (at least for some small interval of x's) and is unique.

Note that the problems $y' = 1/x$, $y(0) = 0$ and $y' = 2\sqrt{|y|}$, $y(0) = 0$ do not satisfy the hypothesis of the theorem. Even if we can use the theorem, we ought to be careful about this existence business. It is quite possible that the solution only exists for a short while.

Example 1.2.3: For some constant A, solve:

$$y' = y^2, \qquad y(0) = A.$$

We know how to solve this equation. First assume that $A \neq 0$, so y is not equal to zero at least for some x near 0. So $x' = 1/y^2$, so $x = -1/y + C$, so $y = \frac{1}{C-x}$. If $y(0) = A$, then $C = 1/A$ so

$$y = \frac{1}{1/A - x}.$$

If $A = 0$, then $y = 0$ is a solution.

For example, when $A = 1$ the solution "blows up" at $x = 1$. Hence, the solution does not exist for all x even if the equation is nice everywhere. The equation $y' = y^2$ certainly looks nice.

For most of this course we will be interested in equations where existence and uniqueness holds, and in fact holds "globally" unlike for the equation $y' = y^2$.

1.2.3 Exercises

***Exercise* 1.2.1:** *Sketch slope field for $y' = e^{x-y}$. How do the solutions behave as x grows? Can you guess a particular solution by looking at the slope field?*

***Exercise* 1.2.2:** *Sketch slope field for $y' = x^2$.*

***Exercise* 1.2.3:** *Sketch slope field for $y' = y^2$.*

---

[*]Named after the French mathematician Charles Émile Picard (1856–1941)

Exercise 1.2.4: *Is it possible to solve the equation $y' = \frac{xy}{\cos x}$ for $y(0) = 1$? Justify.*

Exercise 1.2.5: *Is it possible to solve the equation $y' = y\sqrt{|x|}$ for $y(0) = 0$? Is the solution unique? Justify.*

Exercise 1.2.6: *Match equations $y' = 1 - x$, $y' = x - 2y$, $y' = x(1 - y)$ to slope fields. Justify.*

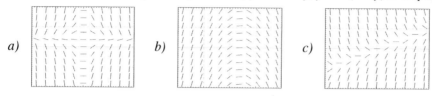

a)   b)   c)

Exercise 1.2.7 (challenging)*: Take $y' = f(x, y)$, $y(0) = 0$, where $f(x, y) > 1$ for all x and y. If the solution exists for all x, can you say what happens to $y(x)$ as x goes to positive infinity? Explain.*

Exercise 1.2.8 (challenging)*: Take $(y - x)y' = 0$, $y(0) = 0$. a) Find two distinct solutions. b) Explain why this does not violate Picard's theorem.*

Exercise 1.2.9: *Suppose $y' = f(x, y)$. What will the slope field look like, explain and sketch an example, if you have the following about $f(x, y)$. a) f does not depend on y. b) f does not depend on x. c) $f(t, t) = 0$ for any number t. d) $f(x, 0) = 0$ and $f(x, 1) = 1$ for all x.*

Exercise 1.2.10: *Find a solution to $y' = |y|$, $y(0) = 0$. Does Picard's theorem apply?*

Exercise 1.2.11: *Take an equation $y' - (y - 2x)g(x, y) + 2$ for some function $g(x, y)$. Can you solve the problem for the initial condition $y(0) = 0$, and if so what is the solution?*

Exercise 1.2.101: *Sketch the slope field of $y' = y^3$. Can you visually find the solution that satisfies $y(0) = 0$?*

Exercise 1.2.102: *Is it possible to solve $y' = xy$ for $y(0) = 0$? Is the solution unique?*

Exercise 1.2.103: *Is it possible to solve $y' = \frac{x}{x^2 - 1}$ for $y(1) = 0$?*

Exercise 1.2.104: *Match equations $y' = \sin x$, $y' = \cos y$, $y' = y\cos(x)$ to slope fields. Justify.*

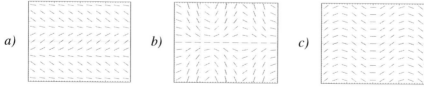

a)   b)   c)

Exercise 1.2.105 (tricky)*: Suppose*

$$f(y) = \begin{cases} 0 & \text{if } y > 0, \\ 1 & \text{if } y \le 0. \end{cases}$$

Does $y' = f(y)$, $y(0) = 0$ have a continuously differentiable solution? Does Picard apply, why, or why not?

1.3 Separable equations

Note: 1 lecture, §1.4 in [EP], §2.2 in [BD]

When a differential equation is of the form $y' = f(x)$, we can just integrate: $y = \int f(x)\,dx + C$. Unfortunately this method no longer works for the general form of the equation $y' = f(x, y)$. Integrating both sides yields

$$y = \int f(x, y)\,dx + C.$$

Notice the dependence on y in the integral.

1.3.1 Separable equations

Let us suppose that the equation is *separable*. That is, let us consider

$$y' = f(x)g(y),$$

for some functions $f(x)$ and $g(y)$. Let us write the equation in the Leibniz notation

$$\frac{dy}{dx} = f(x)g(y).$$

Then we rewrite the equation as

$$\frac{dy}{g(y)} = f(x)\,dx.$$

Now both sides look like something we can integrate. We obtain

$$\int \frac{dy}{g(y)} = \int f(x)\,dx + C.$$

If we can find closed form expressions for these two integrals, we can, perhaps, solve for y.

Example 1.3.1: Take the equation

$$y' = xy.$$

First note that $y = 0$ is a solution, so assume $y \neq 0$ from now on. Write the equation as $\frac{dy}{dx} = xy$, then

$$\int \frac{dy}{y} = \int x\,dx + C.$$

We compute the antiderivatives to get

$$\ln |y| = \frac{x^2}{2} + C.$$

Or

$$|y| = e^{\frac{x^2}{2}+C} = e^{\frac{x^2}{2}} e^C = De^{\frac{x^2}{2}},$$

where $D > 0$ is some constant. Because $y = 0$ is a solution and because of the absolute value we actually can write:

$$y = De^{\frac{x^2}{2}},$$

for any number D (including zero or negative).

We check:

$$y' = Dxe^{\frac{x^2}{2}} = x\left(De^{\frac{x^2}{2}}\right) = xy.$$

Yay!

We should be a little bit more careful with this method. You may be worried that we were integrating in two different variables. We seemed to be doing a different operation to each side. Let us work this method out more rigorously. Take

$$\frac{dy}{dx} = f(x)g(y).$$

We rewrite the equation as follows. Note that $y = y(x)$ is a function of x and so is $\frac{dy}{dx}$!

$$\frac{1}{g(y)} \frac{dy}{dx} = f(x).$$

We integrate both sides with respect to x.

$$\int \frac{1}{g(y)} \frac{dy}{dx} \, dx = \int f(x) \, dx + C.$$

We use the change of variables formula.

$$\int \frac{1}{g(y)} \, dy = \int f(x) \, dx + C.$$

And we are done.

1.3.2   Implicit solutions

It is clear that we might sometimes get stuck even if we can do the integration. For example, take the separable equation

$$y' = \frac{xy}{y^2 + 1}.$$

We separate variables,

$$\frac{y^2 + 1}{y} \, dy = \left(y + \frac{1}{y}\right) dy = x \, dx.$$

We integrate to get

$$\frac{y^2}{2} + \ln|y| = \frac{x^2}{2} + C,$$

or perhaps the easier looking expression (where $D = 2C$)

$$y^2 + 2\ln|y| = x^2 + D.$$

It is not easy to find the solution explicitly as it is hard to solve for y. We, therefore, leave the solution in this form and call it an *implicit solution*. It is still easy to check that an implicit solution satisfies the differential equation. In this case, we differentiate with respect to x to get

$$y'\left(2y + \frac{2}{y}\right) = 2x.$$

It is simple to see that the differential equation holds. If you want to compute values for y, you might have to be tricky. For example, you can graph x as a function of y, and then flip your paper. Computers are also good at some of these tricks.

We note that the above equation also has the solution $y = 0$. The general solution is $y^2 + 2\ln|y| = x^2 + C$ together with $y = 0$. These outlying solutions such as $y = 0$ are sometimes called *singular solutions*.

1.3.3 Examples

Example 1.3.2: Solve $x^2 y' = 1 - x^2 + y^2 - x^2 y^2$, $y(1) = 0$.

First factor the right hand side to obtain

$$x^2 y' = (1 - x^2)(1 + y^2).$$

Separate variables, integrate, and solve for y.

$$\frac{y'}{1 + y^2} = \frac{1 - x^2}{x^2},$$

$$\frac{y'}{1 + y^2} = \frac{1}{x^2} - 1,$$

$$\arctan(y) = \frac{-1}{x} - x + C,$$

$$y = \tan\left(\frac{-1}{x} - x + C\right).$$

Now solve for the initial condition, $0 = \tan(-2 + C)$ to get $C = 2$ (or $2 + \pi$, etc...). The solution we are seeking is, therefore,

$$y = \tan\left(\frac{-1}{x} - x + 2\right).$$

Example 1.3.3: Bob made a cup of coffee, and Bob likes to drink coffee only once it will not burn him at 60 degrees. Initially at time $t = 0$ minutes, Bob measured the temperature and the coffee was 89 degrees Celsius. One minute later, Bob measured the coffee again and it had 85 degrees. The temperature of the room (the ambient temperature) is 22 degrees. When should Bob start drinking?

Let T be the temperature of the coffee, and let A be the ambient (room) temperature. Newton's law of cooling states that the rate at which the temperature of the coffee is changing is proportional to the difference between the ambient temperature and the temperature of the coffee. That is,

$$\frac{dT}{dt} = k(A - T),$$

for some constant k. For our setup $A = 22$, $T(0) = 89$, $T(1) = 85$. We separate variables and integrate (let C and D denote arbitrary constants)

$$\frac{1}{T - A}\frac{dT}{dt} = -k,$$
$$\ln(T - A) = -kt + C, \qquad \text{(note that } T - A > 0\text{)}$$
$$T - A = D\,e^{-kt},$$
$$T = A + D\,e^{-kt}.$$

That is, $T = 22 + D\,e^{-kt}$. We plug in the first condition: $89 = T(0) = 22 + D$, and hence $D = 67$. So $T = 22 + 67\,e^{-kt}$. The second condition says $85 = T(1) = 22 + 67\,e^{-k}$. Solving for k we get $k = -\ln\frac{85-22}{67} \approx 0.0616$. Now we solve for the time t that gives us a temperature of 60 degrees. That is, we solve $60 = 22 + 67e^{-0.0616t}$ to get $t = -\frac{\ln\frac{60-22}{67}}{0.0616} \approx 9.21$ minutes. So Bob can begin to drink the coffee at just over 9 minutes from the time Bob made it. That is probably about the amount of time it took us to calculate how long it would take.

Example 1.3.4: Find the general solution to $y' = \frac{-xy^2}{3}$ (including singular solutions).

First note that $y = 0$ is a solution (a singular solution). So assume that $y \neq 0$ and write

$$\frac{-3}{y^2}y' = x,$$
$$\frac{3}{y} = \frac{x^2}{2} + C,$$
$$y = \frac{3}{x^2/2 + C} = \frac{6}{x^2 + 2C}.$$

1.3.4   Exercises

Exercise 1.3.1: Solve $y' = x/y$.

Exercise 1.3.2: Solve $y' = x^2 y$.

Exercise 1.3.3: *Solve* $\dfrac{dx}{dt} = (x^2 - 1)\,t$, *for* $x(0) = 0$.

Exercise 1.3.4: *Solve* $\dfrac{dx}{dt} = x\,\sin(t)$, *for* $x(0) = 1$.

Exercise 1.3.5: *Solve* $\dfrac{dy}{dx} = xy + x + y + 1$. *Hint: Factor the right hand side.*

Exercise 1.3.6: *Solve* $xy' = y + 2x^2 y$, *where* $y(1) = 1$.

Exercise 1.3.7: *Solve* $\dfrac{dy}{dx} = \dfrac{y^2 + 1}{x^2 + 1}$, *for* $y(0) = 1$.

Exercise 1.3.8: *Find an implicit solution for* $\dfrac{dy}{dx} = \dfrac{x^2 + 1}{y^2 + 1}$, *for* $y(0) = 1$.

Exercise 1.3.9: *Find an explicit solution for* $y' = xe^{-y}$, $y(0) = 1$.

Exercise 1.3.10: *Find an explicit solution for* $xy' = e^{-y}$, *for* $y(1) = 1$.

Exercise 1.3.11: *Find an explicit solution for* $y' = ye^{-x^2}$, $y(0) = 1$. *It is alright to leave a definite integral in your answer.*

Exercise 1.3.12: *Suppose a cup of coffee is at 100 degrees Celsius at time $t = 0$, it is at 70 degrees at $t = 10$ minutes, and it is at 50 degrees at $t = 20$ minutes. Compute the ambient temperature.*

Exercise 1.3.101: *Solve* $y' = 2xy$.

Exercise 1.3.102: *Solve* $x' = 3xt^2 - 3t^2$, $x(0) = 2$.

Exercise 1.3.103: *Find an implicit solution for* $x' = \frac{1}{3x^2+1}$, $x(0) = 1$.

Exercise 1.3.104: *Find an explicit solution to* $xy' = y^2$, $y(1) = 1$.

Exercise 1.3.105: *Find an implicit solution to* $y' = \frac{\sin(x)}{\cos(y)}$.

Exercise 1.3.106: *Take Example 1.3.3 with the same numbers: 89 degrees at $t = 0$, 85 degrees at $t = 1$, and ambient temperature of 22 degrees. Suppose these temperatures were measured with precision of ± 0.5 degrees. Given this imprecision, the time it takes the coffee to cool to (exactly) 60 degrees is also only known in a certain range. Find this range. Hint: Think about what kind of error makes the cooling time longer and what shorter.*

Exercise 1.3.107: *A population x of rabbits on an island is modelled by $x' = x - (1/1000)x^2$, where the independent variable is time in months. At time $t = 0$, there are 40 rabbits on the island. a) Find the solution to the equation with the initial condition. b) How many rabbits are on the island in 1 month, 5 months, 10 months, 15 months (round to the nearest integer).*

1.4   Linear equations and the integrating factor

Note: 1 lecture, §1.5 in [EP], §2.1 in [BD]

One of the most important types of equations we will learn how to solve are the so-called *linear equations*. In fact, the majority of the course is about linear equations. In this section we focus on the *first order linear equation*. A first order equation is linear if we can put it into the form:

$$y' + p(x)y = f(x). \tag{1.3}$$

The word "linear" means linear in y and y'; no higher powers nor functions of y or y' appear. The dependence on x can be more complicated.

Solutions of linear equations have nice properties. For example, the solution exists wherever $p(x)$ and $f(x)$ are defined, and has the same regularity (read: it is just as nice). But most importantly for us right now, there is a method for solving linear first order equations.

The trick is to rewrite the left hand side of (1.3) as a derivative of a product of y with another function. To this end we find a function $r(x)$ such that

$$r(x)y' + r(x)p(x)y = \frac{d}{dx}\big[r(x)y\big].$$

This is the left hand side of (1.3) multiplied by $r(x)$. So if we multiply (1.3) by $r(x)$, we obtain

$$\frac{d}{dx}\big[r(x)y\big] = r(x)f(x).$$

Now we integrate both sides. The right hand side does not depend on y and the left hand side is written as a derivative of a function. Afterwards, we solve for y. The function $r(x)$ is called the *integrating factor* and the method is called the *integrating factor method*.

We are looking for a function $r(x)$, such that if we differentiate it, we get the same function back multiplied by $p(x)$. That seems like a job for the exponential function! Let

$$r(x) = e^{\int p(x)\,dx}.$$

We compute:

$$y' + p(x)y = f(x),$$
$$e^{\int p(x)\,dx}y' + e^{\int p(x)\,dx}p(x)y = e^{\int p(x)\,dx}f(x),$$
$$\frac{d}{dx}\Big[e^{\int p(x)\,dx}y\Big] = e^{\int p(x)\,dx}f(x),$$
$$e^{\int p(x)\,dx}y = \int e^{\int p(x)\,dx}f(x)\,dx + C,$$
$$y = e^{-\int p(x)\,dx}\left(\int e^{\int p(x)\,dx}f(x)\,dx + C\right).$$

Of course, to get a closed form formula for y, we need to be able to find a closed form formula for the integrals appearing above.

Example 1.4.1: Solve

$$y' + 2xy = e^{x-x^2}, \qquad y(0) = -1.$$

First note that $p(x) = 2x$ and $f(x) = e^{x-x^2}$. The integrating factor is $r(x) = e^{\int p(x)\,dx} = e^{x^2}$. We multiply both sides of the equation by $r(x)$ to get

$$e^{x^2} y' + 2xe^{x^2} y = e^{x-x^2} e^{x^2},$$

$$\frac{d}{dx}\left[e^{x^2} y\right] = e^x.$$

We integrate

$$e^{x^2} y = e^x + C,$$

$$y = e^{x-x^2} + Ce^{-x^2}.$$

Next, we solve for the initial condition $-1 = y(0) = 1 + C$, so $C = -2$. The solution is

$$y = e^{x-x^2} - 2e^{-x^2}.$$

Note that we do not care which antiderivative we take when computing $e^{\int p(x)dx}$. You can always add a constant of integration, but those constants will not matter in the end.

Exercise **1.4.1:** *Try it! Add a constant of integration to the integral in the integrating factor and show that the solution you get in the end is the same as what we got above.*

An advice: Do not try to remember the formula itself, that is way too hard. It is easier to remember the process and repeat it.

Since we cannot always evaluate the integrals in closed form, it is useful to know how to write the solution in definite integral form. A definite integral is something that you can plug into a computer or a calculator. Suppose we are given

$$y' + p(x)y = f(x), \qquad y(x_0) = y_0.$$

Look at the solution and write the integrals as definite integrals.

$$y(x) = e^{-\int_{x_0}^{x} p(s)\,ds}\left(\int_{x_0}^{x} e^{\int_{x_0}^{t} p(s)\,ds} f(t)\,dt + y_0\right). \tag{1.4}$$

You should be careful to properly use dummy variables here. If you now plug such a formula into a computer or a calculator, it will be happy to give you numerical answers.

***Exercise* 1.4.2:** *Check that $y(x_0) = y_0$ in formula (1.4).*

***Exercise* 1.4.3:** *Write the solution of the following problem as a definite integral, but try to simplify as far as you can. You will not be able to find the solution in closed form.*

$$y' + y = e^{x^2 - x}, \qquad y(0) = 10.$$

Remark 1.4.1: Before we move on, we should note some interesting properties of linear equations. First, for the linear initial value problem $y' + p(x)y = f(x)$, $y(x_0) = y_0$, there is always an explicit formula (1.4) for the solution. Second, it follows from the formula (1.4) that if $p(x)$ and $f(x)$ are continuous on some interval (a, b), then the solution $y(x)$ exists and is differentiable on (a, b). Compare with the simple nonlinear example we have seen previously, $y' = y^2$, and compare to Theorem 1.2.1.

Example 1.4.2: Let us discuss a common simple application of linear equations. This type of problem is used often in real life. For example, linear equations are used in figuring out the concentration of chemicals in bodies of water (rivers and lakes).

A 100 liter tank contains 10 kilograms of salt dissolved in 60 liters of water. Solution of water and salt (brine) with concentration of 0.1 kilograms per liter is flowing in at the rate of 5 liters a minute. The solution in the tank is well stirred and flows out at a rate of 3 liters a minute. How much salt is in the tank when the tank is full?

Let us come up with the equation. Let x denote the kilograms of salt in the tank, let t denote the time in minutes. For a small change Δt in time, the change in x (denoted Δx) is approximately

$$\Delta x \approx (\text{rate in} \times \text{concentration in})\Delta t - (\text{rate out} \times \text{concentration out})\Delta t.$$

Dividing through by Δt and taking the limit $\Delta t \to 0$ we see that

$$\frac{dx}{dt} = (\text{rate in} \times \text{concentration in}) - (\text{rate out} \times \text{concentration out}).$$

In our example, we have

$$\text{rate in} = 5,$$
$$\text{concentration in} = 0.1,$$
$$\text{rate out} = 3,$$
$$\text{concentration out} = \frac{x}{\text{volume}} = \frac{x}{60 + (5 - 3)t}.$$

Our equation is, therefore,

$$\frac{dx}{dt} = (5 \times 0.1) - \left(3\frac{x}{60 + 2t}\right).$$

Or in the form (1.3)

$$\frac{dx}{dt} + \frac{3}{60 + 2t}x = 0.5.$$

Let us solve. The integrating factor is

$$r(t) = \exp\left(\int \frac{3}{60 + 2t}dt\right) = \exp\left(\frac{3}{2}\ln(60 + 2t)\right) = (60 + 2t)^{3/2}.$$

We multiply both sides of the equation to get

$$(60 + 2t)^{3/2}\frac{dx}{dt} + (60 + 2t)^{3/2}\frac{3}{60 + 2t}x = 0.5(60 + 2t)^{3/2},$$

$$\frac{d}{dt}\left[(60 + 2t)^{3/2}x\right] = 0.5(60 + 2t)^{3/2},$$

$$(60 + 2t)^{3/2}x = \int 0.5(60 + 2t)^{3/2}dt + C,$$

$$x = (60 + 2t)^{-3/2}\int \frac{(60 + 2t)^{3/2}}{2}dt + C(60 + 2t)^{-3/2},$$

$$x = (60 + 2t)^{-3/2}\frac{1}{10}(60 + 2t)^{5/2} + C(60 + 2t)^{-3/2},$$

$$x = \frac{60 + 2t}{10} + C(60 + 2t)^{-3/2}.$$

We need to find C. We know that at $t = 0$, $x = 10$. So

$$10 = x(0) = \frac{60}{10} + C(60)^{-3/2} = 6 + C(60)^{-3/2},$$

or

$$C = 4(60^{3/2}) \approx 1859.03.$$

We are interested in x when the tank is full. So we note that the tank is full when $60 + 2t = 100$, or when $t = 20$. So

$$x(20) = \frac{60 + 40}{10} + C(60 + 40)^{-3/2}$$

$$\approx 10 + 1859.03(100)^{-3/2} \approx 11.86.$$

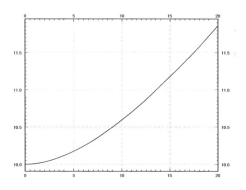

Figure 1.7: Graph of the solution x kilograms of salt in the tank at time t.

See Figure 1.7 for the graph of x over t.

The concentration at the end is approximately 0.1186 kg/liter and we started with 1/6 or 0.167 kg/liter.

1.4.1   Exercises

In the exercises, feel free to leave answer as a definite integral if a closed form solution cannot be found. If you can find a closed form solution, you should give that.

Exercise 1.4.4: *Solve* $y' + xy = x$.

Exercise 1.4.5: *Solve* $y' + 6y = e^x$.

Exercise 1.4.6: *Solve* $y' + 3x^2 y = \sin(x)\, e^{-x^3}$, *with* $y(0) = 1$.

Exercise 1.4.7: *Solve* $y' + \cos(x)y = \cos(x)$.

Exercise 1.4.8: *Solve* $\frac{1}{x^2+1} y' + xy = 3$, *with* $y(0) = 0$.

Exercise 1.4.9: *Suppose there are two lakes located on a stream. Clean water flows into the first lake, then the water from the first lake flows into the second lake, and then water from the second lake flows further downstream. The in and out flow from each lake is 500 liters per hour. The first lake contains 100 thousand liters of water and the second lake contains 200 thousand liters of water. A truck with 500 kg of toxic substance crashes into the first lake. Assume that the water is being continually mixed perfectly by the stream. a) Find the concentration of toxic substance as a function of time in both lakes. b) When will the concentration in the first lake be below 0.001 kg per liter? c) When will the concentration in the second lake be maximal?*

Exercise 1.4.10: *Newton's law of cooling states that* $\frac{dx}{dt} = -k(x - A)$ *where x is the temperature, t is time, A is the ambient temperature, and* $k > 0$ *is a constant. Suppose that* $A = A_0 \cos(\omega t)$ *for some constants* A_0 *and* ω. *That is, the ambient temperature oscillates (for example night and day temperatures). a) Find the general solution. b) In the long term, will the initial conditions make much of a difference? Why or why not?*

Exercise 1.4.11: *Initially 5 grams of salt are dissolved in 20 liters of water. Brine with concentration of salt 2 grams of salt per liter is added at a rate of 3 liters a minute. The tank is mixed well and is drained at 3 liters a minute. How long does the process have to continue until there are 20 grams of salt in the tank?*

Exercise 1.4.12: *Initially a tank contains 10 liters of pure water. Brine of unknown (but constant) concentration of salt is flowing in at 1 liter per minute. The water is mixed well and drained at 1 liter per minute. In 20 minutes there are 15 grams of salt in the tank. What is the concentration of salt in the incoming brine?*

Exercise 1.4.101: *Solve* $y' + 3x^2 y = x^2$.

Exercise 1.4.102: *Solve* $y' + 2\sin(2x)y = 2\sin(2x)$, $y(\pi/2) = 3$.

***Exercise* 1.4.103:** *Suppose a water tank is being pumped out at 3 ℓ/min. The water tank starts at 10 L of clean water. Water with toxic substance is flowing into the tank at 2 ℓ/min, with concentration 20t g/L at time t. When the tank is half empty, how many grams of toxic substance are in the tank (assuming perfect mixing)?*

***Exercise* 1.4.104:** *Suppose we have bacteria on a plate and suppose that we are slowly adding a toxic substance such that the rate of growth is slowing down. That is, suppose that $\frac{dP}{dt} = (2 - 0.1t)P$. If $P(0) = 1000$, find the population at $t = 5$.*

***Exercise* 1.4.105:** *A cylindrical water tank has water flowing in at I cubic meters per second. Let A be the area of the cross section of the tank in meters. Suppose water is flowing from the bottom of the tank at a rate proportional to the height of the water level. Set up the differential equation for h, the height of the water, introducing and naming constants that you need. You should also give the units for your constants.*

1.5   Substitution

Note: 1 lecture, §1.6 in [EP], not in [BD]

Just as when solving integrals, one method to try is to change variables to end up with a simpler equation to solve.

1.5.1   Substitution

The equation

$$y' = (x - y + 1)^2$$

is neither separable nor linear. What can we do? How about trying to change variables, so that in the new variables the equation is simpler. We use another variable v, which we treat as a function of x. Let us try

$$v = x - y + 1.$$

We need to figure out y' in terms of v', v and x. We differentiate (in x) to obtain $v' = 1 - y'$. So $y' = 1 - v'$. We plug this into the equation to get

$$1 - v' = v^2.$$

In other words, $v' = 1 - v^2$. Such an equation we know how to solve by separating variables:

$$\frac{1}{1 - v^2}\, dv = dx.$$

So

$$\frac{1}{2} \ln \left| \frac{v + 1}{v - 1} \right| = x + C,$$

$$\left| \frac{v + 1}{v - 1} \right| = e^{2x + 2C},$$

or $\frac{v+1}{v-1} = De^{2x}$ for some constant D. Note that $v = 1$ and $v = -1$ are also solutions.

Now we need to "unsubstitute" to obtain

$$\frac{x - y + 2}{x - y} = De^{2x},$$

and also the two solutions $x - y + 1 = 1$ or $y = x$, and $x - y + 1 = -1$ or $y = x + 2$. We solve the first equation for y.

$$x - y + 2 = (x - y)De^{2x},$$
$$x - y + 2 = Dxe^{2x} - yDe^{2x},$$
$$-y + yDe^{2x} = Dxe^{2x} - x - 2,$$
$$y(-1 + De^{2x}) = Dxe^{2x} - x - 2,$$
$$y = \frac{Dxe^{2x} - x - 2}{De^{2x} - 1}.$$

Note that $D = 0$ gives $y = x + 2$, but no value of D gives the solution $y = x$.

Substitution in differential equations is applied in much the same way that it is applied in calculus. You guess. Several different substitutions might work. There are some general patterns to look for. We summarize a few of these in a table.

When you see	Try substituting
yy'	$v = y^2$
$y^2 y'$	$v = y^3$
$(\cos y)y'$	$v = \sin y$
$(\sin y)y'$	$v = \cos y$
$y'e^y$	$v = e^y$

Usually you try to substitute in the "most complicated" part of the equation with the hopes of simplifying it. The above table is just a rule of thumb. You might have to modify your guesses. If a substitution does not work (it does not make the equation any simpler), try a different one.

1.5.2 Bernoulli equations

There are some forms of equations where there is a general rule for substitution that always works. One such example is the so-called *Bernoulli equation*[*]:

$$y' + p(x)y = q(x)y^n.$$

This equation looks a lot like a linear equation except for the y^n. If $n = 0$ or $n = 1$, then the equation is linear and we can solve it. Otherwise, the substitution $v = y^{1-n}$ transforms the Bernoulli equation into a linear equation. Note that n need not be an integer.

Example 1.5.1: Solve
$$xy' + y(x + 1) + xy^5 = 0, \qquad y(1) = 1.$$

First, the equation is Bernoulli ($p(x) = (x + 1)/x$ and $q(x) = -1$). We substitute
$$v = y^{1-5} = y^{-4}, \qquad v' = -4y^{-5}y'.$$

In other words, $(-1/4)\, y^5 v' = y'$. So

$$xy' + y(x + 1) + xy^5 = 0,$$

$$\frac{-xy^5}{4}v' + y(x + 1) + xy^5 = 0,$$

$$\frac{-x}{4}v' + y^{-4}(x + 1) + x = 0,$$

$$\frac{-x}{4}v' + v(x + 1) + x = 0,$$

---

[*]There are several things called Bernoulli equations, this is just one of them. The Bernoullis were a prominent Swiss family of mathematicians. These particular equations are named for Jacob Bernoulli (1654–1705).

and finally

$$v' - \frac{4(x+1)}{x}v = 4.$$

The equation is now linear. We can use the integrating factor method. In particular, we use formula (1.4). Let us assume that $x > 0$ so $|x| = x$. This assumption is OK, as our initial condition is $x = 1$. Let us compute the integrating factor. Here $p(s)$ from formula (1.4) is $\frac{-4(s+1)}{s}$.

$$e^{\int_1^x p(s)\,ds} = \exp\left(\int_1^x \frac{-4(s+1)}{s}\,ds\right) = e^{-4x-4\ln(x)+4} = e^{-4x+4}x^{-4} = \frac{e^{-4x+4}}{x^4},$$

$$e^{-\int_1^x p(s)\,ds} = e^{4x+4\ln(x)-4} = e^{4x-4}x^4.$$

We now plug in to (1.4)

$$v(x) = e^{-\int_1^x p(s)\,ds}\left(\int_1^x e^{\int_1^t p(s)\,ds}4\,dt + 1\right)$$

$$= e^{4x-4}x^4\left(\int_1^x 4\frac{e^{-4t+4}}{t^4}\,dt + 1\right).$$

The integral in this expression is not possible to find in closed form. As we said before, it is perfectly fine to have a definite integral in our solution. Now "unsubstitute"

$$y^{-4} = e^{4x-4}x^4\left(4\int_1^x \frac{e^{-4t+4}}{t^4}\,dt + 1\right),$$

$$y = \frac{e^{-x+1}}{x\left(4\int_1^x \frac{e^{-4t+4}}{t^4}\,dt + 1\right)^{1/4}}.$$

1.5.3 Homogeneous equations

Another type of equations we can solve by substitution are the so-called *homogeneous equations*. Suppose that we can write the differential equation as

$$y' = F\left(\frac{y}{x}\right).$$

Here we try the substitutions

$$v = \frac{y}{x} \qquad \text{and therefore} \qquad y' = v + xv'.$$

We note that the equation is transformed into

$$v + xv' = F(v) \qquad \text{or} \qquad xv' = F(v) - v \qquad \text{or} \qquad \frac{v'}{F(v) - v} = \frac{1}{x}.$$

Hence an implicit solution is

$$\int \frac{1}{F(v) - v} \, dv = \ln|x| + C.$$

Example 1.5.2: Solve

$$x^2 y' = y^2 + xy, \qquad y(1) = 1.$$

We put the equation into the form $y' = (y/x)^2 + y/x$. We substitute $v = y/x$ to get the separable equation

$$xv' = v^2 + v - v = v^2,$$

which has a solution

$$\int \frac{1}{v^2} \, dv = \ln|x| + C,$$

$$\frac{-1}{v} = \ln|x| + C,$$

$$v = \frac{-1}{\ln|x| + C}.$$

We unsubstitute

$$\frac{y}{x} = \frac{-1}{\ln|x| + C},$$

$$y = \frac{-x}{\ln|x| + C}.$$

We want $y(1) = 1$, so

$$1 = y(1) = \frac{-1}{\ln|1| + C} = \frac{-1}{C}.$$

Thus $C = -1$ and the solution we are looking for is

$$y = \frac{-x}{\ln|x| - 1}.$$

1.5.4 Exercises

Hint: Answers need not always be in closed form.

Exercise 1.5.1: *Solve $y' + y(x^2 - 1) + xy^6 = 0$, with $y(1) = 1$.*

Exercise 1.5.2: *Solve $2yy' + 1 = y^2 + x$, with $y(0) = 1$.*

Exercise 1.5.3: *Solve $y' + xy = y^4$, with $y(0) = 1$.*

Exercise 1.5.4: *Solve $yy' + x = \sqrt{x^2 + y^2}$.*

Exercise 1.5.5: *Solve $y' = (x + y - 1)^2$.*

Exercise 1.5.6: *Solve $y' = \frac{x^2 - y^2}{xy}$, with $y(1) = 2$.*

Exercise 1.5.101: *Solve $xy' + y + y^2 = 0$, $y(1) = 2$.*

Exercise 1.5.102: *Solve $xy' + y + x = 0$, $y(1) = 1$.*

Exercise 1.5.103: *Solve $y^2 y' = y^3 - 3x$, $y(0) = 2$.*

Exercise 1.5.104: *Solve $2yy' = e^{y^2 - x^2} + 2x$.*

1.6 Autonomous equations

Note: 1 lecture, §2.2 in [EP], §2.5 in [BD]

Let us consider problems of the form

$$\frac{dx}{dt} = f(x),$$

where the derivative of solutions depends only on x (the dependent variable). Such equations are called *autonomous equations*. If we think of t as time, the naming comes from the fact that the equation is independent of time.

Let us return to the cooling coffee problem (see Example 1.3.3). Newton's law of cooling says

$$\frac{dx}{dt} = -k(x - A),$$

where x is the temperature, t is time, k is some constant, and A is the ambient temperature. See Figure 1.8 for an example with $k = 0.3$ and $A = 5$.

Note the solution $x = A$ (in the figure $x = 5$). We call these constant solutions the *equilibrium solutions*. The points on the x axis where $f(x) = 0$ are called *critical points*. The point $x = A$ is a critical point. In fact, each critical point corresponds to an equilibrium solution. Note also, by looking at the graph, that the solution $x = A$ is "stable" in that small perturbations in x do not lead to substantially different solutions as t grows. If we change the initial condition a little bit, then as $t \to \infty$ we get $x(t) \to A$. We call such critical points *stable*. In this simple example it turns out that all solutions in fact go to A as $t \to \infty$. If a critical point is not stable we would say it is *unstable*.

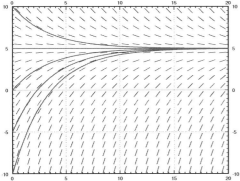
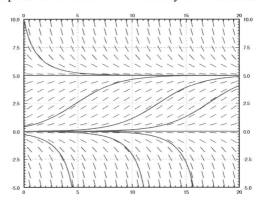
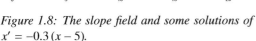

Figure 1.8: The slope field and some solutions of $x' = -0.3\,(x - 5)$.

Figure 1.9: The slope field and some solutions of $x' = 0.1\,x\,(5 - x)$.

Let us consider the *logistic equation*

$$\frac{dx}{dt} = kx(M - x),$$

for some positive k and M. This equation is commonly used to model population if we know the
limiting population M, that is the maximum sustainable population. The logistic equation leads to
less catastrophic predictions on world population than $x' = kx$. In the real world there is no such
thing as negative population, but we will still consider negative x for the purposes of the math.

See Figure 1.9 on the preceding page for an example. Note two critical points, $x = 0$ and $x = 5$.
The critical point at $x = 5$ is stable. On the other hand the critical point at $x = 0$ is unstable.

It is not really necessary to find the exact solutions to talk about the long term behavior of the
solutions. For example, from the above we can easily see that

$$\lim_{t \to \infty} x(t) = \begin{cases} 5 & \text{if } x(0) > 0, \\ 0 & \text{if } x(0) = 0, \\ \text{DNE or } -\infty & \text{if } x(0) < 0. \end{cases}$$

Where DNE means "does not exist." From just looking at the slope field we cannot quite decide
what happens if $x(0) < 0$. It could be that the solution does not exist for t all the way to ∞. Think of
the equation $x' = x^2$; we have seen that solutions only exist for some finite period of time. Same
can happen here. In our example equation above it will actually turn out that the solution does not
exist for all time, but to see that we would have to solve the equation. In any case, the solution does
go to $-\infty$, but it may get there rather quickly.

Often we are interested only in the long term behavior of the solution and we would be doing
unnecessary work if we solved the equation exactly. It is easier to just look at the *phase diagram* or
phase portrait, which is a simple way to visualize the behavior of autonomous equations. In this
case there is one dependent variable x. We draw the x axis, we mark all the critical points, and then
we draw arrows in between. If $f(x) > 0$, we draw an up arrow. If $f(x) < 0$, we draw a down arrow.

Armed with the phase diagram, it is easy to sketch the solutions approximately.

Exercise 1.6.1: *Try sketching a few solutions simply from looking at the phase diagram. Check
with the preceding graphs if you are getting the type of curves.*

Once we draw the phase diagram, we can easily classify critical points as stable or unstable*.

---

*The unstable points that have one of the arrows pointing towards the critical point are sometimes called *semistable*.

Since any mathematical model we cook up will only be an approximation to the real world, unstable points are generally bad news.

Let us think about the logistic equation with harvesting. Suppose an alien race really likes to eat humans. They keep a planet with humans on it and harvest the humans at a rate of h million humans per year. Suppose x is the number of humans in millions on the planet and t is time in years. Let M be the limiting population when no harvesting is done. The number $k > 0$ is a constant depending on how fast humans multiply. Our equation becomes

$$\frac{dx}{dt} = kx(M - x) - h.$$

We expand the right hand side and set it to zero

$$kx(M - x) - h = -kx^2 + kMx - h = 0.$$

Solving for the critical points, let us call them A and B, we get

$$A = \frac{kM + \sqrt{(kM)^2 - 4hk}}{2k}, \qquad B = \frac{kM - \sqrt{(kM)^2 - 4hk}}{2k}.$$

Exercise **1.6.2:** *Draw the phase diagram for different possibilities. Note that these possibilities are $A > B$, or $A = B$, or A and B both complex (i.e. no real solutions). Hint: Fix some simple k and M and then vary h.*

For example, let $M = 8$ and $k = 0.1$. When $h = 1$, then A and B are distinct and positive. The graph we will get is given in Figure 1.10 on the next page. As long as the population starts above B, which is approximately 1.55 million, then the population will not die out. It will in fact tend towards $A \approx 6.45$ million. If ever some catastrophe happens and the population drops below B, humans will die out, and the fast food restaurant serving them will go out of business.

When $h = 1.6$, then $A = B = 4$. There is only one critical point and it is unstable. When the population starts above 4 million it will tend towards 4 million. If it ever drops below 4 million, humans will die out on the planet. This scenario is not one that we (as the human fast food proprietor) want to be in. A small perturbation of the equilibrium state and we are out of business. There is no room for error. See Figure 1.11 on the following page.

Finally if we are harvesting at 2 million humans per year, there are no critical points. The population will always plummet towards zero, no matter how well stocked the planet starts. See Figure 1.12 on the next page.

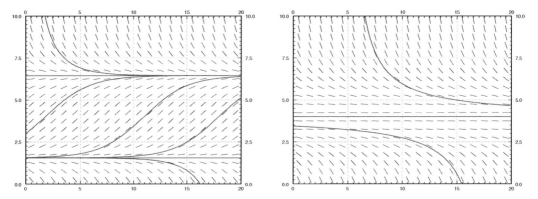

Figure 1.10: The slope field and some solutions of $x' = 0.1\,x\,(8-x) - 1$.

Figure 1.11: The slope field and some solutions of $x' = 0.1\,x\,(8-x) - 1.6$.

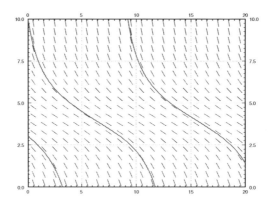

Figure 1.12: The slope field and some solutions of $x' = 0.1\,x\,(8-x) - 2$.

1.6.1 Exercises

Exercise **1.6.3:** *Take $x' = x^2$. a) Draw the phase diagram, find the critical points, and mark them stable or unstable. b) Sketch typical solutions of the equation. c) Find $\lim_{t\to\infty} x(t)$ for the solution with the initial condition $x(0) = -1$.*

Exercise **1.6.4:** *Take $x' = \sin x$. a) Draw the phase diagram for $-4\pi \le x \le 4\pi$. On this interval mark the critical points stable or unstable. b) Sketch typical solutions of the equation. c) Find $\lim_{t\to\infty} x(t)$ for the solution with the initial condition $x(0) = 1$.*

Exercise 1.6.5: Suppose $f(x)$ is positive for $0 < x < 1$, it is zero when $x = 0$ and $x = 1$, and it is negative for all other x. a) Draw the phase diagram for $x' = f(x)$, find the critical points, and mark them stable or unstable. b) Sketch typical solutions of the equation. c) Find $\lim_{t \to \infty} x(t)$ for the solution with the initial condition $x(0) = 0.5$.

Exercise 1.6.6: Start with the logistic equation $\frac{dx}{dt} = kx(M - x)$. Suppose we modify our harvesting. That is we will only harvest an amount proportional to current population. In other words, we harvest hx per unit of time for some $h > 0$ (Similar to earlier example with h replaced with hx). a) Construct the differential equation. b) Show that if $kM > h$, then the equation is still logistic. c) What happens when $kM < h$?

Exercise 1.6.7: A disease is spreading through the country. Let x be the number of people infected. Let the constant S be the number of people susceptible to infection. The infection rate $\frac{dx}{dt}$ is proportional to the product of already infected people, x, and the number of susceptible but uninfected people, $S - x$. a) Write down the differential equation. b) Supposing $x(0) > 0$, that is, some people are infected at time $t = 0$, what is $\lim_{t \to \infty} x(t)$. c) Does the solution to part b) agree with your intuition? Why or why not?

Exercise 1.6.101: Let $x' = (x - 1)(x - 2)x^2$. a) Sketch the phase diagram and find critical points. b) Classify the critical points. c) If $x(0) = 0.5$ then find $\lim_{t \to \infty} x(t)$.

Exercise 1.6.102: Let $x' = e^{-x}$. a) Find and classify all critical points. b) Find $\lim_{t \to \infty} x(t)$ given any initial condition.

Exercise 1.6.103: Assume that a population of fish in a lake satisfies $\frac{dx}{dt} = kx(M - x)$. Now suppose that fish are continually added at A fish per unit of time. a) Find the differential equation for x. b) What is the new limiting population?

Exercise 1.6.104: Suppose $\frac{dx}{dt} = (x - \alpha)(x - \beta)$ for two numbers $\alpha < \beta$.
a) Find the critical points, and classify them.
For b), c), d), find $\lim_{t \to \infty} x(t)$ based on the phase diagram.
b) $x(0) < \alpha$,   c) $\alpha < x(0) < \beta$,   d) $\beta < x(0)$.

1.7   Numerical methods: Euler's method

Note: 1 lecture, §2.4 in [EP], §8.1 in [BD]

As we mentioned before, unless $f(x, y)$ is of a special form, it is generally very hard if not impossible to get a nice formula for the solution of the problem

$$y' = f(x, y), \qquad y(x_0) = y_0.$$

If the equation can be solved in closed form, we should do that. But what if we have an equation that cannot be solved in closed form? What if we want to find the value of the solution at some particular x? Or perhaps we want to produce a graph of the solution to inspect the behavior. In this section we will learn about the basics of numerical approximation of solutions.

The simplest method for approximating a solution is *Euler's method*[*]. It works as follows: We take x_0 and compute the slope $k = f(x_0, y_0)$. The slope is the change in y per unit change in x. We follow the line for an interval of length h on the x axis. Hence if $y = y_0$ at x_0, then we will say that y_1 (the approximate value of y at $x_1 = x_0 + h$) will be $y_1 = y_0 + hk$. Rinse, repeat! That is, compute x_2 and y_2 using x_1 and y_1. For an example of the first two steps of the method see Figure 1.13.

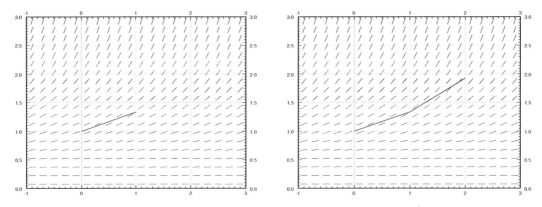

Figure 1.13: First two steps of Euler's method with $h = 1$ for the equation $y' = \frac{y^2}{3}$ with initial conditions $y(0) = 1$.

More abstractly, for any $i = 1, 2, 3, \ldots$, we compute

$$x_{i+1} = x_i + h, \qquad y_{i+1} = y_i + h\, f(x_i, y_i).$$

The line segments we get are an approximate graph of the solution. Generally it is not exactly the solution. See Figure 1.14 on the next page for the plot of the real solution and the approximation.

---

[*]Named after the Swiss mathematician Leonhard Paul Euler (1707–1783). Do note the correct pronunciation of the name sounds more like "oiler."

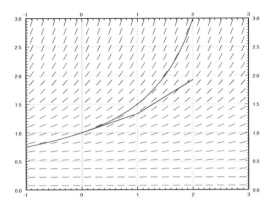

Figure 1.14: Two steps of Euler's method (step size 1) and the exact solution for the equation $y' = \frac{y^2}{3}$ with initial conditions $y(0) = 1$.

Let us see what happens with the equation $y' = y^2/3$, $y(0) = 1$. Let us try to approximate $y(2)$ using Euler's method. In Figures 1.13 and 1.14 we have graphically approximated $y(2)$ with step size 1. With step size 1 we have $y(2) \approx 1.926$. The real answer is 3. So we are approximately 1.074 off. Let us halve the step size. Computing y_4 with $h = 0.5$, we find that $y(2) \approx 2.209$, so an error of about 0.791. Table 1.1 on the following page gives the values computed for various parameters.

***Exercise* 1.7.1:** *Solve this equation exactly and show that $y(2) = 3$.*

The difference between the actual solution and the approximate solution is called the error. We usually talk about just the size of the error and we do not care much about its sign. The point is, we usually do not know the real solution, so we only have a vague understanding of the error. If we knew the error exactly ... what is the point of doing the approximation?

We notice that except for the first few times, every time we halved the interval the error approximately halved. This halving of the error is a general feature of Euler's method as it is a *first order method*. There exists an improved Euler method, see the exercises, that is a second order method reduces the error to approximately one quarter every time we halve the interval. The meaning of "second" order is the squaring in $1/4 = 1/2 \times 1/2 = (1/2)^2$.

To get the error to be within 0.1 of the answer we had to already do 64 steps. To get it to within 0.01 we would have to halve another three or four times, meaning doing 512 to 1024 steps. That is quite a bit to do by hand. The improved Euler method from the exercises should quarter the error every time we halve the interval, so we would have to approximately do half as many "halvings" to get the same error. This reduction can be a big deal. With 10 halvings (starting at $h = 1$) we have 1024 steps, whereas with 5 halvings we only have to do 32 steps, assuming that the error was comparable to start with. A computer may not care about this difference for a problem this simple, but suppose each step would take a second to compute (the function may be substantially more

h	Approximate $y(2)$	Error	$\frac{\text{Error}}{\text{Previous error}}$
1	1.92593	1.07407	
0.5	2.20861	0.79139	0.73681
0.25	2.47250	0.52751	0.66656
0.125	2.68034	0.31966	0.60599
0.0625	2.82040	0.17960	0.56184
0.03125	2.90412	0.09588	0.53385
0.015625	2.95035	0.04965	0.51779
0.0078125	2.97472	0.02528	0.50913

Table 1.1: Euler's method approximation of $y(2)$ where of $y' = y^2/3$, $y(0) = 1$.

difficult to compute than $y^2/3$). Then the difference is 32 seconds versus about 17 minutes. We are not being altogether fair, a second order method would probably double the time to do each step. Even so, it is 1 minute versus 17 minutes. Next, suppose that we have to repeat such a calculation for different parameters a thousand times. You get the idea.

Note that in practice we do not know how large the error is! How do we know what is the right step size? Well, essentially we keep halving the interval, and if we are lucky, we can estimate the error from a few of these calculations and the assumption that the error goes down by a factor of one half each time (if we are using standard Euler).

***Exercise* 1.7.2:** *In the table above, suppose you do not know the error. Take the approximate values of the function in the last two lines, assume that the error goes down by a factor of 2. Can you estimate the error in the last time from this? Does it (approximately) agree with the table? Now do it for the first two rows. Does this agree with the table?*

Let us talk a little bit more about the example $y' = y^2/3$, $y(0) = 1$. Suppose that instead of the value $y(2)$ we wish to find $y(3)$. The results of this effort are listed in Table 1.2 on the next page for successive halvings of h. What is going on here? Well, you should solve the equation exactly and you will notice that the solution does not exist at $x = 3$. In fact, the solution goes to infinity when you approach $x = 3$.

Another case where things go bad is if the solution oscillates wildly near some point. The solution may exist at all points, but even a much better numerical method than Euler would need an insanely small step size to approximate the solution with reasonable precision. And computers might not be able to easily handle such a small step size.

In real applications we would not use a simple method such as Euler's. The simplest method that would probably be used in a real application is the standard Runge-Kutta method (see exercises). That is a fourth order method, meaning that if we halve the interval, the error generally goes down by a factor of 16 (it is fourth order as $1/16 = 1/2 \times 1/2 \times 1/2 \times 1/2$).

h	Approximate $y(3)$
1	3.16232
0.5	4.54329
0.25	6.86079
0.125	10.80321
0.0625	17.59893
0.03125	29.46004
0.015625	50.40121
0.0078125	87.75769

Table 1.2: Attempts to use Euler's to approximate $y(3)$ where of $y' = y^2/3$, $y(0) = 1$.

Choosing the right method to use and the right step size can be very tricky. There are several competing factors to consider.

- Computational time: Each step takes computer time. Even if the function f is simple to compute, we do it many times over. Large step size means faster computation, but perhaps not the right precision.

- Roundoff errors: Computers only compute with a certain number of significant digits. Errors introduced by rounding numbers off during our computations become noticeable when the step size becomes too small relative to the quantities we are working with. So reducing step size may in fact make errors worse. There is a certain optimum step size such that the precision increases as we approach it, but then starts getting worse as we make our step size smaller still. Trouble is: this optimum may be hard to find.

- Stability: Certain equations may be numerically unstable. What may happen is that the numbers never seem to stabilize no matter how many times we halve the interval. We may need a ridiculously small interval size, which may not be practical due to roundoff errors or computational time considerations. Such problems are sometimes called *stiff*. In the worst case, the numerical computations might be giving us bogus numbers that look like a correct answer. Just because the numbers seem to have stabilized after successive halving, does not mean that we must have the right answer.

We have seen just the beginnings of the challenges that appear in real applications. Numerical approximation of solutions to differential equations is an active research area for engineers and mathematicians. For example, the general purpose method used for the ODE solver in Matlab and Octave (as of this writing) is a method that appeared in the literature only in the 1980s.

1.7.1 Exercises

Exercise 1.7.3: *Consider* $\dfrac{dx}{dt} = (2t - x)^2$, $x(0) = 2$. *Use Euler's method with step size* $h = 0.5$ *to approximate* $x(1)$.

Exercise 1.7.4: *Consider* $\dfrac{dx}{dt} = t - x$, $x(0) = 1$. *a) Use Euler's method with step sizes* $h = 1, 1/2, 1/4, 1/8$ *to approximate* $x(1)$. *b) Solve the equation exactly. c) Describe what happens to the errors for each h you used. That is, find the factor by which the error changed each time you halved the interval.*

Exercise 1.7.5: *Approximate the value of e by looking at the initial value problem* $y' = y$ *with* $y(0) = 1$ *and approximating* $y(1)$ *using Euler's method with a step size of* 0.2.

Exercise 1.7.6: *Example of numerical instability: Take* $y' = -5y$, $y(0) = 1$. *We know that the solution should decay to zero as x grows. Using Euler's method, start with* $h = 1$ *and compute* y_1, y_2, y_3, y_4 *to try to approximate* $y(4)$. *What happened? Now halve the interval. Keep halving the interval and approximating* $y(4)$ *until the numbers you are getting start to stabilize (that is, until they start going towards zero). Note: You might want to use a calculator.*

The simplest method used in practice is the *Runge-Kutta method*. Consider $\frac{dy}{dx} = f(x, y)$, $y(x_0) = y_0$, and a step size h. Everything is the same as in Euler's method, except the computation of y_{i+1} and x_{i+1}.

$$k_1 = f(x_i, y_i),$$
$$k_2 = f(x_i + {}^h\!/_2, y_i + k_1({}^h\!/_2)), \qquad x_{i+1} = x_i + h,$$
$$k_3 = f(x_i + {}^h\!/_2, y_i + k_2({}^h\!/_2)), \qquad y_{i+1} = y_i + \frac{k_1 + 2k_2 + 2k_3 + k_4}{6}\, h,$$
$$k_4 = f(x_i + h, y_i + k_3 h).$$

Exercise 1.7.7: *Consider* $\dfrac{dy}{dx} = yx^2$, $y(0) = 1$. *a) Use Runge-Kutta (see above) with step sizes* $h = 1$ *and* $h = 1/2$ *to approximate* $y(1)$. *b) Use Euler's method with* $h = 1$ *and* $h = 1/2$. *c) Solve exactly, find the exact value of* $y(1)$, *and compare.*

Exercise 1.7.101: *Let* $x' = \sin(xt)$, *and* $x(0) = 1$. *Approximate* $x(1)$ *using Euler's method with step sizes* $1, 0.5, 0.25$. *Use a calculator and compute up to 4 decimal digits.*

Exercise 1.7.102: *Let* $x' = 2t$, *and* $x(0) = 0$. *a) Approximate* $x(4)$ *using Euler's method with step sizes* $4, 2$, *and* 1. *b) Solve exactly, and compute the errors. c) Compute the factor by which the errors changed.*

Exercise 1.7.103: *Let* $x' = xe^{xt+1}$, *and* $x(0) = 0$. *a) Approximate* $x(4)$ *using Euler's method with step sizes* $4, 2$, *and* 1. *b) Guess an exact solution based on part a) and compute the errors.*

There is a simple way to improve Euler's method to make it a second order method by doing just one extra step. Consider $\frac{dy}{dx} = f(x, y)$, $y(x_0) = y_0$, and a step size h. What we do is to pretend we compute the next step as in Euler, that is, we start with (x_i, y_i), we compute a slope $k_1 = f(x_i, y_i)$, and then look at the point $(x_i + h, y_i + k_1 h)$. Instead of letting our new point be $(x_i + h, y_i + k_1 h)$, we compute the slope at that point, call it k_2, and then take the average of k_1 and k_2, hoping that the average is going to be closer to the actual slope on the interval from x_i to $x_i + h$. And we are correct, if we halve the step, the error should go down by a factor of $2^2 = 4$. To summarize, the setup is the same as for regular Euler, except the computation of y_{i+1} and x_{i+1}.

$$k_1 = f(x_i, y_i), \qquad\qquad x_{i+1} = x_i + h,$$

$$k_2 = f(x_i + h, y_i + k_1 h), \qquad\qquad y_{i+1} = y_i + \frac{k_1 + k_2}{2} h.$$

Exercise 1.7.104: *Consider* $\dfrac{dy}{dx} = x + y$, $y(0) = 1$. *a) Use the improved Euler's method with step sizes* $h = 1/4$ *and* $h = 1/8$ *to approximate* $y(1)$. *b) Use Euler's method with* $h = 1/4$ *and* $h = 1/8$. *c) Solve exactly, find the exact value of* $y(1)$. *d) Compute the errors, and the factors by which the errors changed.*

1.8   Exact equations

Note: 1–2 lectures, §1.6 in [EP], §2.6 in [BD]

Another type of equation that comes up quite often in physics and engineering is an *exact equation*. Suppose $F(x, y)$ is a function of two variables, which we call the *potential function*. The naming should suggest potential energy, or electric potential. Exact equations and potential functions appear when there is a conservation law at play, such as conservation of energy. Let us make up a simple example. Let

$$F(x, y) = x^2 + y^2.$$

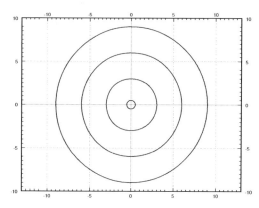

Figure 1.15: Solutions to $F(x, y) = x^2 + y^2 = C$ for various C.

We are interested in the lines of constant energy, that is lines where the energy is conserved; we want curves where $F(x, y) = C$, for some constant C. In our example, the curves $x^2 + y^2 = C$ are circles. See Figure 1.15.

We take the *total derivative* of F:

$$dF = \frac{\partial F}{\partial x} dx + \frac{\partial F}{\partial y} dy.$$

For convenience, we will make use of the notation of $F_x = \frac{\partial F}{\partial x}$ and $F_y = \frac{\partial F}{\partial y}$. In our example,

$$dF = 2x\,dx + 2y\,dy.$$

We apply the total derivative to $F(x, y) = C$, to find the differential equation $dF = 0$. The differential equation we obtain in such a way has the form

$$M\,dx + N\,dy = 0, \qquad \text{or} \qquad M + N\frac{dy}{dx} = 0.$$

An equation of this form is called *exact* if it was obtained as $dF = 0$ for some potential function F. In our toy example, we obtain the equation

$$2x\,dx + 2y\,dy = 0, \qquad \text{or} \qquad 2x + 2y\frac{dy}{dx} = 0.$$

Since we obtained this equation by differentiating $x^2 + y^2 = C$, the equation is exact. We often wish to solve for y in terms of x. In our example,

$$y = \pm \sqrt{C^2 - x^2}.$$

An interpretation of the setup is that at each point $\vec{v} = (M, N)$ is a vector in the plane, that is, a direction and a magnitude. As M and N are functions of (x, y), we have a *vector field*. The particular vector field $\vec{v}$ that comes from an exact equation is a so-called *conservative vector field*, that is, a vector field that comes with a potential function $F(x, y)$, such that

$$\vec{v} = \left(\frac{\partial F}{\partial x}, \frac{\partial F}{\partial y} \right).$$

Let γ be a path in the plane starting at (x_1, y_1) and ending at (x_2, y_2). If we think of $\vec{v}$ as force, then the work required to move along γ is

$$\int_\gamma \vec{v}(\vec{r}) \cdot d\vec{r} = \int_\gamma M \, dx + N \, dy = F(x_2, y_2) - F(x_1, y_1).$$

That is, the work done only depends on endpoints, that is where we start and where we end. For example, suppose F is gravitational potential. The derivative of F given by $\vec{v}$ is the gravitational force. What we are saying is that the work required to move a heavy box from the ground floor to the roof, only depends on the change in potential energy. That is, the work done is the same no matter what path we took; if we took the stairs or the elevator. Although if we took the elevator, the elevator is doing the work for us. The curves $F(x, y) = C$ are those where no work need be done, such as the heavy box sliding along without accelerating or breaking on a perfectly flat roof, on a cart with incredibly well oiled wheels.

An exact equation is a conservative vector field, and the implicit solution of this equation is the potential function.

1.8.1 Solving exact equations

Now you, the reader, should ask: Where did we solve a differential equation? Well, in applications we generally know M and N, but we do not know F. That is, we may have just started with $2x + 2y\frac{dy}{dx} = 0$, or perhaps even

$$x + y\frac{dy}{dx} = 0.$$

It is up to us to find some potential F that works. Many different F will work; adding a constant to F does not change the equation. Once we have a potential function F, the equation $F(x, y(x)) = C$ gives an implicit solution of the ODE.

Example 1.8.1: Let us find the general solution to $2x + 2y\frac{dy}{dx} = 0$. Forget we knew what F was.

If we know that this is an exact equation, we start looking for a potential function F. We have $M = 2x$ and $N = 2y$. If F exists, it must be such that $F_x(x, y) = 2x$. Integrate in the x variable to find

$$F(x, y) = x^2 + A(y), \tag{1.5}$$

for some function $A(y)$. The function A is the "constant of integration", though it is only constant as far as x is concerned, and may still depend on y. Now differentiate (1.5) in y and set it equal to N, which is what F_y is supposed to be:

$$2y = F_y(x, y) = A'(y).$$

Integrating, we find $A(y) = y^2$. We could add a constant of integration if we wanted to, but there is no need. We found $F(x, y) = x^2 + y^2$. Next for a constant C, we solve

$$F(x, y(x)) = C.$$

for y in terms of x. In this case, we obtain $y = \pm \sqrt{C^2 - x^2}$ as we did before.

***Exercise* 1.8.1**: *Why did we not need to add a constant of integration when integrating $A'(y) = 2y$? Add a constant of integration, say 3, and see what F you get. What is the difference from what we got above, and why does it not matter?*

The procedure, once we know that the equation is exact, is:

(i) Integrate $F_x = M$ in x resulting in $F(x, y) =$ something $+ A(y)$.

(ii) Differentiate this F in y, and set that equal to N, so that we may find $A(y)$ by integration.

The procedure can also be done by first integrating in y and then differentiating in x. Pretty easy huh? Let's try this again.

Example 1.8.2: Consider now $2x + y + xy\frac{dy}{dx} = 0$.

OK, so $M = 2x + y$ and $N = xy$. We try to proceed as before. Suppose F exists. Then $F_x(x, y) = 2x + y$. We integrate:

$$F(x, y) = x^2 + xy + A(y)$$

for some function $A(y)$. Differentiate in y and set equal to N:

$$N = xy = F_y(x, y) = x + A'(y).$$

But there is no way to satisfy this requirement! The function xy cannot be written as x plus a function of y. The equation is not exact; no potential function F exists.

Is there an easier way to check for the existence of F, other than failing in trying to find it? Turns out there is. Suppose $M = F_x$ and $N = F_y$. Then as long as the second derivatives are continuous,

$$\frac{\partial M}{\partial y} = \frac{\partial^2 F}{\partial y \partial x} = \frac{\partial^2 F}{\partial x \partial y} = \frac{\partial N}{\partial x}.$$

Let us state it as a theorem. Usually this is called the Poincarè Lemma[*].

Theorem 1.8.1 (Poincarè). *If M and N are continuously differentiable functions of (x, y), and $\frac{\partial M}{\partial y} = \frac{\partial N}{\partial x}$. Then near any point there is a function $F(x, y)$ such that $M = \frac{\partial F}{\partial x}$ and $N = \frac{\partial F}{\partial y}$.*

---

[*]Named for the French polymath Jules Henri Poincarè (1854–1912).

The theorem doesn't give us a global F defined everywhere. In general, we can only find the potential locally, near some initial point. By this time, we have come to expect this from differential equations.

Let us return to the example above where $M = 2x + y$ and $N = xy$. Notice $M_y = 1$ and $N_x = y$, which are clearly not equal. The equation is not exact.

Example 1.8.3: Solve

$$\frac{dy}{dx} = \frac{-2x - y}{x - 1}, \qquad y(0) = 1.$$

We write the equation as

$$(2x + y) + (x - 1)\frac{dy}{dx} = 0,$$

so $M = 2x + y$ and $N = x - 1$. Then

$$M_y = 1 = N_x.$$

The equation is exact. Integrating M in x, we find

$$F(x, y) = x^2 + xy + A(y).$$

Differentiating in y and setting to N, we find

$$x - 1 = x + A'(y).$$

So $A'(y) = -1$, and $A(y) = -y$ will work. Take $F(x, y) = x^2 + xy - y$. We wish to solve $x^2 + xy - y = C$. First let us find C. As $y(0) = 1$ then $F(0, 1) = C$. Therefore $0^2 + 0 \times 1 - 1 = C$, so $C = -1$. Now we solve $x^2 + xy - y = -1$ for y to get

$$y = \frac{-x^2 - 1}{x - 1}.$$

Example 1.8.4: Solve

$$-\frac{y}{x^2 + y^2}dx + \frac{x}{x^2 + y^2}dy = 0, \qquad y(1) = 2.$$

We leave to the reader to check that $M_y = N_x$.

This vector field (M, N) is not conservative if considered as a vector field of the entire plane minus the origin. The problem is that if a curve γ was a circle around the origin, say starting at $(1, 0)$ and ending at $(1, 0)$ going counterclockwise, then if F existed we would expect

$$0 = F(1, 0) - F(1, 0) = \int_\gamma F_x\, dx + F_y\, dy = \int_\gamma \frac{-y}{x^2 + y^2}\, dx + \frac{x}{x^2 + y^2}\, dy = 2\pi.$$

That is nonsense! We leave the computation of the path integral to the interested reader, or you can consult your multivariable calculus textbook. So there is no potential function F defined everywhere outside the origin $(0, 0)$.

If we think back to the theorem, it did not guarantee such a function anyway. It only guaranteed a potential function locally. As $y(1) = 2$ we start at the point $(1, 2)$. Considering $x > 0$ and integrating M in x or N in y we find

$$F(x, y) = \arctan(y/x).$$

So the implicit solution is $\arctan(y/x) = C$. Solving, $y = \tan(C)x$. That is, the solution is a straight line. Solving $y(1) = 2$ gives us that $\tan(C) = 2$, and so $y = 2x$ is the desired solution. See Figure 1.16, and note that the solution only exists for $x > 0$.

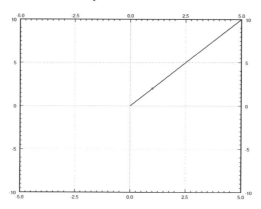

Figure 1.16: Solution to $-\frac{y}{x^2+y^2}dx + \frac{x}{x^2+y^2}dy = 0$, $y(1) = 2$, with initial point marked.

Example 1.8.5: Solve

$$x^2 + y^2 + 2y(x + 1)\frac{dy}{dx} = 0.$$

The reader should check that this equation is exact. Let $M = x^2 + y^2$ and $N = 2y(x + 1)$. We follow the procdure for exact equations

$$F(x, y) = \frac{1}{3}x^3 + xy^2 + A(y),$$

and

$$2y(x + 1) = 2xy + A'(y).$$

Therefore $A'(y) = 2y$ or $A(y) = y^2$ and $F(x, y) = \frac{1}{3}x^3 + xy^2 + y^2$. We try to solve $F(x, y) = C$. We easily solve for y^2 and then just take the square root:

$$y^2 = \frac{C - (1/3)x^3}{x + 1}, \qquad \text{so} \qquad y = \pm\sqrt{\frac{C - (1/3)x^3}{x + 1}}.$$

When $x = -1$, the term in front of $\frac{dy}{dx}$ vanishes. You can also see that our solution is not valid in that case. However, one could in that case try to solve for x in terms of y starting from the implicit solution $\frac{1}{3}x^3 + xy^2 + y^2 = C$. In this case the solution is somewhat messy and we leave it as implicit.

1.8.2 Integrating factors

Sometimes an equation $M\,dx + N\,dy = 0$ is not exact, but it can be made exact by multiplying with a function $u(x, y)$. That is, perhaps for some nonzero function $u(x, y)$,

$$u(x, y)M(x, y)\,dx + u(x, y)N(x, y)\,dy = 0$$

is exact. Any solution to this new equation is also a solution to $M\,dx + N\,dy = 0$.

In fact, a linear equation

$$\frac{dy}{dx} + p(x)y = f(x), \qquad \text{or} \qquad (p(x)y - f(x))\,dx + dy = 0$$

is always such an equation. Let $r(x) = e^{\int p(x)\,dx}$ be the integrating factor for a linear equation. Multiply the equation by $r(x)$ and write it in the form of $M + N\frac{dy}{dx} = 0$.

$$r(x)p(x)y - r(x)f(x) + r(x)\frac{dy}{dx} = 0.$$

Then $M = r(x)p(x)y - r(x)f(x)$, so $M_y = r(x)p(x)$, while $N = r(x)$, so $N_x = r'(x) = r(x)p(x)$. In other words, we have an exact equation. So integrating factors for linear functions are just a special case of integrating factors for exact equations.

But how do we find the integrating factor u? Well, given an equation

$$M\,dx + N\,dy = 0,$$

u should be a function such that

$$\frac{\partial}{\partial y}[uM] = u_y M + uM_y = \frac{\partial}{\partial x}[uN] = u_x N + uN_x.$$

Therefore,

$$(M_y - N_x)u = u_x N - u_y M.$$

At first it may seem we replaced one differential equation by another one. True, but hope is not lost.

A strategy that often works is to look for a u that is a function of x alone, or a function of y alone. If u is a function of x alone, that is $u(x)$, then we write $u'(x)$ instead of u_x, and u_y is just zero. Then

$$\frac{M_y - N_x}{N}u = u'.$$

In particular, $\frac{M_y - N_x}{N}$ ought to be a function of x alone (not depend on y). If so, then we have a linear equation

$$u' - \frac{M_y - N_x}{N}u = 0.$$

Letting $P(x) = \frac{M_y - N_x}{N}$, we solve using the standard integrating factor method, to find $u(x) = Ce^{\int P(x)\,dx}$. The constant in the solution is not relevant, we need any nonzero solution, we take $C = 1$. So $u(x) = e^{\int P(x)\,dx}$ is the integrating factor.

Similarly we could try a function of the form $u(y)$. Then

$$\frac{M_y - N_x}{M} u = -u'.$$

In particular $\frac{M_y - N_x}{M}$ ought to be a function of y alone. If so, then we have a linear equation

$$u' + \frac{M_y - N_x}{M} u = 0.$$

Letting $Q(y) = \frac{M_y - N_x}{M}$ we find $u(y) = Ce^{-\int Q(y)\,dy}$, and we can take $C = 1$. So $u(y) = e^{-\int Q(y)\,dy}$ is the integrating factor.

Example 1.8.6: Solve

$$\frac{x^2 + y^2}{x + 1} + 2y\frac{dy}{dx} = 0.$$

Let $M = \frac{x^2 + y^2}{x+1}$ and $N = 2y$. Compute

$$M_y - N_x = \frac{2y}{x + 1} - 0 = \frac{2y}{x + 1}.$$

As this is not zero, the equation is not exact. We notice

$$P(x) = \frac{M_y - N_x}{N} = \frac{2y}{x + 1}\frac{1}{2y} = \frac{1}{x + 1}$$

is a function of x alone. We compute the integrating factor

$$e^{\int P(x)\,dx} = e^{\ln(x+1)} = x + 1.$$

We multiply our given equation by $(x + 1)$ to obtain

$$x^2 + y^2 + 2y(x + 1)\frac{dy}{dx} = 0,$$

which is an exact equation that we solved in Example 1.8.5. The solution was

$$y = \pm\sqrt{\frac{C - (^1/_3)x^3}{x + 1}}.$$

Example 1.8.7: Solve

$$y^2 + (xy + 1)\frac{dy}{dx} = 0.$$

First compute

$$M_y - N_x = 2y - y = y.$$

As this is not zero, the equation is not exact. We observe

$$Q(y) = \frac{M_y - N_x}{M} = \frac{y}{y^2} = \frac{1}{y}$$

is a function of y alone. We compute the integrating factor

$$e^{-\int Q(y)\,dy} = e^{-\ln y} = \frac{1}{y}.$$

Therefore we look at the exact equation

$$y + \frac{xy + 1}{y}\frac{dy}{dx} = 0.$$

The reader should double check that this equation is exact. We follow the procdure for exact equations

$$F(x, y) = xy + A(y),$$

and

$$\frac{xy + 1}{y} = x + \frac{1}{y} = x + A'(y). \tag{1.6}$$

Consequently $A'(y) = \frac{1}{y}$ or $A(y) = \ln y$. Thus $F(x, y) = xy + \ln y$. It is not possible to solve $F(x, y) = C$ for y in terms of elementary functions, so let us be content with the implicit solution:

$$xy + \ln y = C.$$

We are looking for the general solution and we divided by y above. We should check what happens when $y = 0$, as the equation itself makes perfect sense in that case. We plug in $y = 0$ to find the equation is satisfied. So $y = 0$ is also a solution.

1.8.3   Exercises

Exercise **1.8.2:** *Solve the following exact equations, implicit general solutions will suffice:*
a) $(2xy + x^2)\,dx + (x^2 + y^2 + 1)\,dy = 0$
b) $x^5 + y^5\frac{dy}{dx} = 0$
c) $e^x + y^3 + 3xy^2\frac{dy}{dx} = 0$
d) $(x + y)\cos(x) + \sin(x) + \sin(x)y' = 0$

Exercise 1.8.3: *Find the integrating factor for the following equations making them into exact equations:*
a) $e^{xy} dx + \frac{y}{x} e^{xy} dy = 0$
b) $\frac{e^x + y^3}{y^2} dx + 3x dy = 0$
c) $4(y^2 + x) dx + \frac{2x + 2y^2}{y} dy = 0$
d) $2 \sin(y) dx + x \cos(y) dy = 0$

Exercise 1.8.4: *Suppose you have an equation of the form:* $f(x) + g(y)\frac{dy}{dx} = 0$.
a) Show it is exact.
b) Find the form of the potential function in terms of f and g.

Exercise 1.8.5: *Suppose that we have the equation* $f(x) dx - dy = 0$.
a) Is this equation exact?
b) Find the general solution using a definite integral.

Exercise 1.8.6: *Find the potential function* $F(x, y)$ *of the exact equation* $\frac{1+xy}{x} dx + (1/y + x) dy = 0$ *in two different ways.*
a) Integrate M in terms of x and then differentiate in y and set to N.
b) Integrate N in terms of y and then differentiate in x and set to M.

Exercise 1.8.7: *A function* $u(x, y)$ *is said to be* harmonic function *if* $u_{xx} + u_{yy} = 0$.
a) Show that $-u_y dx + u_x dy = 0$ *is an exact equation. Therefore there exists (at least locally) the so-called* harmonic conjugate *function* $v(x, y)$ *such that* $v_x = -u_y$ *and* $v_y = u_x$.
Verify that the following u are harmonic and find the corresponding harmonic conjugates v:
b) $u = 2xy$
c) $u = e^x \cos y$
d) $u = x^3 - 3xy^2$

Exercise 1.8.101: *Solve the following exact equations, implicit general solutions will suffice:*
a) $\cos(x) + ye^{xy} + xe^{xy}y' = 0$
b) $(2x + y) dx + (x - 4y) dy = 0$
c) $e^x + e^y \frac{dy}{dx} = 0$
d) $(3x^2 + 3y) dx + (3y^2 + 3x) dy = 0$

Exercise 1.8.102: *Find the integrating factor for the following equations making them into exact equations:*
a) $\frac{1}{y} dx + 3y dy = 0$
b) $dx - e^{-x-y} dy = 0$
c) $\left(\frac{\cos(x)}{y^2} + \frac{1}{y}\right) dx + \frac{1}{y^2} dy = 0$
d) $\left(2y + \frac{y^2}{x}\right) dx + (2y + x) dy = 0$

Exercise 1.8.103: *a) Show that every separable equation* $y' = f(x)g(y)$ *can be written as an exact equation, and verify that it is indeed exact. b) Using this rewrite* $y' = xy$ *as an exact equation, solve it and verify that the solution is the same as it was in Example 1.3.1.*

Chapter 2

Higher order linear ODEs

2.1 Second order linear ODEs

Note: less than 1 lecture, first part of §3.1 in [EP], parts of §3.1 and §3.2 in [BD]

Let us consider the general *second order linear differential equation*

$$A(x)y'' + B(x)y' + C(x)y = F(x).$$

We usually divide through by $A(x)$ to get

$$y'' + p(x)y' + q(x)y = f(x), \tag{2.1}$$

where $p(x) = {}^{B(x)}/_{A(x)}$, $q(x) = {}^{C(x)}/_{A(x)}$, and $f(x) = {}^{F(x)}/_{A(x)}$. The word *linear* means that the equation contains no powers nor functions of y, y', and y''.

In the special case when $f(x) = 0$, we have a so-called *homogeneous* equation

$$y'' + p(x)y' + q(x)y = 0. \tag{2.2}$$

We have already seen some second order linear homogeneous equations.

$$y'' + k^2 y = 0 \qquad \text{Two solutions are:} \quad y_1 = \cos(kx), \quad y_2 = \sin(kx).$$
$$y'' - k^2 y = 0 \qquad \text{Two solutions are:} \quad y_1 = e^{kx}, \quad y_2 = e^{-kx}.$$

If we know two solutions of a linear homogeneous equation, we know a lot more of them.

Theorem 2.1.1 (Superposition). *Suppose y_1 and y_2 are two solutions of the homogeneous equation* (2.2). *Then*

$$y(x) = C_1 y_1(x) + C_2 y_2(x),$$

also solves (2.2) *for arbitrary constants C_1 and C_2.*

That is, we can add solutions together and multiply them by constants to obtain new and different solutions. We call the expression $C_1y_1 + C_2y_2$ a *linear combination* of y_1 and y_2. Let us prove this theorem; the proof is very enlightening and illustrates how linear equations work.

Proof: Let $y = C_1y_1 + C_2y_2$. Then

$$
\begin{aligned}
y'' + py' + qy &= (C_1y_1 + C_2y_2)'' + p(C_1y_1 + C_2y_2)' + q(C_1y_1 + C_2y_2) \\
&= C_1y_1'' + C_2y_2'' + C_1py_1' + C_2py_2' + C_1qy_1 + C_2qy_2 \\
&= C_1(y_1'' + py_1' + qy_1) + C_2(y_2'' + py_2' + qy_2) \\
&= C_1 \cdot 0 + C_2 \cdot 0 = 0.
\end{aligned}
$$

The proof becomes even simpler to state if we use the operator notation. An *operator* is an object that eats functions and spits out functions (kind of like what a function is, but a function eats numbers and spits out numbers). Define the operator L by

$$ Ly = y'' + py' + qy. $$

The differential equation now becomes $Ly = 0$. The operator (and the equation) L being *linear* means that $L(C_1y_1 + C_2y_2) = C_1Ly_1 + C_2Ly_2$. The proof above becomes

$$ Ly = L(C_1y_1 + C_2y_2) = C_1Ly_1 + C_2Ly_2 = C_1 \cdot 0 + C_2 \cdot 0 = 0. $$

Two different solutions to the second equation $y'' - k^2y = 0$ are $y_1 = \cosh(kx)$ and $y_2 = \sinh(kx)$. Let us remind ourselves of the definition, $\cosh x = \frac{e^x + e^{-x}}{2}$ and $\sinh x = \frac{e^x - e^{-x}}{2}$. Therefore, these are solutions by superposition as they are linear combinations of the two exponential solutions.

The functions sinh and cosh are sometimes more convenient to use than the exponential. Let us review some of their properties.

$$ \cosh 0 = 1 \qquad\qquad\qquad \sinh 0 = 0 $$

$$ \frac{d}{dx} \cosh x = \sinh x \qquad\qquad \frac{d}{dx} \sinh x = \cosh x $$

$$ \cosh^2 x - \sinh^2 x = 1 $$

Exercise 2.1.1: *Derive these properties using the definitions of* sinh *and* cosh *in terms of exponentials.*

Linear equations have nice and simple answers to the existence and uniqueness question.

Theorem 2.1.2 (Existence and uniqueness). *Suppose p, q, f are continuous functions on some interval I, a is a number in I, and a, b_0, b_1 are constants. The equation*

$$ y'' + p(x)y' + q(x)y = f(x), $$

has exactly one solution $y(x)$ defined on the same interval I satisfying the initial conditions

$$ y(a) = b_0, \qquad y'(a) = b_1. $$

For example, the equation $y'' + k^2 y = 0$ with $y(0) = b_0$ and $y'(0) = b_1$ has the solution

$$y(x) = b_0 \cos(kx) + \frac{b_1}{k} \sin(kx).$$

The equation $y'' - k^2 y = 0$ with $y(0) = b_0$ and $y'(0) = b_1$ has the solution

$$y(x) = b_0 \cosh(kx) + \frac{b_1}{k} \sinh(kx).$$

Using cosh and sinh in this solution allows us to solve for the initial conditions in a cleaner way than if we have used the exponentials.

The initial conditions for a second order ODE consist of two equations. Common sense tells us that if we have two arbitrary constants and two equations, then we should be able to solve for the constants and find a solution to the differential equation satisfying the initial conditions.

Question: Suppose we find two different solutions y_1 and y_2 to the homogeneous equation (2.2). Can every solution be written (using superposition) in the form $y = C_1 y_1 + C_2 y_2$?

Answer is affirmative! Provided that y_1 and y_2 are different enough in the following sense. We say y_1 and y_2 are *linearly independent* if one is not a constant multiple of the other.

Theorem 2.1.3. *Let p, q be continuous functions. Let y_1 and y_2 be two linearly independent solutions to the homogeneous equation (2.2). Then every other solution is of the form*

$$y = C_1 y_1 + C_2 y_2.$$

That is, $y = C_1 y_1 + C_2 y_2$ is the general solution.

For example, we found the solutions $y_1 = \sin x$ and $y_2 = \cos x$ for the equation $y'' + y = 0$. It is not hard to see that sine and cosine are not constant multiples of each other. If $\sin x = A \cos x$ for some constant A, we let $x = 0$ and this would imply $A = 0$. But then $\sin x = 0$ for all x, which is preposterous. So y_1 and y_2 are linearly independent. Hence

$$y = C_1 \cos x + C_2 \sin x$$

is the general solution to $y'' + y = 0$.

We will study the solution of nonhomogeneous equations in § 2.5. We will first focus on finding general solutions to homogeneous equations.

2.1.1 Exercises

***Exercise* 2.1.2:** *Show that $y = e^x$ and $y = e^{2x}$ are linearly independent.*

***Exercise* 2.1.3:** *Take $y'' + 5y = 10x + 5$. Find (guess!) a solution.*

Exercise 2.1.4: *Prove the superposition principle for nonhomogeneous equations. Suppose that y_1 is a solution to $Ly_1 = f(x)$ and y_2 is a solution to $Ly_2 = g(x)$ (same linear operator L). Show that $y = y_1 + y_2$ solves $Ly = f(x) + g(x)$.*

Exercise 2.1.5: *For the equation $x^2 y'' - xy' = 0$, find two solutions, show that they are linearly independent and find the general solution. Hint: Try $y = x^r$.*

Equations of the form $ax^2 y'' + bxy' + cy = 0$ are called *Euler's equations* or *Cauchy-Euler equations*. They are solved by trying $y = x^r$ and solving for r (assume that $x \geq 0$ for simplicity).

Exercise 2.1.6: *Suppose that $(b - a)^2 - 4ac > 0$. a) Find a formula for the general solution of $ax^2 y'' + bxy' + cy = 0$. Hint: Try $y = x^r$ and find a formula for r. b) What happens when $(b - a)^2 - 4ac = 0$ or $(b - a)^2 - 4ac < 0$?*

We will revisit the case when $(b - a)^2 - 4ac < 0$ later.

Exercise 2.1.7: *Same equation as in Exercise 2.1.6. Suppose $(b - a)^2 - 4ac = 0$. Find a formula for the general solution of $ax^2 y'' + bxy' + cy = 0$. Hint: Try $y = x^r \ln x$ for the second solution.*

If you have one solution to a second order linear homogeneous equation you can find another one. This is the *reduction of order method*.

Exercise 2.1.8 (reduction of order)**:** *Suppose y_1 is a solution to $y'' + p(x)y' + q(x)y = 0$. Show that*

$$y_2(x) = y_1(x) \int \frac{e^{-\int p(x)\,dx}}{(y_1(x))^2}\, dx$$

is also a solution.

Note: If you wish to come up with the formula for reduction of order yourself, start by trying $y_2(x) = y_1(x)v(x)$. Then plug y_2 into the equation, use the fact that y_1 is a solution, substitute $w = v'$, and you have a first order linear equation in w. Solve for w and then for v. When solving for w, make sure to include a constant of integration. Let us solve some famous equations using the method.

Exercise 2.1.9 (Chebyshev's equation of order 1)**:** *Take $(1 - x^2)y'' - xy' + y = 0$. a) Show that $y = x$ is a solution. b) Use reduction of order to find a second linearly independent solution. c) Write down the general solution.*

Exercise 2.1.10 (Hermite's equation of order 2)**:** *Take $y'' - 2xy' + 4y = 0$. a) Show that $y = 1 - 2x^2$ is a solution. b) Use reduction of order to find a second linearly independent solution. c) Write down the general solution.*

Exercise 2.1.101: *Are $\sin(x)$ and e^x linearly independent? Justify.*

Exercise 2.1.102: *Are e^x and e^{x+2} linearly independent? Justify.*

Exercise **2.1.103**: *Guess a solution to* $y'' + y' + y = 5$.

Exercise **2.1.104**: *Find the general solution to* $xy'' + y' = 0$. *Hint: Notice that it is a first order ODE in* y'.

Exercise **2.1.105**: *Write down an equation (guess) for which we have the solutions* e^x *and* e^{2x}. *Hint: Try an equation of the form* $y'' + Ay' + By = 0$ *for constants* A *and* B, *plug in both* e^x *and* e^{2x} *and solve for* A *and* B.

2.2   Constant coefficient second order linear ODEs

Note: more than 1 lecture, second part of §3.1 in [EP], §3.1 in [BD]

Suppose we have the problem

$$y'' - 6y' + 8y = 0, \qquad y(0) = -2, \qquad y'(0) = 6.$$

This is a second order linear homogeneous equation with constant coefficients. *Constant coefficients* means that the functions in front of y'', y', and y are constants, not depending on x.

To guess a solution, think of a function that you know stays essentially the same when we differentiate it, so that we can take the function and its derivatives, add some multiples of these together, and end up with zero.

Let us try a solution of the form $y = e^{rx}$. Then $y' = re^{rx}$ and $y'' = r^2 e^{rx}$. Plug in to get

$$y'' - 6y' + 8y = 0,$$
$$r^2 e^{rx} - 6re^{rx} + 8e^{rx} = 0,$$
$$r^2 - 6r + 8 = 0 \qquad \text{(divide through by } e^{rx}),$$
$$(r - 2)(r - 4) = 0.$$

Hence, if $r = 2$ or $r = 4$, then e^{rx} is a solution. So let $y_1 = e^{2x}$ and $y_2 = e^{4x}$.

Exercise 2.2.1: Check that y_1 and y_2 are solutions.

The functions e^{2x} and e^{4x} are linearly independent. If they were not linearly independent we could write $e^{4x} = Ce^{2x}$ for some constant C, implying that $e^{2x} = C$ for all x, which is clearly not possible. Hence, we can write the general solution as

$$y = C_1 e^{2x} + C_2 e^{4x}.$$

We need to solve for C_1 and C_2. To apply the initial conditions we first find $y' = 2C_1 e^{2x} + 4C_2 e^{4x}$. We plug in $x = 0$ and solve.

$$-2 = y(0) = C_1 + C_2,$$
$$6 = y'(0) = 2C_1 + 4C_2.$$

Either apply some matrix algebra, or just solve these by high school math. For example, divide the second equation by 2 to obtain $3 = C_1 + 2C_2$, and subtract the two equations to get $5 = C_2$. Then $C_1 = -7$ as $-2 = C_1 + 5$. Hence, the solution we are looking for is

$$y = -7e^{2x} + 5e^{4x}.$$

Let us generalize this example into a method. Suppose that we have an equation

$$ay'' + by' + cy = 0, \qquad\qquad\qquad (2.3)$$

where a, b, c are constants. Try the solution $y = e^{rx}$ to obtain

$$ar^2 e^{rx} + br e^{rx} + c e^{rx} = 0,$$
$$ar^2 + br + c = 0.$$

The equation $ar^2 + br + c = 0$ is called the *characteristic equation* of the ODE. Solve for the r by using the quadratic formula.

$$r_1, r_2 = \frac{-b \pm \sqrt{b^2 - 4ac}}{2a}.$$

Therefore, we have $e^{r_1 x}$ and $e^{r_2 x}$ as solutions. There is still a difficulty if $r_1 = r_2$, but it is not hard to overcome.

Theorem 2.2.1. *Suppose that r_1 and r_2 are the roots of the characteristic equation.*

(i) *If r_1 and r_2 are distinct and real (when $b^2 - 4ac > 0$), then (2.3) has the general solution*

$$y = C_1 e^{r_1 x} + C_2 e^{r_2 x}.$$

(ii) *If $r_1 = r_2$ (happens when $b^2 - 4ac = 0$), then (2.3) has the general solution*

$$y = (C_1 + C_2 x) e^{r_1 x}.$$

For another example of the first case, take the equation $y'' - k^2 y = 0$. Here the characteristic equation is $r^2 - k^2 = 0$ or $(r - k)(r + k) = 0$. Consequently, e^{-kx} and e^{kx} are the two linearly independent solutions.

Example 2.2.1: Find the general solution of

$$y'' - 8y' + 16y = 0.$$

The characteristic equation is $r^2 - 8r + 16 = (r - 4)^2 = 0$. The equation has a double root $r_1 = r_2 = 4$. The general solution is, therefore,

$$y = (C_1 + C_2 x) e^{4x} = C_1 e^{4x} + C_2 x e^{4x}.$$

Exercise 2.2.2: Check that e^{4x} and xe^{4x} are linearly independent.

That e^{4x} solves the equation is clear. If xe^{4x} solves the equation, then we know we are done. Let us compute $y' = e^{4x} + 4xe^{4x}$ and $y'' = 8e^{4x} + 16xe^{4x}$. Plug in

$$y'' - 8y' + 16y = 8e^{4x} + 16xe^{4x} - 8(e^{4x} + 4xe^{4x}) + 16xe^{4x} = 0.$$

We should note that in practice, doubled root rarely happens. If coefficients are picked truly randomly we are very unlikely to get a doubled root.

Let us give a short proof for why the solution xe^{rx} works when the root is doubled. This case is really a limiting case of when the two roots are distinct and very close. Note that $\frac{e^{r_2 x} - e^{r_1 x}}{r_2 - r_1}$ is a solution when the roots are distinct. When we take the limit as r_1 goes to r_2, we are really taking the derivative of e^{rx} using r as the variable. Therefore, the limit is xe^{rx}, and hence this is a solution in the doubled root case.

2.2.1   Complex numbers and Euler's formula

It may happen that a polynomial has some complex roots. For example, the equation $r^2 + 1 = 0$ has no real roots, but it does have two complex roots. Here we review some properties of complex numbers.

Complex numbers may seem a strange concept, especially because of the terminology. There is nothing imaginary or really complicated about complex numbers. A complex number is simply a pair of real numbers, (a, b). We can think of a complex number as a point in the plane. We add complex numbers in the straightforward way: $(a, b) + (c, d) = (a + c, b + d)$. We define multiplication by

$$(a, b) \times (c, d) \stackrel{\text{def}}{=} (ac - bd, ad + bc).$$

It turns out that with this multiplication rule, all the standard properties of arithmetic hold. Further, and most importantly $(0, 1) \times (0, 1) = (-1, 0)$.

Generally we just write (a, b) as $a + ib$, and we treat i as if it were an unknown. When b is zero, then $(a, 0)$ is just the number a. We do arithmetic with complex numbers just as we would with polynomials. The property we just mentioned becomes $i^2 = -1$. So whenever we see i^2, we replace it by -1. The numbers i and $-i$ are the two roots of $r^2 + 1 = 0$.

Note that engineers often use the letter j instead of i for the square root of -1. We will use the mathematicians' convention and use i.

Exercise 2.2.3: Make sure you understand (that you can justify) the following identities:

* $i^2 = -1,\ i^3 = -i,\ i^4 = 1,$

* $\dfrac{1}{i} = -i,$

* $(3 - 7i)(-2 - 9i) = \cdots = -69 - 13i,$

* $(3 - 2i)(3 + 2i) = 3^2 - (2i)^2 = 3^2 + 2^2 = 13,$

* $\dfrac{1}{3-2i} = \dfrac{1}{3-2i}\dfrac{3+2i}{3+2i} = \dfrac{3+2i}{13} = \dfrac{3}{13} + \dfrac{2}{13}i.$

We also define the exponential e^{a+ib} of a complex number. We do this by writing down the Taylor series and plugging in the complex number. Because most properties of the exponential can be proved by looking at the Taylor series, these properties still hold for the complex exponential. For example the very important property: $e^{x+y} = e^x e^y$. This means that $e^{a+ib} = e^a e^{ib}$. Hence if we can compute e^{ib}, we can compute e^{a+ib}. For e^{ib} we use the so-called *Euler's formula*.

Theorem 2.2.2 (Euler's formula).

$$\boxed{e^{i\theta} = \cos\theta + i\sin\theta \qquad and \qquad e^{-i\theta} = \cos\theta - i\sin\theta.}$$

In other words, $e^{a+ib} = e^a(\cos(b) + i\sin(b)) = e^a\cos(b) + ie^a\sin(b)$.

Exercise 2.2.4: *Using Euler's formula, check the identities:*

$$\cos\theta = \frac{e^{i\theta} + e^{-i\theta}}{2} \qquad and \qquad \sin\theta = \frac{e^{i\theta} - e^{-i\theta}}{2i}.$$

Exercise 2.2.5: *Double angle identities: Start with $e^{i(2\theta)} = (e^{i\theta})^2$. Use Euler on each side and deduce:*

$$\cos(2\theta) = \cos^2\theta - \sin^2\theta \qquad and \qquad \sin(2\theta) = 2\sin\theta\cos\theta.$$

For a complex number $a + ib$ we call a the *real part* and b the *imaginary part* of the number. Often the following notation is used,

$$\text{Re}(a + ib) = a \qquad and \qquad \text{Im}(a + ib) = b.$$

2.2.2 Complex roots

Suppose the equation $ay'' + by' + cy = 0$ has the characteristic equation $ar^2 + br + c = 0$ that has complex roots. By the quadratic formula, the roots are $\frac{-b \pm \sqrt{b^2 - 4ac}}{2a}$. These roots are complex if $b^2 - 4ac < 0$. In this case the roots are

$$r_1, r_2 = \frac{-b}{2a} \pm i\frac{\sqrt{4ac - b^2}}{2a}.$$

As you can see, we always get a pair of roots of the form $\alpha \pm i\beta$. In this case we can still write the solution as

$$y = C_1 e^{(\alpha + i\beta)x} + C_2 e^{(\alpha - i\beta)x}.$$

However, the exponential is now complex valued. We would need to allow C_1 and C_2 to be complex numbers to obtain a real-valued solution (which is what we are after). While there is nothing particularly wrong with this approach, it can make calculations harder and it is generally preferred to find two real-valued solutions.

Here we can use Euler's formula. Let

$$y_1 = e^{(\alpha + i\beta)x} \qquad and \qquad y_2 = e^{(\alpha - i\beta)x}.$$

Then

$$y_1 = e^{\alpha x}\cos(\beta x) + ie^{\alpha x}\sin(\beta x),$$
$$y_2 = e^{\alpha x}\cos(\beta x) - ie^{\alpha x}\sin(\beta x).$$

Linear combinations of solutions are also solutions. Hence,

$$y_3 = \frac{y_1 + y_2}{2} = e^{\alpha x}\cos(\beta x),$$
$$y_4 = \frac{y_1 - y_2}{2i} = e^{\alpha x}\sin(\beta x),$$

are also solutions. Furthermore, they are real-valued. It is not hard to see that they are linearly independent (not multiples of each other). Therefore, we have the following theorem.

Theorem 2.2.3. *Take the equation*

$$ay'' + by' + cy = 0.$$

If the characteristic equation has the roots $\alpha \pm i\beta$ (when $b^2 - 4ac < 0$), then the general solution is

$$y = C_1 e^{\alpha x} \cos(\beta x) + C_2 e^{\alpha x} \sin(\beta x).$$

Example 2.2.2: Find the general solution of $y'' + k^2 y = 0$, for a constant $k > 0$.

The characteristic equation is $r^2 + k^2 = 0$. Therefore, the roots are $r = \pm ik$, and by the theorem, we have the general solution

$$y = C_1 \cos(kx) + C_2 \sin(kx).$$

Example 2.2.3: Find the solution of $y'' - 6y' + 13y = 0$, $y(0) = 0$, $y'(0) = 10$.

The characteristic equation is $r^2 - 6r + 13 = 0$. By completing the square we get $(r - 3)^2 + 2^2 = 0$ and hence the roots are $r = 3 \pm 2i$. By the theorem we have the general solution

$$y = C_1 e^{3x} \cos(2x) + C_2 e^{3x} \sin(2x).$$

To find the solution satisfying the initial conditions, we first plug in zero to get

$$0 = y(0) = C_1 e^0 \cos 0 + C_2 e^0 \sin 0 = C_1.$$

Hence $C_1 = 0$ and $y = C_2 e^{3x} \sin(2x)$. We differentiate

$$y' = 3C_2 e^{3x} \sin(2x) + 2C_2 e^{3x} \cos(2x).$$

We again plug in the initial condition and obtain $10 = y'(0) = 2C_2$, or $C_2 = 5$. Hence the solution we are seeking is

$$y = 5e^{3x} \sin(2x).$$

2.2.3   Exercises

Exercise 2.2.6: *Find the general solution of $2y'' + 2y' - 4y = 0$.*

Exercise 2.2.7: *Find the general solution of $y'' + 9y' - 10y = 0$.*

Exercise 2.2.8: *Solve $y'' - 8y' + 16y = 0$ for $y(0) = 2$, $y'(0) = 0$.*

Exercise 2.2.9: *Solve $y'' + 9y' = 0$ for $y(0) = 1$, $y'(0) = 1$.*

Exercise 2.2.10: *Find the general solution of $2y'' + 50y = 0$.*

Exercise 2.2.11: *Find the general solution of $y'' + 6y' + 13y = 0$.*

Exercise 2.2.12: *Find the general solution of $y'' = 0$ using the methods of this section.*

Exercise 2.2.13: *The method of this section applies to equations of other orders than two. We will see higher orders later. Try to solve the first order equation $2y' + 3y = 0$ using the methods of this section.*

Exercise 2.2.14: *Let us revisit the Cauchy-Euler equations of Exercise 2.1.6 on page 66. Suppose now that $(b - a)^2 - 4ac < 0$. Find a formula for the general solution of $ax^2y'' + bxy' + cy = 0$. Hint: Note that $x^r = e^{r \ln x}$.*

Exercise 2.2.15: *Find the solution to $y'' - (2\alpha)y' + \alpha^2 y = 0$, $y(0) = a$, $y'(0) = b$, where α, a, and b are real numbers.*

Exercise 2.2.16: *Construct an equation such that $y = C_1 e^{-2x} \cos(3x) + C_2 e^{-2x} \sin(3x)$ is the general solution.*

Exercise 2.2.101: *Find the general solution to $y'' + 4y' + 2y = 0$.*

Exercise 2.2.102: *Find the general solution to $y'' - 6y' + 9y = 0$.*

Exercise 2.2.103: *Find the solution to $2y'' + y' + y = 0$, $y(0) = 1$, $y'(0) = -2$.*

Exercise 2.2.104: *Find the solution to $2y'' + y' - 3y = 0$, $y(0) = a$, $y'(0) = b$.*

Exercise 2.2.105: *Find the solution to $z''(t) = -2z'(t) - 2z(t)$, $z(0) = 2$, $z'(0) = -2$.*

Exercise 2.2.106: *Find the solution to $y'' - (\alpha + \beta)y' + \alpha\beta y = 0$, $y(0) = a$, $y'(0) = b$, where α, β, a, and b are real numbers, and $\alpha \neq \beta$.*

Exercise 2.2.107: *Construct an equation such that $y = C_1 e^{3x} + C_2 e^{-2x}$ is the general solution.*

2.3   Higher order linear ODEs

Note: somewhat more than 1 lecture, §3.2 and §3.3 in [EP], §4.1 and §4.2 in [BD]

We briefly study higher order equations. Equations appearing in applications tend to be second order. Higher order equations do appear from time to time, but generally the world around us is "second order."

The basic results about linear ODEs of higher order are essentially the same as for second order equations, with 2 replaced by n. The important concept of linear independence is somewhat more complicated when more than two functions are involved. For higher order constant coefficient ODEs, the methods developed are also somewhat harder to apply, but we will not dwell on these complications. It is also possible to use the methods for systems of linear equations from chapter 3 to solve higher order constant coefficient equations.

Let us start with a general homogeneous linear equation

$$y^{(n)} + p_{n-1}(x)y^{(n-1)} + \cdots + p_1(x)y' + p_0(x)y = 0. \tag{2.4}$$

Theorem 2.3.1 (Superposition). *Suppose $y_1, y_2, \ldots, y_n$ are solutions of the homogeneous equation* (2.4). *Then*

$$y(x) = C_1 y_1(x) + C_2 y_2(x) + \cdots + C_n y_n(x)$$

also solves (2.4) *for arbitrary constants $C_1, C_2, \ldots, C_n$.*

In other words, a *linear combination* of solutions to (2.4) is also a solution to (2.4). We also have the existence and uniqueness theorem for nonhomogeneous linear equations.

Theorem 2.3.2 (Existence and uniqueness). *Suppose p_0 through p_{n-1}, and f are continuous functions on some interval I, a is a number in I, and $b_0, b_1, \ldots, b_{n-1}$ are constants. The equation*

$$y^{(n)} + p_{n-1}(x)y^{(n-1)} + \cdots + p_1(x)y' + p_0(x)y = f(x)$$

has exactly one solution $y(x)$ defined on the same interval I satisfying the initial conditions

$$y(a) = b_0, \quad y'(a) = b_1, \quad \ldots, \quad y^{(n-1)}(a) = b_{n-1}.$$

2.3.1   Linear independence

When we had two functions y_1 and y_2 we said they were linearly independent if one was not the multiple of the other. Same idea holds for n functions. In this case it is easier to state as follows. The functions $y_1, y_2, \ldots, y_n$ are *linearly independent* if

$$c_1 y_1 + c_2 y_2 + \cdots + c_n y_n = 0$$

has only the trivial solution $c_1 = c_2 = \cdots = c_n = 0$, where the equation must hold for all x. If we can solve equation with some constants where for example $c_1 \neq 0$, then we can solve for y_1 as a linear combination of the others. If the functions are not linearly independent, they are *linearly dependent*.

Example 2.3.1: Show that e^x, e^{2x}, e^{3x} are linearly independent.

Let us give several ways to show this fact. Many textbooks (including [EP] and [F]) introduce Wronskians, but it is difficult to see why they work and they are not really necessary here.

Let us write down

$$c_1 e^x + c_2 e^{2x} + c_3 e^{3x} = 0.$$

We use rules of exponentials and write $z = e^x$. Hence $z^2 = e^{2x}$ and $z^3 = e^{3x}$. Then we have

$$c_1 z + c_2 z^2 + c_3 z^3 = 0.$$

The left hand side is a third degree polynomial in z. It is either identically zero, or it has at most 3 zeros. Therefore, it is identically zero, $c_1 = c_2 = c_3 = 0$, and the functions are linearly independent.

Let us try another way. As before we write

$$c_1 e^x + c_2 e^{2x} + c_3 e^{3x} = 0.$$

This equation has to hold for all x. We divide through by e^{3x} to get

$$c_1 e^{-2x} + c_2 e^{-x} + c_3 = 0.$$

As the equation is true for all x, let $x \to \infty$. After taking the limit we see that $c_3 = 0$. Hence our equation becomes

$$c_1 e^x + c_2 e^{2x} = 0.$$

Rinse, repeat!

How about yet another way. We again write

$$c_1 e^x + c_2 e^{2x} + c_3 e^{3x} = 0.$$

We can evaluate the equation and its derivatives at different values of x to obtain equations for c_1, c_2, and c_3. Let us first divide by e^x for simplicity.

$$c_1 + c_2 e^x + c_3 e^{2x} = 0.$$

We set $x = 0$ to get the equation $c_1 + c_2 + c_3 = 0$. Now differentiate both sides

$$c_2 e^x + 2c_3 e^{2x} = 0.$$

We set $x = 0$ to get $c_2 + 2c_3 = 0$. We divide by e^x again and differentiate to get $2c_3 e^x = 0$. It is clear that c_3 is zero. Then c_2 must be zero as $c_2 = -2c_3$, and c_1 must be zero because $c_1 + c_2 + c_3 = 0$.

There is no one best way to do it. All of these methods are perfectly valid. The important thing is to understand why the functions are linearly independent.

Example 2.3.2: On the other hand, the functions e^x, e^{-x}, and $\cosh x$ are linearly dependent. Simply apply definition of the hyperbolic cosine:

$$\cosh x = \frac{e^x + e^{-x}}{2} \qquad \text{or} \qquad 2\cosh x - e^x - e^{-x} = 0.$$

2.3.2 Constant coefficient higher order ODEs

When we have a higher order constant coefficient homogeneous linear equation, the song and dance is exactly the same as it was for second order. We just need to find more solutions. If the equation is n^{th} order we need to find n linearly independent solutions. It is best seen by example.

Example 2.3.3: Find the general solution to

$$y''' - 3y'' - y' + 3y = 0. \tag{2.5}$$

Try: $y = e^{rx}$. We plug in and get

$$r^3 e^{rx} - 3r^2 e^{rx} - r e^{rx} + 3 e^{rx} = 0.$$

We divide through by e^{rx}. Then

$$r^3 - 3r^2 - r + 3 = 0.$$

The trick now is to find the roots. There is a formula for the roots of degree 3 and 4 polynomials but it is very complicated. There is no formula for higher degree polynomials. That does not mean that the roots do not exist. There are always n roots for an nth degree polynomial. They may be repeated and they may be complex. Computers are pretty good at finding roots approximately for reasonable size polynomials.

A good place to start is to plot the polynomial and check where it is zero. We can also simply try plugging in. We just start plugging in numbers $r = -2, -1, 0, 1, 2, \ldots$ and see if we get a hit (we can also try complex numbers). Even if we do not get a hit, we may get an indication of where the root is. For example, we plug $r = -2$ into our polynomial and get -15; we plug in $r = 0$ and get 3. That means there is a root between $r = -2$ and $r = 0$, because the sign changed. If we find one root, say r_1, then we know $(r - r_1)$ is a factor of our polynomial. Polynomial long division can then be used.

A good strategy is to begin with $r = -1$, 1, or 0. These are easy to compute. Our polynomial happens to have two such roots, $r_1 = -1$ and $r_2 = 1$. There should be 3 roots and the last root is reasonably easy to find. The constant term in a monic polynomial such as this is the multiple of the negations of all the roots because $r^3 - 3r^2 - r + 3 = (r - r_1)(r - r_2)(r - r_3)$. So

$$3 = (-r_1)(-r_2)(-r_3) = (1)(-1)(-r_3) = r_3.$$

You should check that $r_3 = 3$ really is a root. Hence we know that e^{-x}, e^x and e^{3x} are solutions to (2.5). They are linearly independent as can easily be checked, and there are 3 of them, which happens to be exactly the number we need. Hence the general solution is

$$y = C_1 e^{-x} + C_2 e^x + C_3 e^{3x}.$$

Suppose we were given some initial conditions $y(0) = 1$, $y'(0) = 2$, and $y''(0) = 3$. Then

$$1 = y(0) = C_1 + C_2 + C_3,$$
$$2 = y'(0) = -C_1 + C_2 + 3C_3,$$
$$3 = y''(0) = C_1 + C_2 + 9C_3.$$

It is possible to find the solution by high school algebra, but it would be a pain. The sensible way to solve a system of equations such as this is to use matrix algebra, see § 3.2. For now we note that the solution is $C_1 = -1/4$, $C_2 = 1$, and $C_3 = 1/4$. The specific solution to the ODE is

$$y = \frac{-1}{4} e^{-x} + e^x + \frac{1}{4} e^{3x}.$$

Next, suppose that we have real roots, but they are repeated. Let us say we have a root r repeated k times. In the spirit of the second order solution, and for the same reasons, we have the solutions

$$e^{rx}, \quad xe^{rx}, \quad x^2 e^{rx}, \quad \ldots, \quad x^{k-1} e^{rx}.$$

We take a linear combination of these solutions to find the general solution.

Example 2.3.4: Solve

$$y^{(4)} - 3y''' + 3y'' - y' = 0.$$

We note that the characteristic equation is

$$r^4 - 3r^3 + 3r^2 - r = 0.$$

By inspection we note that $r^4 - 3r^3 + 3r^2 - r = r(r-1)^3$. Hence the roots given with multiplicity are $r = 0, 1, 1, 1$. Thus the general solution is

$$y = \underbrace{(C_1 + C_2 x + C_3 x^2) e^x}_{\text{terms coming from } r=1} + \underbrace{C_4}_{\text{from } r=0}.$$

The case of complex roots is similar to second order equations. Complex roots always come in pairs $r = \alpha \pm i\beta$. Suppose we have two such complex roots, each repeated k times. The corresponding solution is

$$(C_0 + C_1 x + \cdots + C_{k-1} x^{k-1}) e^{\alpha x} \cos(\beta x) + (D_0 + D_1 x + \cdots + D_{k-1} x^{k-1}) e^{\alpha x} \sin(\beta x).$$

where $C_0, \ldots, C_{k-1}, D_0, \ldots, D_{k-1}$ are arbitrary constants.

Example 2.3.5: Solve

$$y^{(4)} - 4y''' + 8y'' - 8y' + 4y = 0.$$

The characteristic equation is

$$r^4 - 4r^3 + 8r^2 - 8r + 4 = 0,$$
$$(r^2 - 2r + 2)^2 = 0,$$
$$((r-1)^2 + 1)^2 = 0.$$

Hence the roots are $1 \pm i$, both with multiplicity 2. Hence the general solution to the ODE is

$$y = (C_1 + C_2 x) e^x \cos x + (C_3 + C_4 x) e^x \sin x.$$

The way we solved the characteristic equation above is really by guessing or by inspection. It is not so easy in general. We could also have asked a computer or an advanced calculator for the roots.

2.3.3 Exercises

***Exercise* 2.3.1:** *Find the general solution for* $y''' - y'' + y' - y = 0$.

***Exercise* 2.3.2:** *Find the general solution for* $y^{(4)} - 5y''' + 6y'' = 0$.

***Exercise* 2.3.3:** *Find the general solution for* $y''' + 2y'' + 2y' = 0$.

***Exercise* 2.3.4:** *Suppose the characteristic equation for a differential equation is* $(r - 1)^2(r - 2)^2 = 0$. *a) Find such a differential equation. b) Find its general solution.*

***Exercise* 2.3.5:** *Suppose that a fourth order equation has a solution* $y = 2e^{4x}x \cos x$. *a) Find such an equation. b) Find the initial conditions that the given solution satisfies.*

***Exercise* 2.3.6:** *Find the general solution for the equation of Exercise 2.3.5.*

***Exercise* 2.3.7:** *Let* $f(x) = e^x - \cos x$, $g(x) = e^x + \cos x$, *and* $h(x) = \cos x$. *Are* $f(x)$, $g(x)$, *and* $h(x)$ *linearly independent? If so, show it, if not, find a linear combination that works.*

***Exercise* 2.3.8:** *Let* $f(x) = 0$, $g(x) = \cos x$, *and* $h(x) = \sin x$. *Are* $f(x)$, $g(x)$, *and* $h(x)$ *linearly independent? If so, show it, if not, find a linear combination that works.*

***Exercise* 2.3.9:** *Are* x, x^2, *and* x^4 *linearly independent? If so, show it, if not, find a linear combination that works.*

***Exercise* 2.3.10:** *Are* e^x, xe^x, *and* $x^2 e^x$ *linearly independent? If so, show it, if not, find a linear combination that works.*

***Exercise* 2.3.11:** *Find an equation such that* $y = xe^{-2x} \sin(3x)$ *is a solution.*

***Exercise* 2.3.101:** *Find the general solution of* $y^{(5)} - y^{(4)} = 0$

***Exercise* 2.3.102:** *Suppose that the characteristic equation of a third order differential equation has roots* $3, \pm 2i$. *a) What is the characteristic equation? b) Find the corresponding differential equation. c) Find the general solution.*

***Exercise* 2.3.103:** *Solve* $1001y''' + 3.2y'' + \pi y' - \sqrt{4}y = 0$, $y(0) = 0$, $y'(0) = 0$, $y''(0) = 0$.

***Exercise* 2.3.104:** *Are* e^x, e^{x+1}, e^{2x}, $\sin(x)$ *linearly independent? If so, show it, if not find a linear combination that works.*

***Exercise* 2.3.105:** *Are* $\sin(x)$, x, $x \sin(x)$ *linearly independent? If so, show it, if not find a linear combination that works.*

***Exercise* 2.3.106:** *Find an equation such that* $y = \cos(x)$, $y = \sin(x)$, $y = e^x$ *are solutions.*

2.4 Mechanical vibrations

Note: 2 lectures, §3.4 in [EP], §3.7 in [BD]

Let us look at some applications of linear second order constant coefficient equations.

2.4.1 Some examples

Our first example is a mass on a spring. Suppose we have a mass $m > 0$ (in kilograms) connected by a spring with spring constant $k > 0$ (in newtons per meter) to a fixed wall. There may be some external force $F(t)$ (in newtons) acting on the mass. Finally, there is some friction measured by $c \geq 0$ (in newton-seconds per meter) as the mass slides along the floor (or perhaps there is a damper connected).

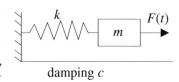

damping c

Let x be the displacement of the mass ($x = 0$ is the rest position), with x growing to the right (away from the wall). The force exerted by the spring is proportional to the compression of the spring by Hooke's law. Therefore, it is kx in the negative direction. Similarly the amount of force exerted by friction is proportional to the velocity of the mass. By Newton's second law we know that force equals mass times acceleration and hence $mx'' = F(t) - cx' - kx$ or

$$mx'' + cx' + kx = F(t).$$

This is a linear second order constant coefficient ODE. We set up some terminology about this equation. We say the motion is

(i) *forced*, if $F \not\equiv 0$ (if F is not identically zero),

(ii) *unforced* or *free*, if $F \equiv 0$ (if F is identically zero),

(iii) *damped*, if $c > 0$, and

(iv) *undamped*, if $c = 0$.

This system appears in lots of applications even if it does not at first seem like it. Many real world scenarios can be simplified to a mass on a spring. For example, a bungee jump setup is essentially a mass and spring system (you are the mass). It would be good if someone did the math before you jump off the bridge, right? Let us give two other examples.

Here is an example for electrical engineers. Suppose that we have the pictured RLC circuit. There is a resistor with a resistance of R ohms, an inductor with an inductance of L henries, and a capacitor with a capacitance of C farads. There is also an electric source (such as a battery) giving a voltage of $E(t)$ volts at time t (measured in seconds). Let $Q(t)$ be the charge

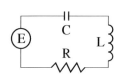

in coulombs on the capacitor and $I(t)$ be the current in the circuit. The relation between the two is $Q' = I$. By elementary principles we have that $LI' + RI + Q/c = E$. If we differentiate we get

$$LI''(t) + RI'(t) + \frac{1}{C}I(t) = E'(t).$$

This is an nonhomogeneous second order constant coefficient linear equation. Further, as L, R, and C are all positive, this system behaves just like the mass and spring system. The position of the mass is replaced by the current. Mass is replaced by the inductance, damping is replaced by resistance and the spring constant is replaced by one over the capacitance. The change in voltage becomes the forcing function. Hence for constant voltage this is an unforced motion.

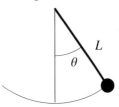 Our next example behaves like a mass and spring system only approximately. Suppose we have a mass m on a pendulum of length L. We wish to find an equation for the angle $\theta(t)$. Let g be the force of gravity. Elementary physics mandates that the equation is of the form

$$\theta'' + \frac{g}{L}\sin\theta = 0.$$

Let us derive this equation using Newton's second law; force equals mass times acceleration. The acceleration is $L\theta''$ and mass is m. So $mL\theta''$ has to be equal to the tangential component of the force given by the gravity, that is $mg\sin\theta$ in the opposite direction. So $mL\theta'' = -mg\sin\theta$. The m curiously cancels from the equation.

Now we make our approximation. For small θ we have that approximately $\sin\theta \approx \theta$. This can be seen by looking at the graph. In Figure 2.1 we can see that for approximately $-0.5 < \theta < 0.5$ (in radians) the graphs of $\sin\theta$ and θ are almost the same.

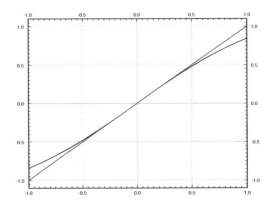

Figure 2.1: The graphs of $\sin\theta$ *and* θ *(in radians).*

Therefore, when the swings are small, θ is always small and we can model the behavior by the simpler linear equation

$$\theta'' + \frac{g}{L}\theta = 0.$$

Note that the errors that we get from the approximation build up. So after a very long time, the behavior of the real system might be substantially different from our solution. Also we will see that in a mass-spring system, the amplitude is independent of the period. This is not true for a pendulum. Nevertheless, for reasonably short periods of time and small swings (for example if the pendulum is very long), the approximation is reasonably good.

In real world problems it is often necessary to make these types of simplifications. We must understand both the mathematics and the physics of the situation to see if the simplification is valid in the context of the questions we are trying to answer.

2.4.2 Free undamped motion

In this section we will only consider free or unforced motion, as we cannot yet solve nonhomogeneous equations. Let us start with undamped motion where $c = 0$. We have the equation

$$mx'' + kx = 0.$$

If we divide by m and let $\omega_0 = \sqrt{k/m}$, then we can write the equation as

$$x'' + \omega_0^2 x = 0.$$

The general solution to this equation is

$$x(t) = A \cos(\omega_0 t) + B \sin(\omega_0 t).$$

By a trigonometric identity, we have that for two different constants C and γ, we have

$$A \cos(\omega_0 t) + B \sin(\omega_0 t) = C \cos(\omega_0 t - \gamma).$$

It is not hard to compute that $C = \sqrt{A^2 + B^2}$ and $\tan \gamma = B/A$. Therefore, we let C and γ be our arbitrary constants and write $x(t) = C \cos(\omega_0 t - \gamma)$.

Exercise 2.4.1: *Justify the above identity and verify the equations for C and γ. Hint: Start with* $\cos(\alpha - \beta) = \cos(\alpha)\cos(\beta) + \sin(\alpha)\sin(\beta)$ *and multiply by C. Then think what should α and β be.*

While it is generally easier to use the first form with A and B to solve for the initial conditions, the second form is much more natural. The constants C and γ have very nice interpretation. We look at the form of the solution

$$x(t) = C \cos(\omega_0 t - \gamma).$$

We can see that the *amplitude* is C, ω_0 is the (angular) frequency, and γ is the so-called *phase shift*. The phase shift just shifts the graph left or right. We call ω_0 the *natural (angular) frequency*. This entire setup is usually called *simple harmonic motion*.

Let us pause to explain the word *angular* before the word *frequency*. The units of ω_0 are radians per unit time, not cycles per unit time as is the usual measure of frequency. Because one cycle is 2π

radians, the usual frequency is given by $\frac{\omega_0}{2\pi}$. It is simply a matter of where we put the constant 2π, and that is a matter of taste.

The *period* of the motion is one over the frequency (in cycles per unit time) and hence $\frac{2\pi}{\omega_0}$. That is the amount of time it takes to complete one full cycle.

Example 2.4.1: Suppose that $m = 2\,\text{kg}$ and $k = 8\,\text{N/m}$. The whole mass and spring setup is sitting on a truck that was traveling at $1\,\text{m/s}$. The truck crashes and hence stops. The mass was held in place 0.5 meters forward from the rest position. During the crash the mass gets loose. That is, the mass is now moving forward at $1\,\text{m/s}$, while the other end of the spring is held in place. The mass therefore starts oscillating. What is the frequency of the resulting oscillation and what is the amplitude. The units are the mks units (meters-kilograms-seconds).

The setup means that the mass was at half a meter in the positive direction during the crash and relative to the wall the spring is mounted to, the mass was moving forward (in the positive direction) at $1\,\text{m/s}$. This gives us the initial conditions.

So the equation with initial conditions is

$$2x'' + 8x = 0, \qquad x(0) = 0.5, \qquad x'(0) = 1.$$

We can directly compute $\omega_0 = \sqrt{k/m} = \sqrt{4} = 2$. Hence the angular frequency is 2. The usual frequency in Hertz (cycles per second) is $2/2\pi = 1/\pi \approx 0.318$.

The general solution is

$$x(t) = A\cos(2t) + B\sin(2t).$$

Letting $x(0) = 0.5$ means $A = 0.5$. Then $x'(t) = -2(0.5)\sin(2t) + 2B\cos(2t)$. Letting $x'(0) = 1$ we get $B = 0.5$. Therefore, the amplitude is $C = \sqrt{A^2 + B^2} = \sqrt{0.25 + 0.25} = \sqrt{0.5} \approx 0.707$. The solution is

$$x(t) = 0.5\cos(2t) + 0.5\sin(2t).$$

A plot of $x(t)$ is shown in Figure 2.2 on the facing page.

In general, for free undamped motion, a solution of the form

$$x(t) = A\cos(\omega_0 t) + B\sin(\omega_0 t),$$

corresponds to the initial conditions $x(0) = A$ and $x'(0) = \omega_0 B$. Therefore, it is easy to figure out A and B from the initial conditions. The amplitude and the phase shift can then be computed from A and B. In the example, we have already found the amplitude C. Let us compute the phase shift. We know that $\tan\gamma = B/A = 1$. We take the arctangent of 1 and get $\pi/4$ or approximately 0.785. We still need to check if this γ is in the correct quadrant (and add π to γ if it is not). Since both A and B are positive, then γ should be in the first quadrant, $\pi/4$ radians is in the first quadrant, so $\gamma = \pi/4$.

Note: Many calculators and computer software do not only have the `atan` function for arctangent, but also what is sometimes called `atan2`. This function takes two arguments, B and A, and returns a γ in the correct quadrant for you.

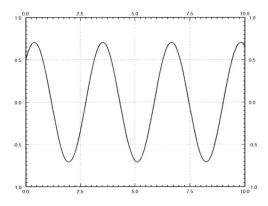

Figure 2.2: Simple undamped oscillation.

2.4.3 Free damped motion

Let us now focus on damped motion. Let us rewrite the equation

$$mx'' + cx' + kx = 0,$$

as

$$x'' + 2px' + \omega_0^2 x = 0,$$

where

$$\omega_0 = \sqrt{\frac{k}{m}}, \qquad p = \frac{c}{2m}.$$

The characteristic equation is

$$r^2 + 2pr + \omega_0^2 = 0.$$

Using the quadratic formula we get that the roots are

$$r = -p \pm \sqrt{p^2 - \omega_0^2}.$$

The form of the solution depends on whether we get complex or real roots. We get real roots if and only if the following number is nonnegative:

$$p^2 - \omega_0^2 = \left(\frac{c}{2m}\right)^2 - \frac{k}{m} = \frac{c^2 - 4km}{4m^2}.$$

The sign of $p^2 - \omega_0^2$ is the same as the sign of $c^2 - 4km$. Thus we get real roots if and only if $c^2 - 4km$ is nonnegative, or in other words if $c^2 \geq 4km$.

Overdamping

When $c^2 - 4km > 0$, we say the system is *overdamped*. In this case, there are two distinct real roots r_1 and r_2. Both roots are negative: As $\sqrt{p^2 - \omega_0^2}$ is always less than p, then $-p \pm \sqrt{p^2 - \omega_0^2}$ is negative in either case.

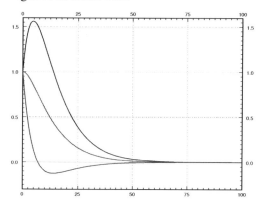

Figure 2.3: Overdamped motion for several different initial conditions.

The solution is

$$x(t) = C_1 e^{r_1 t} + C_2 e^{r_2 t}.$$

Since r_1, r_2 are negative, $x(t) \to 0$ as $t \to \infty$. Thus the mass will tend towards the rest position as time goes to infinity. For a few sample plots for different initial conditions, see Figure 2.3.

Do note that no oscillation happens. In fact, the graph will cross the x axis at most once. To see why, we try to solve $0 = C_1 e^{r_1 t} + C_2 e^{r_2 t}$. Therefore, $C_1 e^{r_1 t} = -C_2 e^{r_2 t}$ and using laws of exponents we obtain

$$\frac{-C_1}{C_2} = e^{(r_2 - r_1)t}.$$

This equation has at most one solution $t \geq 0$. For some initial conditions the graph will never cross the x axis, as is evident from the sample graphs.

Example 2.4.2: Suppose the mass is released from rest. That is $x(0) = x_0$ and $x'(0) = 0$. Then

$$x(t) = \frac{x_0}{r_1 - r_2} \left(r_1 e^{r_2 t} - r_2 e^{r_1 t} \right).$$

It is not hard to see that this satisfies the initial conditions.

Critical damping

When $c^2 - 4km = 0$, we say the system is *critically damped*. In this case, there is one root of multiplicity 2 and this root is $-p$. Our solution is

$$x(t) = C_1 e^{-pt} + C_2 t e^{-pt}.$$

The behavior of a critically damped system is very similar to an overdamped system. After all a critically damped system is in some sense a limit of overdamped systems. Since these equations are really only an approximation to the real world, in reality we are never critically damped, it is a place we can only reach in theory. We are always a little bit underdamped or a little bit overdamped. It is better not to dwell on critical damping.

Underdamping

When $c^2 - 4km < 0$, we say the system is *underdamped*. In this case, the roots are complex.

$$r = -p \pm \sqrt{p^2 - \omega_0^2}$$
$$= -p \pm \sqrt{-1}\sqrt{\omega_0^2 - p^2}$$
$$= -p \pm i\omega_1,$$

where $\omega_1 = \sqrt{\omega_0^2 - p^2}$. Our solution is

$$x(t) = e^{-pt}(A\cos(\omega_1 t) + B\sin(\omega_1 t)),$$

or

$$x(t) = Ce^{-pt}\cos(\omega_1 t - \gamma).$$

An example plot is given in Figure 2.4. Note that we still have that $x(t) \to 0$ as $t \to \infty$.

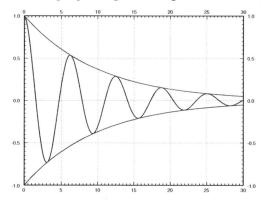

Figure 2.4: Underdamped motion with the envelope curves shown.

In the figure we also show the *envelope curves* Ce^{-pt} and $-Ce^{-pt}$. The solution is the oscillating line between the two envelope curves. The envelope curves give the maximum amplitude of the oscillation at any given point in time. For example if you are bungee jumping, you are really interested in computing the envelope curve so that you do not hit the concrete with your head.

The phase shift γ just shifts the graph left or right but within the envelope curves (the envelope curves do not change if γ changes).

Finally note that the angular *pseudo-frequency* (we do not call it a frequency since the solution is not really a periodic function) ω_1 becomes smaller when the damping c (and hence p) becomes larger. This makes sense. When we change the damping just a little bit, we do not expect the behavior of the solution to change dramatically. If we keep making c larger, then at some point the solution should start looking like the solution for critical damping or overdamping, where no oscillation happens. So if c^2 approaches $4km$, we want ω_1 to approach 0.

On the other hand when c becomes smaller, ω_1 approaches ω_0 (ω_1 is always smaller than ω_0), and the solution looks more and more like the steady periodic motion of the undamped case. The envelope curves become flatter and flatter as c (and hence p) goes to 0.

2.4.4   Exercises

Exercise 2.4.2: *Consider a mass and spring system with a mass m = 2, spring constant k = 3, and damping constant c = 1. a) Set up and find the general solution of the system. b) Is the system underdamped, overdamped or critically damped? c) If the system is not critically damped, find a c that makes the system critically damped.*

Exercise 2.4.3: *Do Exercise 2.4.2 for m = 3, k = 12, and c = 12.*

Exercise 2.4.4: *Using the mks units (meters-kilograms-seconds), suppose you have a spring with spring constant 4 N/m. You want to use it to weigh items. Assume no friction. You place the mass on the spring and put it in motion. a) You count and find that the frequency is 0.8 Hz (cycles per second). What is the mass? b) Find a formula for the mass m given the frequency ω in Hz.*

Exercise 2.4.5: *Suppose we add possible friction to Exercise 2.4.4. Further, suppose you do not know the spring constant, but you have two reference weights 1 kg and 2 kg to calibrate your setup. You put each in motion on your spring and measure the frequency. For the 1 kg weight you measured 1.1 Hz, for the 2 kg weight you measured 0.8 Hz. a) Find k (spring constant) and c (damping constant). b) Find a formula for the mass in terms of the frequency in Hz.* Note that there may be more than one possible mass for a given frequency. *c) For an unknown object you measured 0.2 Hz, what is the mass of the object? Suppose that you know that the mass of the unknown object is more than a kilogram.*

Exercise 2.4.6: *Suppose you wish to measure the friction a mass of 0.1 kg experiences as it slides along a floor (you wish to find c). You have a spring with spring constant k = 5 N/m. You take the spring, you attach it to the mass and fix it to a wall. Then you pull on the spring and let the mass go. You find that the mass oscillates with frequency 1 Hz. What is the friction?*

Exercise 2.4.101: *A mass of 2 kilograms is on a spring with spring constant k newtons per meter with no damping. Suppose the system is at rest and at time t = 0 the mass is kicked and starts traveling at 2 meters per second. How large does k have to be to so that the mass does not go further than 3 meters from the rest position?*

Exercise 2.4.102: *Suppose we have an RLC circuit with a resistor of 100 miliohms (0.1 ohms), inductor of inductance of 50 millihenries (0.05 henries), and a capacitor of 5 farads, with constant voltage. a) Set up the ODE equation for the current I. b) Find the general solution. c) Solve for I(0) = 10 and I'(0) = 0.*

Exercise 2.4.103: *A 5000 kg railcar hits a bumper (a spring) at 1 m/s, and the spring compresses by 0.1 m. Assume no damping. a) Find k. b) Find out how far does the spring compress when a 10000 kg railcar hits the spring at the same speed. c) If the spring would break if it compresses further than 0.3 m, what is the maximum mass of a railcar that can hit it at 1 m/s? d) What is the maximum mass of a railcar that can hit the spring without breaking at 2 m/s?*

***Exercise* 2.4.104:** *A mass of m kg is on a spring with $k = 3$ N/m and $c = 2$ Ns/m. Find the mass m_0 for which there is critical damping. If $m < m_0$, does the system oscillate or not, that is, is it underdamped or overdamped.*

2.5   Nonhomogeneous equations

Note: 2 lectures, §3.5 in [EP], §3.5 and §3.6 in [BD]

2.5.1   Solving nonhomogeneous equations

We have solved linear constant coefficient homogeneous equations. What about nonhomogeneous linear ODEs? For example, the equations for forced mechanical vibrations. That is, suppose we have an equation such as

$$y'' + 5y' + 6y = 2x + 1. \tag{2.6}$$

We will write $Ly = 2x + 1$ when the exact form of the operator is not important. We solve (2.6) in the following manner. First, we find the general solution y_c to the *associated homogeneous equation*

$$y'' + 5y' + 6y = 0. \tag{2.7}$$

We call y_c the *complementary solution*. Next, we find a single *particular solution* y_p to (2.6) in some way. Then

$$y = y_c + y_p$$

is the general solution to (2.6). We have $Ly_c = 0$ and $Ly_p = 2x + 1$. As L is a *linear operator* we verify that y is a solution, $Ly = L(y_c + y_p) = Ly_c + Ly_p = 0 + (2x + 1)$. Let us see why we obtain the *general* solution.

Let y_p and $\tilde{y}_p$ be two different particular solutions to (2.6). Write the difference as $w = y_p - \tilde{y}_p$. Then plug w into the left hand side of the equation to get

$$w'' + 5w' + 6w = (y_p'' + 5y_p' + 6y_p) - (\tilde{y}_p'' + 5\tilde{y}_p' + 6\tilde{y}_p) = (2x + 1) - (2x + 1) = 0.$$

Using the operator notation the calculation becomes simpler. As L is a linear operator we write

$$Lw = L(y_p - \tilde{y}_p) = Ly_p - L\tilde{y}_p = (2x + 1) - (2x + 1) = 0.$$

So $w = y_p - \tilde{y}_p$ is a solution to (2.7), that is $Lw = 0$. Any two solutions of (2.6) differ by a solution to the homogeneous equation (2.7). The solution $y = y_c + y_p$ includes *all* solutions to (2.6), since y_c is the general solution to the associated homogeneous equation.

Theorem 2.5.1. *Let $Ly = f(x)$ be a linear ODE (not necessarily constant coefficient). Let y_c be the complementary solution (the general solution to the associated homogeneous equation $Ly = 0$) and let y_p be any particular solution to $Ly = f(x)$. Then the general solution to $Ly = f(x)$ is*

$$y = y_c + y_p.$$

The moral of the story is that we can find the particular solution in any old way. If we find a different particular solution (by a different method, or simply by guessing), then we still get the same general solution. The formula may look different, and the constants we will have to choose to satisfy the initial conditions may be different, but it is the same solution.

2.5.2 Undetermined coefficients

The trick is to somehow, in a smart way, guess one particular solution to (2.6). Note that $2x + 1$ is a polynomial, and the left hand side of the equation will be a polynomial if we let y be a polynomial of the same degree. Let us try

$$y_p = Ax + B.$$

We plug in to obtain

$$y_p'' + 5y_p' + 6y_p = (Ax + B)'' + 5(Ax + B)' + 6(Ax + B) = 0 + 5A + 6Ax + 6B = 6Ax + (5A + 6B).$$

So $6Ax + (5A + 6B) = 2x + 1$. Therefore, $A = 1/3$ and $B = -1/9$. That means $y_p = \frac{1}{3} x - \frac{1}{9} = \frac{3x-1}{9}$. Solving the complementary problem (exercise!) we get

$$y_c = C_1 e^{-2x} + C_2 e^{-3x}.$$

Hence the general solution to (2.6) is

$$y = C_1 e^{-2x} + C_2 e^{-3x} + \frac{3x - 1}{9}.$$

Now suppose we are further given some initial conditions. For example, $y(0) = 0$ and $y'(0) = 1/3$. First find $y' = -2C_1 e^{-2x} - 3C_2 e^{-3x} + 1/3$. Then

$$0 = y(0) = C_1 + C_2 - \frac{1}{9}, \qquad \frac{1}{3} = y'(0) = -2C_1 - 3C_2 + \frac{1}{3}.$$

We solve to get $C_1 = 1/3$ and $C_2 = -2/9$. The particular solution we want is

$$y(x) = \frac{1}{3}e^{-2x} - \frac{2}{9}e^{-3x} + \frac{3x - 1}{9} = \frac{3e^{-2x} - 2e^{-3x} + 3x - 1}{9}.$$

Exercise 2.5.1: *Check that y really solves the equation* (2.6) *and the given initial conditions.*

Note: A common mistake is to solve for constants using the initial conditions with y_c and only add the particular solution y_p after that. That will *not* work. You need to first compute $y = y_c + y_p$ and *only then* solve for the constants using the initial conditions.

A right hand side consisting of exponentials, sines, and cosines can be handled similarly. For example,

$$y'' + 2y' + 2y = \cos(2x).$$

Let us find some y_p. We start by guessing the solution includes some multiple of $\cos(2x)$. We may have to also add a multiple of $\sin(2x)$ to our guess since derivatives of cosine are sines. We try

$$y_p = A\cos(2x) + B\sin(2x).$$

We plug y_p into the equation and we get

$$-4A\cos(2x) - 4B\sin(2x) - 4A\sin(2x) + 4B\cos(2x) + 2A\cos(2x) + 2B\sin(2x) = \cos(2x).$$

The left hand side must equal to right hand side. We group terms and we get that $-4A + 4B + 2A = 1$ and $-4B - 4A + 2B = 0$. So $-2A + 4B = 1$ and $2A + B = 0$ and hence $A = {}^{-1}/_{10}$ and $B = {}^{1}/_{5}$. So

$$y_p = A\cos(2x) + B\sin(2x) = \frac{-\cos(2x) + 2\sin(2x)}{10}.$$

Similarly, if the right hand side contains exponentials we try exponentials. For example, for

$$Ly = e^{3x},$$

we will try $y = Ae^{3x}$ as our guess and try to solve for A.

When the right hand side is a multiple of sines, cosines, exponentials, and polynomials, we can use the product rule for differentiation to come up with a guess. We need to guess a form for y_p such that Ly_p is of the same form, and has all the terms needed to for the right hand side. For example,

$$Ly = (1 + 3x^2)\,e^{-x}\cos(\pi x).$$

For this equation, we will guess

$$y_p = (A + Bx + Cx^2)\,e^{-x}\cos(\pi x) + (D + Ex + Fx^2)\,e^{-x}\sin(\pi x).$$

We will plug in and then hopefully get equations that we can solve for $A, B, C, D, E,$ and F. As you can see this can make for a very long and tedious calculation very quickly. C'est la vie!

There is one hiccup in all this. It could be that our guess actually solves the associated homogeneous equation. That is, suppose we have

$$y'' - 9y = e^{3x}.$$

We would love to guess $y = Ae^{3x}$, but if we plug this into the left hand side of the equation we get

$$y'' - 9y = 9Ae^{3x} - 9Ae^{3x} = 0 \neq e^{3x}.$$

There is no way we can choose A to make the left hand side be e^{3x}. The trick in this case is to multiply our guess by x to get rid of duplication with the complementary solution. That is first we compute y_c (solution to $Ly = 0$)

$$y_c = C_1 e^{-3x} + C_2 e^{3x}$$

and we note that the e^{3x} term is a duplicate with our desired guess. We modify our guess to $y = Axe^{3x}$ and notice there is no duplication anymore. Let us try. Note that $y' = Ae^{3x} + 3Axe^{3x}$ and $y'' = 6Ae^{3x} + 9Axe^{3x}$. So

$$y'' - 9y = 6Ae^{3x} + 9Axe^{3x} - 9Axe^{3x} = 6Ae^{3x}.$$

Thus $6Ae^{3x}$ is supposed to equal e^{3x}. Hence, $6A = 1$ and so $A = \frac{1}{6}$. We can now write the general solution as

$$y = y_c + y_p = C_1 e^{-3x} + C_2 e^{3x} + \frac{1}{6} x e^{3x}.$$

It is possible that multiplying by x does not get rid of all duplication. For example,

$$y'' - 6y' + 9y = e^{3x}.$$

The complementary solution is $y_c = C_1 e^{3x} + C_2 x e^{3x}$. Guessing $y = A x e^{3x}$ would not get us anywhere. In this case we want to guess $y_p = A x^2 e^{3x}$. Basically, we want to multiply our guess by x until all duplication is gone. *But no more!* Multiplying too many times will not work.

Finally, what if the right hand side has several terms, such as

$$Ly = e^{2x} + \cos x.$$

In this case we find u that solves $Lu = e^{2x}$ and v that solves $Lv = \cos x$ (that is, do each term separately). Then note that if $y = u + v$, then $Ly = e^{2x} + \cos x$. This is because L is linear; we have $Ly = L(u + v) = Lu + Lv = e^{2x} + \cos x$.

2.5.3   Variation of parameters

The method of undetermined coefficients will work for many basic problems that crop up. But it does not work all the time. It only works when the right hand side of the equation $Ly = f(x)$ has only finitely many linearly independent derivatives, so that we can write a guess that consists of them all. Some equations are a bit tougher. Consider

$$y'' + y = \tan x.$$

Note that each new derivative of $\tan x$ looks completely different and cannot be written as a linear combination of the previous derivatives. We get $\sec^2 x$, $2 \sec^2 x \tan x$, etc. . . .

This equation calls for a different method. We present the method of *variation of parameters*, which will handle any equation of the form $Ly = f(x)$, provided we can solve certain integrals. For simplicity, we restrict ourselves to second order constant coefficient equations, but the method works for higher order equations just as well (the computations become more tedious). The method also works for equations with nonconstant coefficients, provided we can solve the associated homogeneous equation.

Perhaps it is best to explain this method by example. Let us try to solve the equation

$$Ly = y'' + y = \tan x.$$

First we find the complementary solution (solution to $Ly_c = 0$). We get $y_c = C_1 y_1 + C_2 y_2$, where $y_1 = \cos x$ and $y_2 = \sin x$. To find a particular solution to the nonhomogeneous equation we try

$$y_p = y = u_1 y_1 + u_2 y_2,$$

where u_1 and u_2 are *functions* and not constants. We are trying to satisfy $Ly = \tan x$. That gives us one condition on the functions u_1 and u_2. Compute (note the product rule!)

$$y' = (u_1' y_1 + u_2' y_2) + (u_1 y_1' + u_2 y_2').$$

We can still impose one more condition at our discretion to simplify computations (we have two unknown functions, so we should be allowed two conditions). We require that $(u_1' y_1 + u_2' y_2) = 0$. This makes computing the second derivative easier.

$$y' = u_1 y_1' + u_2 y_2',$$
$$y'' = (u_1' y_1' + u_2' y_2') + (u_1 y_1'' + u_2 y_2'').$$

Since y_1 and y_2 are solutions to $y'' + y = 0$, we know that $y_1'' = -y_1$ and $y_2'' = -y_2$. (Note: If the equation was instead $y'' + p(x)y' + q(x)y = 0$ we would have $y_i'' = -p(x)y_i' - q(x)y_i$.) So

$$y'' = (u_1' y_1' + u_2' y_2') - (u_1 y_1 + u_2 y_2).$$

We have $(u_1 y_1 + u_2 y_2) = y$ and so

$$y'' = (u_1' y_1' + u_2' y_2') - y,$$

and hence

$$y'' + y = Ly = u_1' y_1' + u_2' y_2'.$$

For y to satisfy $Ly = f(x)$ we must have $f(x) = u_1' y_1' + u_2' y_2'$.

So what we need to solve are the two equations (conditions) we imposed on u_1 and u_2:

$$\boxed{\begin{aligned} u_1' y_1 + u_2' y_2 &= 0, \\ u_1' y_1' + u_2' y_2' &= f(x). \end{aligned}}$$

We now solve for u_1' and u_2' in terms of $f(x)$, y_1 and y_2. We always get these formulas for any $Ly = f(x)$, where $Ly = y'' + p(x)y' + q(x)y$. There is a general formula for the solution we can just plug into, but it is better to just repeat what we do below. In our case the two equations become

$$u_1' \cos(x) + u_2' \sin(x) = 0,$$
$$-u_1' \sin(x) + u_2' \cos(x) = \tan(x).$$

Hence

$$u_1' \cos(x) \sin(x) + u_2' \sin^2(x) = 0,$$
$$-u_1' \sin(x) \cos(x) + u_2' \cos^2(x) = \tan(x) \cos(x) = \sin(x).$$

And thus

$$u_2'(\sin^2(x) + \cos^2(x)) = \sin(x),$$
$$u_2' = \sin(x),$$
$$u_1' = \frac{-\sin^2(x)}{\cos(x)} = -\tan(x)\sin(x).$$

Now we need to integrate u_1' and u_2' to get u_1 and u_2.

$$u_1 = \int u_1' \, dx = \int -\tan(x)\sin(x) \, dx = \frac{1}{2}\ln\left|\frac{\sin(x)-1}{\sin(x)+1}\right| + \sin(x),$$
$$u_2 = \int u_2' \, dx = \int \sin(x) \, dx = -\cos(x).$$

So our particular solution is

$$y_p = u_1 y_1 + u_2 y_2 = \frac{1}{2}\cos(x)\ln\left|\frac{\sin(x)-1}{\sin(x)+1}\right| + \cos(x)\sin(x) - \cos(x)\sin(x) =$$
$$= \frac{1}{2}\cos(x)\ln\left|\frac{\sin(x)-1}{\sin(x)+1}\right|.$$

The general solution to $y'' + y = \tan x$ is, therefore,

$$y = C_1\cos(x) + C_2\sin(x) + \frac{1}{2}\cos(x)\ln\left|\frac{\sin(x)-1}{\sin(x)+1}\right|.$$

2.5.4 Exercises

Exercise 2.5.2: *Find a particular solution of $y'' - y' - 6y = e^{2x}$.*

Exercise 2.5.3: *Find a particular solution of $y'' - 4y' + 4y = e^{2x}$.*

Exercise 2.5.4: *Solve the initial value problem $y'' + 9y = \cos(3x) + \sin(3x)$ for $y(0) = 2$, $y'(0) = 1$.*

Exercise 2.5.5: *Set up the form of the particular solution but do not solve for the coefficients for $y^{(4)} - 2y''' + y'' = e^x$.*

Exercise 2.5.6: *Set up the form of the particular solution but do not solve for the coefficients for $y^{(4)} - 2y''' + y'' = e^x + x + \sin x$.*

Exercise 2.5.7: *a) Using variation of parameters find a particular solution of $y'' - 2y' + y = e^x$. b) Find a particular solution using undetermined coefficients. c) Are the two solutions you found the same? See also Exercise 2.5.10.*

Exercise 2.5.8: *Find a particular solution of $y'' - 2y' + y = \sin(x^2)$. It is OK to leave the answer as a definite integral.*

Exercise 2.5.9: *For an arbitrary constant c find a particular solution to $y'' - y = e^{cx}$. Hint: Make sure to handle every possible real c.*

Exercise 2.5.10: *a) Using variation of parameters find a particular solution of $y'' - y = e^x$. b) Find a particular solution using undetermined coefficients. c) Are the two solutions you found the same? What is going on?*

Exercise 2.5.101: *Find a particular solution to $y'' - y' + y = 2\sin(3x)$*

Exercise 2.5.102: *a) Find a particular solution to $y'' + 2y = e^x + x^3$. b) Find the general solution.*

Exercise 2.5.103: *Solve $y'' + 2y' + y = x^2$, $y(0) = 1$, $y'(0) = 2$.*

Exercise 2.5.104: *Use variation of parameters to find a particular solution of $y'' - y = \frac{1}{e^x + e^{-x}}$.*

Exercise 2.5.105: *For an arbitrary constant c find the general solution to $y'' - 2y = \sin(x + c)$.*

2.6 Forced oscillations and resonance

Note: 2 lectures, §3.6 in [EP], §3.8 in [BD]

Let us return back to the example of a mass on a spring. We now examine the case of forced oscillations, which we did not yet handle. That is, we consider the equation

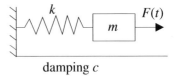

damping c

$$mx'' + cx' + kx = F(t),$$

for some nonzero $F(t)$. The setup is again: m is mass, c is friction, k is the spring constant and $F(t)$ is an external force acting on the mass.

We are interested in periodic forcing, such as noncentered rotating parts, or perhaps loud sounds, or other sources of periodic force. Once we learn about Fourier series in chapter 4, we will see that we cover all periodic functions by simply considering $F(t) = F_0 \cos(\omega t)$ (or sine instead of cosine, the calculations are essentially the same).

2.6.1 Undamped forced motion and resonance

First let us consider undamped ($c = 0$) motion for simplicity. We have the equation

$$mx'' + kx = F_0 \cos(\omega t).$$

This equation has the complementary solution (solution to the associated homogeneous equation)

$$x_c = C_1 \cos(\omega_0 t) + C_2 \sin(\omega_0 t),$$

where $\omega_0 = \sqrt{k/m}$ is the *natural frequency* (angular). It is the frequency at which the system "wants to oscillate" without external interference.

Let us suppose that $\omega_0 \neq \omega$. We try the solution $x_p = A \cos(\omega t)$ and solve for A. Note that we do not need a sine in our trial solution as after plugging in we only have cosines. If you include a sine, it is fine; you will find that its coefficient is zero (I could not find a second rhyme).

We solve using the method of undetermined coefficients. We find that

$$x_p = \frac{F_0}{m(\omega_0^2 - \omega^2)} \cos(\omega t).$$

We leave it as an exercise to do the algebra required.

The general solution is

$$\boxed{x = C_1 \cos(\omega_0 t) + C_2 \sin(\omega_0 t) + \frac{F_0}{m(\omega_0^2 - \omega^2)} \cos(\omega t).}$$

Written another way

$$x = C\cos(\omega_0 t - \gamma) + \frac{F_0}{m(\omega_0^2 - \omega^2)}\cos(\omega t).$$

The solution is a superposition of two cosine waves at different frequencies.

Example 2.6.1: Take

$$0.5x'' + 8x = 10\cos(\pi t), \qquad x(0) = 0, \qquad x'(0) = 0.$$

Let us compute. First we read off the parameters: $\omega = \pi$, $\omega_0 = \sqrt{8/0.5} = 4$, $F_0 = 10$, $m = 0.5$. The general solution is

$$x = C_1\cos(4t) + C_2\sin(4t) + \frac{20}{16 - \pi^2}\cos(\pi t).$$

Solve for C_1 and C_2 using the initial conditions. It is easy to see that $C_1 = \frac{-20}{16-\pi^2}$ and $C_2 = 0$. Hence

$$x = \frac{20}{16 - \pi^2}(\cos(\pi t) - \cos(4t)).$$

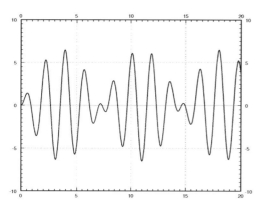

Notice the "beating" behavior in Figure 2.5. First use the trigonometric identity

$$2\sin\left(\frac{A - B}{2}\right)\sin\left(\frac{A + B}{2}\right) = \cos B - \cos A$$

to get

$$x = \frac{20}{16 - \pi^2}\left(2\sin\left(\frac{4 - \pi}{2}t\right)\sin\left(\frac{4 + \pi}{2}t\right)\right).$$

The function x is a high frequency wave modulated by a low frequency wave.

Figure 2.5: Graph of $\frac{20}{16-\pi^2}(\cos(\pi t) - \cos(4t))$.

Now suppose $\omega_0 = \omega$. Obviously, we cannot try the solution $A\cos(\omega t)$ and then use the method of undetermined coefficients. We notice that $\cos(\omega t)$ solves the associated homogeneous equation. Therefore, we try $x_p = At\cos(\omega t) + Bt\sin(\omega t)$. This time we need the sine term, since the second derivative of $t\cos(\omega t)$ contains sines. We write the equation

$$x'' + \omega^2 x = \frac{F_0}{m}\cos(\omega t).$$

Plugging x_p into the left hand side we get

$$2B\omega\cos(\omega t) - 2A\omega\sin(\omega t) = \frac{F_0}{m}\cos(\omega t).$$

Hence $A = 0$ and $B = \frac{F_0}{2m\omega}$. Our particular solution is $\frac{F_0}{2m\omega} t \sin(\omega t)$ and our general solution is

$$x = C_1 \cos(\omega t) + C_2 \sin(\omega t) + \frac{F_0}{2m\omega} t \sin(\omega t).$$

The important term is the last one (the particular solution we found). This term grows without bound as $t \to \infty$. In fact it oscillates between $\frac{F_0 t}{2m\omega}$ and $\frac{-F_0 t}{2m\omega}$. The first two terms only oscillate between $\pm \sqrt{C_1^2 + C_2^2}$, which becomes smaller and smaller in proportion to the oscillations of the last term as t gets larger. In Figure 2.6 we see the graph with $C_1 = C_2 = 0$, $F_0 = 2$, $m = 1$, $\omega = \pi$.

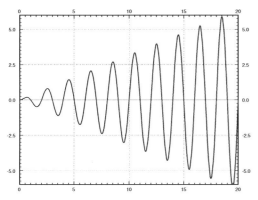

Figure 2.6: Graph of $\frac{1}{\pi} t \sin(\pi t)$.

By forcing the system in just the right frequency we produce very wild oscillations. This kind of behavior is called *resonance* or perhaps *pure resonance*. Sometimes resonance is desired. For example, remember when as a kid you could start swinging by just moving back and forth on the swing seat in the "correct frequency"? You were trying to achieve resonance. The force of each one of your moves was small, but after a while it produced large swings.

On the other hand resonance can be destructive. In an earthquake some buildings collapse while others may be relatively undamaged. This is due to different buildings having different resonance frequencies. So figuring out the resonance frequency can be very important.

A common (but wrong) example of destructive force of resonance is the Tacoma Narrows bridge failure. It turns out there was a different phenomenon at play[*].

2.6.2 Damped forced motion and practical resonance

In real life things are not as simple as they were above. There is, of course, some damping. Our equation becomes

$$mx'' + cx' + kx = F_0 \cos(\omega t), \tag{2.8}$$

for some $c > 0$. We solved the homogeneous problem before. We let

$$p = \frac{c}{2m}, \qquad \omega_0 = \sqrt{\frac{k}{m}}.$$

---

[*]K. Billah and R. Scanlan, *Resonance, Tacoma Narrows Bridge Failure, and Undergraduate Physics Textbooks*, American Journal of Physics, 59(2), 1991, 118–124, http://www.ketchum.org/billah/Billah-Scanlan.pdf

We replace equation (2.8) with

$$x'' + 2px' + \omega_0^2 x = \frac{F_0}{m}\cos(\omega t).$$

The roots of the characteristic equation of the associated homogeneous problem are $r_1, r_2 = -p \pm \sqrt{p^2 - \omega_0^2}$. The form of the general solution of the associated homogeneous equation depends on the sign of $p^2 - \omega_0^2$, or equivalently on the sign of $c^2 - 4km$, as we have seen before. That is,

$$x_c = \begin{cases} C_1 e^{r_1 t} + C_2 e^{r_2 t} & \text{if } c^2 > 4km, \\ C_1 e^{-pt} + C_2 t e^{-pt} & \text{if } c^2 = 4km, \\ e^{-pt}(C_1 \cos(\omega_1 t) + C_2 \sin(\omega_1 t)) & \text{if } c^2 < 4km, \end{cases}$$

where $\omega_1 = \sqrt{\omega_0^2 - p^2}$. In any case, we see that $x_c(t) \to 0$ as $t \to \infty$. Furthermore, there can be no conflicts when trying to solve for the undetermined coefficients by trying $x_p = A\cos(\omega t) + B\sin(\omega t)$. Let us plug in and solve for A and B. We get (the tedious details are left to reader)

$$((\omega_0^2 - \omega^2)B - 2\omega pA)\sin(\omega t) + ((\omega_0^2 - \omega^2)A + 2\omega pB)\cos(\omega t) = \frac{F_0}{m}\cos(\omega t).$$

We solve for A and B:

$$A = \frac{(\omega_0^2 - \omega^2)F_0}{m(2\omega p)^2 + m(\omega_0^2 - \omega^2)^2},$$

$$B = \frac{2\omega p F_0}{m(2\omega p)^2 + m(\omega_0^2 - \omega^2)^2}.$$

We also compute $C = \sqrt{A^2 + B^2}$ to be

$$C = \frac{F_0}{m\sqrt{(2\omega p)^2 + (\omega_0^2 - \omega^2)^2}}.$$

Thus our particular solution is

$$x_p = \frac{(\omega_0^2 - \omega^2)F_0}{m(2\omega p)^2 + m(\omega_0^2 - \omega^2)^2}\cos(\omega t) + \frac{2\omega p F_0}{m(2\omega p)^2 + m(\omega_0^2 - \omega^2)^2}\sin(\omega t).$$

Or in the alternative notation we have amplitude C and phase shift γ where (if $\omega \neq \omega_0$)

$$\tan\gamma = \frac{B}{A} = \frac{2\omega p}{\omega_0^2 - \omega^2}.$$

Hence we have

$$x_p = \frac{F_0}{m\sqrt{(2\omega p)^2 + (\omega_0^2 - \omega^2)^2}} \cos(\omega t - \gamma).$$

If $\omega = \omega_0$ we see that $A = 0$, $B = C = \frac{F_0}{2m\omega p}$, and $\gamma = \pi/2$.

The exact formula is not as important as the idea. Do not memorize the above formula, you should instead remember the ideas involved. For a different forcing function F, you will get a different formula for x_p. So there is no point in memorizing this specific formula. You can always recompute it later or look it up if you really need it.

For reasons we will explain in a moment, we call x_c the *transient solution* and denote it by x_{tr}. We call the x_p we found above the *steady periodic solution* and denote it by x_{sp}. The general solution to our problem is

$$x = x_c + x_p = x_{tr} + x_{sp}.$$

The transient solution $x_c = x_{tr}$ goes to zero as $t \to \infty$, as all the terms involve an exponential with a negative exponent. So for large t, the effect of x_{tr} is negligible and we see essentially only x_{sp}. Hence the name *transient*. Notice that x_{sp} involves no arbitrary constants, and the initial conditions only affect x_{tr}. This means that the effect of the initial conditions is negligible after some period of time. Because of this behavior, we might as well focus on the steady periodic solution and ignore the transient solution. See Figure 2.7 for a graph given several different initial conditions.

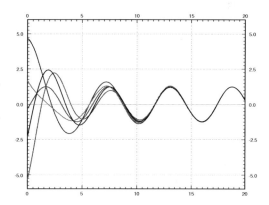

Figure 2.7: Solutions with different initial conditions for parameters $k = 1$, $m = 1$, $F_0 = 1$, $c = 0.7$, and $\omega = 1.1$.

The speed at which x_{tr} goes to zero depends on p (and hence c). The bigger p is (the bigger c is), the "faster" x_{tr} becomes negligible. So the smaller the damping, the longer the "transient region." This agrees with the observation that when $c = 0$, the initial conditions affect the behavior for all time (i.e. an infinite "transient region").

Let us describe what we mean by resonance when damping is present. Since there were no conflicts when solving with undetermined coefficient, there is no term that goes to infinity. We look at the maximum value of the amplitude of the steady periodic solution. Let C be the amplitude of x_{sp}. If we plot C as a function of ω (with all other parameters fixed) we can find its maximum. We call the ω that achieves this maximum the *practical resonance frequency*. We call the maximal amplitude $C(\omega)$ the *practical resonance amplitude*. Thus when damping is present we talk of *practical resonance* rather than pure resonance. A sample plot for three different values of c is given in Figure 2.8. As you can see the practical resonance amplitude grows as damping gets smaller, and practical resonance can disappear altogether when damping is large.

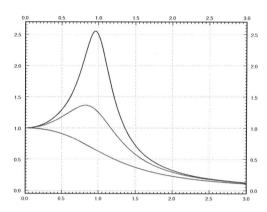

Figure 2.8: Graph of $C(\omega)$ showing practical resonance with parameters $k = 1$, $m = 1$, $F_0 = 1$. The top line is with $c = 0.4$, the middle line with $c = 0.8$, and the bottom line with $c = 1.6$.

To find the maximum we need to find the derivative $C'(\omega)$. Computation shows

$$C'(\omega) = \frac{-2\omega(2p^2 + \omega^2 - \omega_0^2)F_0}{m\left((2\omega p)^2 + (\omega_0^2 - \omega^2)^2\right)^{3/2}}.$$

This is zero either when $\omega = 0$ or when $2p^2 + \omega^2 - \omega_0^2 = 0$. In other words, $C'(\omega) = 0$ when

$$\boxed{\omega = \sqrt{\omega_0^2 - 2p^2} \quad \text{or} \quad \omega = 0.}$$

It can be shown that if $\omega_0^2 - 2p^2$ is positive, then $\sqrt{\omega_0^2 - 2p^2}$ is the practical resonance frequency (that is the point where $C(\omega)$ is maximal, note that in this case $C'(\omega) > 0$ for small ω). If $\omega = 0$ is the maximum, then there is no practical resonance since we assume $\omega > 0$ in our system. In this case the amplitude gets larger as the forcing frequency gets smaller.

If practical resonance occurs, the frequency is smaller than ω_0. As the damping c (and hence p) becomes smaller, the practical resonance frequency goes to ω_0. So when damping is very small, ω_0 is a good estimate of the resonance frequency. This behavior agrees with the observation that when $c = 0$, then ω_0 is the resonance frequency.

The behavior is more complicated if the forcing function is not an exact cosine wave, but for example a square wave. The reader is encouraged to come back to this section once we have learned about the Fourier series.

2.6.3 Exercises

Exercise 2.6.1: *Derive a formula for x_{sp} if the equation is $mx'' + cx' + kx = F_0 \sin(\omega t)$. Assume $c > 0$.*

Exercise 2.6.2: *Derive a formula for x_{sp} if the equation is $mx'' + cx' + kx = F_0 \cos(\omega t) + F_1 \cos(3\omega t)$. Assume $c > 0$.*

Exercise 2.6.3: *Take $mx'' + cx' + kx = F_0 \cos(\omega t)$. Fix $m > 0$, $k > 0$, and $F_0 > 0$. Consider the function $C(\omega)$. For what values of c (solve in terms of m, k, and F_0) will there be no practical resonance (that is, for what values of c is there no maximum of $C(\omega)$ for $\omega > 0$)?*

Exercise 2.6.4: *Take $mx'' + cx' + kx = F_0 \cos(\omega t)$. Fix $c > 0$, $k > 0$, and $F_0 > 0$. Consider the function $C(\omega)$. For what values of m (solve in terms of c, k, and F_0) will there be no practical resonance (that is, for what values of m is there no maximum of $C(\omega)$ for $\omega > 0$)?*

Exercise 2.6.5: *Suppose a water tower in an earthquake acts as a mass-spring system. Assume that the container on top is full and the water does not move around. The container then acts as a mass and the support acts as the spring, where the induced vibrations are horizontal. Suppose that the container with water has a mass of $m = 10,000$ kg. It takes a force of 1000 newtons to displace the container 1 meter. For simplicity assume no friction. When the earthquake hits the water tower is at rest (it is not moving).*

Suppose that an earthquake induces an external force $F(t) = mA\omega^2 \cos(\omega t)$.

a) What is the natural frequency of the water tower?

b) If ω is not the natural frequency, find a formula for the maximal amplitude of the resulting oscillations of the water container (the maximal deviation from the rest position). The motion will be a high frequency wave modulated by a low frequency wave, so simply find the constant in front of the sines.

c) Suppose $A = 1$ and an earthquake with frequency 0.5 cycles per second comes. What is the amplitude of the oscillations? Suppose that if the water tower moves more than 1.5 meter from the rest position, the tower collapses. Will the tower collapse?

Exercise 2.6.101: *A mass of 4 kg on a spring with $k = 4$ N/m and a damping constant $c = 1$ Ns/m. Suppose that $F_0 = 2$ N. Using forcing function $F_0 \cos(\omega t)$, find the ω that causes practical resonance and find the amplitude.*

Exercise 2.6.102: *Derive a formula for x_{sp} for $mx'' + cx' + kx = F_0 \cos(\omega t) + A$, where A is some constant. Assume $c > 0$.*

Exercise 2.6.103: *Suppose there is no damping in a mass and spring system with $m = 5$, $k = 20$, and $F_0 = 5$. Suppose ω is chosen to be precisely the resonance frequency. a) Find ω. b) Find the amplitude of the oscillations at time $t = 100$, given the system is at rest at $t = 0$.*

Chapter 3

Systems of ODEs

3.1 Introduction to systems of ODEs

Note: 1 lecture, §4.1 in [EP], §7.1 in [BD]

Often we do not have just one dependent variable and one equation. And as we will see, we may end up with systems of several equations and several dependent variables even if we start with a single equation.

If we have several dependent variables, suppose $y_1, y_2, \ldots, y_n$, then we can have a differential equation involving all of them and their derivatives. For example, $y_1'' = f(y_1', y_2', y_1, y_2, x)$. Usually, when we have two dependent variables we have two equations such as

$$y_1'' = f_1(y_1', y_2', y_1, y_2, x),$$
$$y_2'' = f_2(y_1', y_2', y_1, y_2, x),$$

for some functions f_1 and f_2. We call the above a *system of differential equations*. More precisely, the above is a second order system of ODEs.

Example 3.1.1: Sometimes a system is easy to solve by solving for one variable and then for the second variable. Take the first order system

$$y_1' = y_1,$$
$$y_2' = y_1 - y_2,$$

with initial conditions of the form $y_1(0) = 1$, $y_2(0) = 2$.

We note that $y_1 = C_1 e^x$ is the general solution of the first equation. We then plug this y_1 into the second equation and get the equation $y_2' = C_1 e^x - y_2$, which is a linear first order equation that is easily solved for y_2. By the method of integrating factor we get

$$e^x y_2 = \frac{C_1}{2} e^{2x} + C_2,$$

or $y_2 = \frac{C_1}{2}e^x + C_2e^{-x}$. The general solution to the system is, therefore,

$$y_1 = C_1e^x,$$

$$y_2 = \frac{C_1}{2}e^x + C_2e^{-x}.$$

We solve for C_1 and C_2 given the initial conditions. We substitute $x = 0$ and find that $C_1 = 1$ and $C_2 = 3/2$. Thus the solution is $y_1 = e^x$, and $y_2 = (1/2)e^x + (3/2)e^{-x}$.

Generally, we will not be so lucky to be able to solve for each variable separately as in the example above, and we will have to solve for all variables at once.

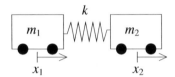

As an example application, let us think of mass and spring systems again. Suppose we have one spring with constant k, but two masses m_1 and m_2. We can think of the masses as carts, and we will suppose that they ride along a straight track with no friction. Let x_1 be the displacement of the first cart and x_2 be the displacement of the second cart. That is, we put the two carts somewhere with no tension on the spring, and we mark the position of the first and second cart and call those the zero positions. Then x_1 measures how far the first cart is from its zero position, and x_2 measures how far the second cart is from its zero position. The force exerted by the spring on the first cart is $k(x_2 - x_1)$, since $x_2 - x_1$ is how far the string is stretched (or compressed) from the rest position. The force exerted on the second cart is the opposite, thus the same thing with a negative sign. Newton's second law states that force equals mass times acceleration. So the system of equations governing the setup is

$$m_1 x_1'' = k(x_2 - x_1),$$
$$m_2 x_2'' = -k(x_2 - x_1).$$

In this system we cannot solve for the x_1 or x_2 variable separately. That we must solve for both x_1 and x_2 at once is intuitively clear, since where the first cart goes depends on exactly where the second cart goes and vice-versa.

Before we talk about how to handle systems, let us note that in some sense we need only consider first order systems. Let us take an nth order differential equation

$$y^{(n)} = F(y^{(n-1)}, \dots, y', y, x).$$

We define new variables $u_1, u_2, \dots, u_n$ and write the system

$$u_1' = u_2,$$
$$u_2' = u_3,$$
$$\vdots$$
$$u_{n-1}' = u_n,$$
$$u_n' = F(u_n, u_{n-1}, \dots, u_2, u_1, x).$$

We solve this system for $u_1, u_2, \ldots, u_n$. Once we have solved for the u's, we can discard u_2 through u_n and let $y = u_1$. We note that this y solves the original equation.

A similar process can be followed for a system of higher order differential equations. For example, a system of k differential equations in k unknowns, all of order n, can be transformed into a first order system of $n \times k$ equations and $n \times k$ unknowns.

Example 3.1.2: We can use this idea in reverse as well. Let us consider the system

$$x' = 2y - x, \qquad y' = x,$$

where the independent variable is t. We wish to solve for the initial conditions $x(0) = 1$, $y(0) = 0$.

If we differentiate the second equation we get $y'' = x'$. We know what x' is in terms of x and y, and we know that $x = y'$. So,

$$y'' = x' = 2y - x = 2y - y'.$$

We now have the equation $y'' + y' - 2y = 0$. We know how to solve this equation and we find that $y = C_1 e^{-2t} + C_2 e^t$. Once we have y we use the equation $y' = x$ to get x.

$$x = y' = -2C_1 e^{-2t} + C_2 e^t.$$

We solve for the initial conditions $1 = x(0) = -2C_1 + C_2$ and $0 = y(0) = C_1 + C_2$. Hence, $C_1 = -C_2$ and $1 = 3C_2$. So $C_1 = -\frac{1}{3}$ and $C_2 = \frac{1}{3}$. Our solution is

$$x = \frac{2e^{-2t} + e^t}{3}, \qquad y = \frac{-e^{-2t} + e^t}{3}.$$

Exercise 3.1.1: Plug in and check that this really is the solution.

It is useful to go back and forth between systems and higher order equations for other reasons. For example, the ODE approximation methods are generally only given as solutions for first order systems. It is not very hard to adapt the code for the Euler method for first order equations to handle first order systems. We essentially just treat the dependent variable not as a number but as a vector. In many mathematical computer languages there is almost no distinction in syntax.

The above example is what we call a *linear first order system*, as none of the dependent variables appear in any functions or with any higher powers than one. It is also *autonomous* as the equations do not depend on the independent variable t.

For autonomous systems we can draw the so-called *direction field* or *vector field*. That is, a plot similar to a slope field, but instead of giving a slope at each point, we give a direction (and a magnitude). The previous example $x' = 2y - x$, $y' = x$ says that at the point (x, y) the direction in which we should travel to satisfy the equations should be the direction of the vector $(2y - x, x)$ with the speed equal to the magnitude of this vector. So we draw the vector $(2y - x, x)$ based at the point (x, y) and we do this for many points on the xy-plane. We may want to scale down the size of our vectors to fit many of them on the same direction field. See Figure 3.1 on the next page.

We can now draw a path of the solution in the plane. That is, suppose the solution is given by $x = f(t)$, $y = g(t)$, then we pick an interval of t (say $0 \le t \le 2$ for our example) and plot all the points $(f(t), g(t))$ for t in the selected range. The resulting picture is called the *phase portrait* (or phase plane portrait). The particular curve obtained is called the *trajectory* or *solution curve*. An example plot is given in Figure 3.2. In this figure the line starts at $(1, 0)$ and travels along the vector field for a distance of 2 units of t. Since we solved this system precisely we can compute $x(2)$ and $y(2)$. We get that $x(2) \approx 2.475$ and $y(2) \approx 2.457$. This point corresponds to the top right end of the plotted solution curve in the figure.

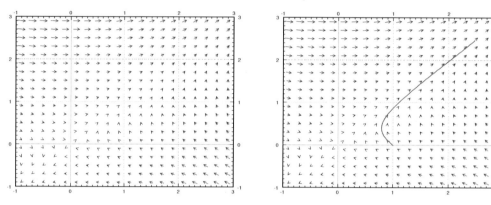

Figure 3.1: The direction field for $x' = 2y-x$, $y' = x$.   Figure 3.2: The direction field for $x' = 2y-x$, $y' = x$ with the trajectory of the solution starting at $(1,0)$ for $0 \le t \le 2$.

Notice the similarity to the diagrams we drew for autonomous systems in one dimension. But now note how much more complicated things became when we allowed just one more dimension.

We can draw phase portraits and trajectories in the xy-plane even if the system is not autonomous. In this case however we cannot draw the direction field, since the field changes as t changes. For each t we would get a different direction field.

3.1.1   Exercises

Exercise 3.1.2: *Find the general solution of $x_1' = x_2 - x_1 + t$, $x_2' = x_2$.*

Exercise 3.1.3: *Find the general solution of $x_1' = 3x_1 - x_2 + e^t$, $x_2' = x_1$.*

Exercise 3.1.4: *Write $ay'' + by' + cy = f(x)$ as a first order system of ODEs.*

Exercise 3.1.5: *Write $x'' + y^2y' - x^3 = \sin(t)$, $y'' + (x' + y')^2 - x = 0$ as a first order system of ODEs.*

Exercise 3.1.101: *Find the general solution to $y_1' = 3y_1$, $y_2' = y_1 + y_2$, $y_3' = y_1 + y_3$.*

Exercise 3.1.102: *Solve $y' = 2x$, $x' = x + y$, $x(0) = 1$, $y(0) = 3$.*

Exercise 3.1.103: *Write $x''' = x + t$ as a first order system.*

Exercise 3.1.104: *Write $y_1'' + y_1 + y_2 = t$, $y_2'' + y_1 - y_2 = t^2$ as a first order system.*

Exercise 3.1.105: *Suppose two masses on carts on frictionless surface are at displacements x_1 and x_2 as in the example of this section. Suppose initial displacement is $x_1(0) = x_2(0) = 0$, and initial velocity is $x_1'(0) = x_2'(0) = a$ for some number a. Use your intuition to solve the system, explain your reasoning.*

3.2   Matrices and linear systems

Note: 1 and a half lectures, first part of §5.1 in [EP], §7.2 and §7.3 in [BD]

3.2.1   Matrices and vectors

Before we can start talking about linear systems of ODEs, we will need to talk about matrices, so let us review these briefly. A *matrix* is an $m \times n$ array of numbers (m rows and n columns). For example, we denote a 3×5 matrix as follows

$$A = \begin{bmatrix} a_{11} & a_{12} & a_{13} & a_{14} & a_{15} \\ a_{21} & a_{22} & a_{23} & a_{24} & a_{25} \\ a_{31} & a_{32} & a_{33} & a_{34} & a_{35} \end{bmatrix}.$$

By a *vector* we will usually mean a *column vector*, that is an $m \times 1$ matrix. If we mean a *row vector* we will explicitly say so (a row vector is a $1 \times n$ matrix). We will usually denote matrices by upper case letters and vectors by lower case letters with an arrow such as $\vec{x}$ or $\vec{b}$. By $\vec{0}$ we will mean the vector of all zeros.

It is easy to define some operations on matrices. Note that we will want 1×1 matrices to really act like numbers, so our operations will have to be compatible with this viewpoint.

First, we can multiply by a *scalar* (a number). This means just multiplying each entry by the same number. For example,

$$2 \begin{bmatrix} 1 & 2 & 3 \\ 4 & 5 & 6 \end{bmatrix} = \begin{bmatrix} 2 & 4 & 6 \\ 8 & 10 & 12 \end{bmatrix}.$$

Matrix addition is also easy. We add matrices element by element. For example,

$$\begin{bmatrix} 1 & 2 & 3 \\ 4 & 5 & 6 \end{bmatrix} + \begin{bmatrix} 1 & 1 & -1 \\ 0 & 2 & 4 \end{bmatrix} = \begin{bmatrix} 2 & 3 & 2 \\ 4 & 7 & 10 \end{bmatrix}.$$

If the sizes do not match, then addition is not defined.

If we denote by 0 the matrix of with all zero entries, by c, d scalars, and by A, B, C matrices, we have the following familiar rules.

$$A + 0 = A = 0 + A,$$
$$A + B = B + A,$$
$$(A + B) + C = A + (B + C),$$
$$c(A + B) = cA + cB,$$
$$(c + d)A = cA + dA.$$

Another useful operation for matrices is the so-called *transpose*. This operation just swaps rows and columns of a matrix. The transpose of A is denoted by A^T. Example:

$$\begin{bmatrix} 1 & 2 & 3 \\ 4 & 5 & 6 \end{bmatrix}^T = \begin{bmatrix} 1 & 4 \\ 2 & 5 \\ 3 & 6 \end{bmatrix}$$

3.2.2 Matrix multiplication

Let us now define matrix multiplication. First we define the so-called *dot product* (or *inner product*) of two vectors. Usually this will be a row vector multiplied with a column vector of the same size. For the dot product we multiply each pair of entries from the first and the second vector and we sum these products. The result is a single number. For example,

$$\begin{bmatrix} a_1 & a_2 & a_3 \end{bmatrix} \cdot \begin{bmatrix} b_1 \\ b_2 \\ b_3 \end{bmatrix} = a_1 b_1 + a_2 b_2 + a_3 b_3.$$

And similarly for larger (or smaller) vectors.

Armed with the dot product we define the *product of matrices*. First let us denote by $\text{row}_i(A)$ the ith row of A and by $\text{column}_j(A)$ the jth column of A. For an $m \times n$ matrix A and an $n \times p$ matrix B we can define the product AB. We let AB be an $m \times p$ matrix whose ij^{th} entry is the dot product

$$\text{row}_i(A) \cdot \text{column}_j(B).$$

Do note how the sizes match up. Example:

$$\begin{bmatrix} 1 & 2 & 3 \\ 4 & 5 & 6 \end{bmatrix} \begin{bmatrix} 1 & 0 & -1 \\ 1 & 1 & 1 \\ 1 & 0 & 0 \end{bmatrix} =$$

$$= \begin{bmatrix} 1 \cdot 1 + 2 \cdot 1 + 3 \cdot 1 & 1 \cdot 0 + 2 \cdot 1 + 3 \cdot 0 & 1 \cdot (-1) + 2 \cdot 1 + 3 \cdot 0 \\ 4 \cdot 1 + 5 \cdot 1 + 6 \cdot 1 & 4 \cdot 0 + 5 \cdot 1 + 6 \cdot 0 & 4 \cdot (-1) + 5 \cdot 1 + 6 \cdot 0 \end{bmatrix} = \begin{bmatrix} 6 & 2 & 1 \\ 15 & 5 & 1 \end{bmatrix}$$

For multiplication we want an analogue of a 1. This analogue is the so-called *identity matrix*. The identity matrix is a square matrix with 1s on the main diagonal and zeros everywhere else. It is usually denoted by I. For each size we have a different identity matrix and so sometimes we may denote the size as a subscript. For example, the I_3 would be the 3×3 identity matrix

$$I = I_3 = \begin{bmatrix} 1 & 0 & 0 \\ 0 & 1 & 0 \\ 0 & 0 & 1 \end{bmatrix}.$$

We have the following rules for matrix multiplication. Suppose that A, B, C are matrices of the correct sizes so that the following make sense. Let α denote a scalar (number).

$$A(BC) = (AB)C,$$
$$A(B + C) = AB + AC,$$
$$(B + C)A = BA + CA,$$
$$\alpha(AB) = (\alpha A)B = A(\alpha B),$$
$$IA = A = AI.$$

A few warnings are in order.

(i) $AB \neq BA$ in general (it may be true by fluke sometimes). That is, matrices do not commute. For example, take $A = \begin{bmatrix} 1 & 1 \\ 1 & 1 \end{bmatrix}$ and $B = \begin{bmatrix} 1 & 0 \\ 0 & 2 \end{bmatrix}$.

(ii) $AB = AC$ does not necessarily imply $B = C$, even if A is not 0.

(iii) $AB = 0$ does not necessarily mean that $A = 0$ or $B = 0$. For example, take $A = B = \begin{bmatrix} 0 & 1 \\ 0 & 0 \end{bmatrix}$.

For the last two items to hold we would need to "divide" by a matrix. This is where the *matrix inverse* comes in. Suppose that A and B are $n \times n$ matrices such that

$$AB = I = BA.$$

Then we call B the inverse of A and we denote B by A^{-1}. If the inverse of A exists, then we call A *invertible*. If A is not invertible we sometimes say A is *singular*.

If A is invertible, then $AB = AC$ does imply that $B = C$ (in particular the inverse of A is unique). We just multiply both sides by A^{-1} to get $A^{-1}AB = A^{-1}AC$ or $IB = IC$ or $B = C$. It is also not hard to see that $\left(A^{-1}\right)^{-1} = A$.

3.2.3 The determinant

Let us now discuss *determinants* of square matrices. We define the determinant of a 1×1 matrix as the value of its only entry. For a 2×2 matrix we define

$$\det\left(\begin{bmatrix} a & b \\ c & d \end{bmatrix}\right) \overset{\text{def}}{=} ad - bc.$$

Before trying to compute the determinant for larger matrices, let us first note the meaning of the determinant. Consider an $n \times n$ matrix as a mapping of the n dimensional euclidean space $\mathbb{R}^n$ to itself, where $\vec{x}$ gets sent to $A\vec{x}$. In particular, a 2×2 matrix A is a mapping of the plane to itself. The determinant of A is the factor by which the area of objects gets changed. If we take the unit square

(square of side 1) in the plane, then A takes the square to a parallelogram of area $|\det(A)|$. The sign of $\det(A)$ denotes changing of orientation (negative if the axes get flipped). For example, let

$$A = \begin{bmatrix} 1 & 1 \\ -1 & 1 \end{bmatrix}.$$

Then $\det(A) = 1 + 1 = 2$. Let us see where the square with vertices $(0,0)$, $(1,0)$, $(0,1)$, and $(1,1)$ gets sent. Clearly $(0,0)$ gets sent to $(0,0)$.

$$\begin{bmatrix} 1 & 1 \\ -1 & 1 \end{bmatrix}\begin{bmatrix} 1 \\ 0 \end{bmatrix} = \begin{bmatrix} 1 \\ -1 \end{bmatrix}, \qquad \begin{bmatrix} 1 & 1 \\ -1 & 1 \end{bmatrix}\begin{bmatrix} 0 \\ 1 \end{bmatrix} = \begin{bmatrix} 1 \\ 1 \end{bmatrix}, \qquad \begin{bmatrix} 1 & 1 \\ -1 & 1 \end{bmatrix}\begin{bmatrix} 1 \\ 1 \end{bmatrix} = \begin{bmatrix} 2 \\ 0 \end{bmatrix}.$$

The image of the square is another square with vertices $(0,0)$, $(1,-1)$, $(1,1)$, and $(2,0)$. The image square has a side of length $\sqrt{2}$ and is therefore of area 2.

If you think back to high school geometry, you may have seen a formula for computing the area of a parallelogram with vertices $(0,0)$, (a,c), (b,d) and $(a+b,c+d)$. And it is precisely

$$\left| \det\left(\begin{bmatrix} a & b \\ c & d \end{bmatrix} \right) \right|.$$

The vertical lines above mean absolute value. The matrix $\begin{bmatrix} a & b \\ c & d \end{bmatrix}$ carries the unit square to the given parallelogram.

Now we define the determinant for larger matrices. We define A_{ij} as the matrix A with the ith row and the jth column deleted. To compute the determinant of a matrix, pick one row, say the ith row and compute:

$$\det(A) = \sum_{j=1}^{n} (-1)^{i+j} a_{ij} \det(A_{ij}).$$

For the first row we get

$$\det(A) = a_{11} \det(A_{11}) - a_{12} \det(A_{12}) + a_{13} \det(A_{13}) - \cdots \begin{cases} +a_{1n} \det(A_{1n}) & \text{if } n \text{ is odd,} \\ -a_{1n} \det(A_{1n}) & \text{if } n \text{ even.} \end{cases}$$

We alternately add and subtract the determinants of the submatrices A_{ij} for a fixed i and all j. For a 3×3 matrix, picking the first row, we get $\det(A) = a_{11} \det(A_{11}) - a_{12} \det(A_{12}) + a_{13} \det(A_{13})$. For example,

$$\det\left(\begin{bmatrix} 1 & 2 & 3 \\ 4 & 5 & 6 \\ 7 & 8 & 9 \end{bmatrix} \right) = 1 \cdot \det\left(\begin{bmatrix} 5 & 6 \\ 8 & 9 \end{bmatrix} \right) - 2 \cdot \det\left(\begin{bmatrix} 4 & 6 \\ 7 & 9 \end{bmatrix} \right) + 3 \cdot \det\left(\begin{bmatrix} 4 & 5 \\ 7 & 8 \end{bmatrix} \right)$$

$$= 1(5 \cdot 9 - 6 \cdot 8) - 2(4 \cdot 9 - 6 \cdot 7) + 3(4 \cdot 8 - 5 \cdot 7) = 0.$$

The numbers $(-1)^{i+j} \det(A_{ij})$ are called *cofactors* of the matrix and this way of computing the determinant is called the *cofactor expansion*. It is also possible to compute the determinant by expanding along columns (picking a column instead of a row above).

Note that a common notation for the determinant is a pair of vertical lines:

$$\begin{vmatrix} a & b \\ c & d \end{vmatrix} = \det\left(\begin{bmatrix} a & b \\ c & d \end{bmatrix}\right).$$

I personally find this notation confusing as vertical lines usually mean a positive quantity, while determinants can be negative. I will not use this notation in this book.

One of the most important properties of determinants (in the context of this course) is the following theorem.

Theorem 3.2.1. *An $n \times n$ matrix A is invertible if and only if $\det(A) \neq 0$.*

In fact, there is a formula for the inverse of a 2×2 matrix

$$\begin{bmatrix} a & b \\ c & d \end{bmatrix}^{-1} = \frac{1}{ad - bc} \begin{bmatrix} d & -b \\ -c & a \end{bmatrix}.$$

Notice the determinant of the matrix in the denominator of the fraction. The formula only works if the determinant is nonzero, otherwise we are dividing by zero.

3.2.4   Solving linear systems

One application of matrices we will need is to solve systems of linear equations. This is best shown by example. Suppose that we have the following system of linear equations

$$\begin{aligned} 2x_1 + 2x_2 + 2x_3 &= 2, \\ x_1 + x_2 + 3x_3 &= 5, \\ x_1 + 4x_2 + x_3 &= 10. \end{aligned}$$

Without changing the solution, we could swap equations in this system, we could multiply any of the equations by a nonzero number, and we could add a multiple of one equation to another equation. It turns out these operations always suffice to find a solution.

It is easier to write the system as a matrix equation. The system above can be written as

$$\begin{bmatrix} 2 & 2 & 2 \\ 1 & 1 & 3 \\ 1 & 4 & 1 \end{bmatrix} \begin{bmatrix} x_1 \\ x_2 \\ x_3 \end{bmatrix} = \begin{bmatrix} 2 \\ 5 \\ 10 \end{bmatrix}.$$

To solve the system we put the coefficient matrix (the matrix on the left hand side of the equation) together with the vector on the right and side and get the so-called *augmented matrix*

$$\left[\begin{array}{ccc|c} 2 & 2 & 2 & 2 \\ 1 & 1 & 3 & 5 \\ 1 & 4 & 1 & 10 \end{array}\right].$$

We apply the following three elementary operations.

(i) Swap two rows.

(ii) Multiply a row by a nonzero number.

(iii) Add a multiple of one row to another row.

We keep doing these operations until we get into a state where it is easy to read off the answer, or until we get into a contradiction indicating no solution, for example if we come up with an equation such as $0 = 1$.

Let us work through the example. First multiply the first row by $1/2$ to obtain

$$\left[\begin{array}{ccc|c} 1 & 1 & 1 & 1 \\ 1 & 1 & 3 & 5 \\ 1 & 4 & 1 & 10 \end{array}\right].$$

Now subtract the first row from the second and third row.

$$\left[\begin{array}{ccc|c} 1 & 1 & 1 & 1 \\ 0 & 0 & 2 & 4 \\ 0 & 3 & 0 & 9 \end{array}\right]$$

Multiply the last row by $1/3$ and the second row by $1/2$.

$$\left[\begin{array}{ccc|c} 1 & 1 & 1 & 1 \\ 0 & 0 & 1 & 2 \\ 0 & 1 & 0 & 3 \end{array}\right]$$

Swap rows 2 and 3.

$$\left[\begin{array}{ccc|c} 1 & 1 & 1 & 1 \\ 0 & 1 & 0 & 3 \\ 0 & 0 & 1 & 2 \end{array}\right]$$

Subtract the last row from the first, then subtract the second row from the first.

$$\left[\begin{array}{ccc|c} 1 & 0 & 0 & -4 \\ 0 & 1 & 0 & 3 \\ 0 & 0 & 1 & 2 \end{array}\right]$$

If we think about what equations this augmented matrix represents, we see that $x_1 = -4$, $x_2 = 3$, and $x_3 = 2$. We try this solution in the original system and, voilà, it works!

Exercise 3.2.1: *Check that the solution above really solves the given equations.*

If we write this equation in matrix notation as

$$A\vec{x} = \vec{b},$$

where A is the matrix $\begin{bmatrix} 2 & 2 & 2 \\ 1 & 1 & 3 \\ 1 & 4 & 1 \end{bmatrix}$ and $\vec{b}$ is the vector $\begin{bmatrix} 2 \\ 5 \\ 10 \end{bmatrix}$. The solution can be also computed via the inverse,

$$\vec{x} = A^{-1}A\vec{x} = A^{-1}\vec{b}.$$

One last note to make about linear systems of equations is that it is possible that the solution is not unique (or that no solution exists). It is easy to tell if a solution does not exist. If during the row reduction you come up with a row where all the entries except the last one are zero (the last entry in a row corresponds to the right hand side of the equation) the system is *inconsistent* and has no solution. For example for a system of 3 equations and 3 unknowns, if you find a row such as [0   0   0 | 1] in the augmented matrix, you know the system is inconsistent.

You generally try to use row operations until the following conditions are satisfied. The first nonzero entry in each row is called the *leading entry.*

(i) There is only one leading entry in each column.

(ii) All the entries above and below a leading entry are zero.

(iii) All leading entries are 1.

Such a matrix is said to be in *reduced row echelon form.* The variables corresponding to columns with no leading entries are said to be *free variables.* Free variables mean that we can pick those variables to be anything we want and then solve for the rest of the unknowns.

Example 3.2.1: The following augmented matrix is in reduced row echelon form.

$$\begin{bmatrix} 1 & 2 & 0 & 3 \\ 0 & 0 & 1 & 1 \\ 0 & 0 & 0 & 0 \end{bmatrix}$$

Suppose the variables are x_1, x_2, and x_3. Then x_2 is the free variable, $x_1 = 3 - 2x_2$, and $x_3 = 1$.

On the other hand if during the row reduction process you come up with the matrix

$$\begin{bmatrix} 1 & 2 & 13 & 3 \\ 0 & 0 & 1 & 1 \\ 0 & 0 & 0 & 3 \end{bmatrix},$$

there is no need to go further. The last row corresponds to the equation $0x_1 + 0x_2 + 0x_3 = 3$, which is preposterous. Hence, no solution exists.

3.2.5 Computing the inverse

If the coefficient matrix is square and there exists a unique solution $\vec{x}$ to $A\vec{x} = \vec{b}$ for any $\vec{b}$, then A is invertible. In fact by multiplying both sides by A^{-1} you can see that $\vec{x} = A^{-1}\vec{b}$. So it is useful to compute the inverse if you want to solve the equation for many different right hand sides $\vec{b}$.

We have a formula for the 2×2 inverse, but it is also not hard to compute inverses of larger matrices. While we will not have too much occasion to compute inverses for larger matrices than 2×2 by hand, let us touch on how to do it. Finding the inverse of A is actually just solving a bunch of linear equations. If we can solve $A\vec{x}_k = \vec{e}_k$ where $\vec{e}_k$ is the vector with all zeros except a 1 at the kth position, then the inverse is the matrix with the columns $\vec{x}_k$ for $k = 1, \ldots, n$ (exercise: why?). Therefore, to find the inverse we write a larger $n \times 2n$ augmented matrix $[A \mid I]$, where I is the identity matrix. We then perform row reduction. The reduced row echelon form of $[A \mid I]$ will be of the form $[I \mid A^{-1}]$ if and only if A is invertible. We then just read off the inverse A^{-1}.

3.2.6 Exercises

***Exercise* 3.2.2:** *Solve* $\left[\begin{smallmatrix} 1 & 2 \\ 3 & 4 \end{smallmatrix}\right] \vec{x} = \left[\begin{smallmatrix} 5 \\ 6 \end{smallmatrix}\right]$ *by using matrix inverse.*

***Exercise* 3.2.3:** *Compute determinant of* $\left[\begin{smallmatrix} 9 & -2 & -6 \\ -8 & 3 & 6 \\ 10 & -2 & -6 \end{smallmatrix}\right]$.

***Exercise* 3.2.4:** *Compute determinant of* $\left[\begin{smallmatrix} 1 & 2 & 3 & 1 \\ 4 & 0 & 5 & 0 \\ 6 & 0 & 7 & 0 \\ 8 & 0 & 10 & 1 \end{smallmatrix}\right]$. *Hint: Expand along the proper row or column to make the calculations simpler.*

***Exercise* 3.2.5:** *Compute inverse of* $\left[\begin{smallmatrix} 1 & 2 & 3 \\ 1 & 1 & 1 \\ 0 & 1 & 0 \end{smallmatrix}\right]$.

***Exercise* 3.2.6:** *For which h is* $\left[\begin{smallmatrix} 1 & 2 & 3 \\ 4 & 5 & 6 \\ 7 & 8 & h \end{smallmatrix}\right]$ *not invertible? Is there only one such h? Are there several? Infinitely many?*

***Exercise* 3.2.7:** *For which h is* $\left[\begin{smallmatrix} h & 1 & 1 \\ 0 & h & 0 \\ 1 & 1 & h \end{smallmatrix}\right]$ *not invertible? Find all such h.*

***Exercise* 3.2.8:** *Solve* $\left[\begin{smallmatrix} 9 & -2 & -6 \\ -8 & 3 & 6 \\ 10 & -2 & -6 \end{smallmatrix}\right] \vec{x} = \left[\begin{smallmatrix} 1 \\ 2 \\ 3 \end{smallmatrix}\right]$.

***Exercise* 3.2.9:** *Solve* $\left[\begin{smallmatrix} 5 & 3 & 7 \\ 8 & 4 & 4 \\ 6 & 3 & 3 \end{smallmatrix}\right] \vec{x} = \left[\begin{smallmatrix} 2 \\ 0 \\ 0 \end{smallmatrix}\right]$.

***Exercise* 3.2.10:** *Solve* $\left[\begin{smallmatrix} 3 & 2 & 3 & 0 \\ 3 & 3 & 3 & 3 \\ 0 & 2 & 4 & 2 \\ 2 & 3 & 4 & 3 \end{smallmatrix}\right] \vec{x} = \left[\begin{smallmatrix} 2 \\ 0 \\ 4 \\ 1 \end{smallmatrix}\right]$.

***Exercise* 3.2.11:** *Find 3 nonzero 2×2 matrices A, B, and C such that $AB = AC$ but $B \neq C$.*

***Exercise* 3.2.101:** *Compute determinant of* $\left[\begin{smallmatrix} 1 & 1 & 1 \\ 2 & 3 & -5 \\ 1 & -1 & 0 \end{smallmatrix}\right]$

***Exercise* 3.2.102:** *Find t such that $\left[\begin{smallmatrix} 1 & t \\ -1 & 2 \end{smallmatrix}\right]$ is not invertible.*

***Exercise* 3.2.103:** *Solve $\left[\begin{smallmatrix} 1 & 1 \\ 1 & -1 \end{smallmatrix}\right] \vec{x} = \left[\begin{smallmatrix} 10 \\ 20 \end{smallmatrix}\right]$.*

***Exercise* 3.2.104:** *Suppose a, b, c are nonzero numbers. Let $M = \left[\begin{smallmatrix} a & 0 \\ 0 & b \end{smallmatrix}\right]$, $N = \left[\begin{smallmatrix} a & 0 & 0 \\ 0 & b & 0 \\ 0 & 0 & c \end{smallmatrix}\right]$. a) Compute M^{-1}. b) Compute N^{-1}.*

3.3 Linear systems of ODEs

Note: less than 1 lecture, second part of §5.1 in [EP], §7.4 in [BD]

First let us talk about matrix or vector valued functions. Such a function is just a matrix whose entries depend on some variable. If t is the independent variable, we write a *vector valued function* $\vec{x}(t)$ as

$$\vec{x}(t) = \begin{bmatrix} x_1(t) \\ x_2(t) \\ \vdots \\ x_n(t) \end{bmatrix}.$$

Similarly a *matrix valued function* $A(t)$ is

$$A(t) = \begin{bmatrix} a_{11}(t) & a_{12}(t) & \cdots & a_{1n}(t) \\ a_{21}(t) & a_{22}(t) & \cdots & a_{2n}(t) \\ \vdots & \vdots & \ddots & \vdots \\ a_{n1}(t) & a_{n2}(t) & \cdots & a_{nn}(t) \end{bmatrix}.$$

We can talk about the derivative $A'(t)$ or $\frac{dA}{dt}$. This is just the matrix valued function whose ij^{th} entry is $a'_{ij}(t)$.

Rules of differentiation of matrix valued functions are similar to rules for normal functions. Let $A(t)$ and $B(t)$ be matrix valued functions. Let c a scalar and let C be a constant matrix. Then

$$(A(t) + B(t))' = A'(t) + B'(t),$$
$$(A(t)B(t))' = A'(t)B(t) + A(t)B'(t),$$
$$(cA(t))' = cA'(t),$$
$$(CA(t))' = CA'(t),$$
$$(A(t)\,C)' = A'(t)\,C.$$

Note the order of the multiplication in the last two expressions.

A *first order linear system of ODEs* is a system that can be written as the vector equation

$$\vec{x}'(t) = P(t)\vec{x}(t) + \vec{f}(t),$$

where $P(t)$ is a matrix valued function, and $\vec{x}(t)$ and $\vec{f}(t)$ are vector valued functions. We will often suppress the dependence on t and only write $\vec{x}' = P\vec{x} + \vec{f}$. A solution of the system is a vector valued function $\vec{x}$ satisfying the vector equation.

For example, the equations

$$x_1' = 2tx_1 + e^t x_2 + t^2,$$
$$x_2' = \frac{x_1}{t} - x_2 + e^t,$$

can be written as

$$\vec{x}' = \begin{bmatrix} 2t & e^t \\ 1/t & -1 \end{bmatrix} \vec{x} + \begin{bmatrix} t^2 \\ e^t \end{bmatrix}.$$

We will mostly concentrate on equations that are not just linear, but are in fact *constant coefficient* equations. That is, the matrix P will be constant; it will not depend on t.

When $\vec{f} = \vec{0}$ (the zero vector), then we say the system is *homogeneous*. For homogeneous linear systems we have the principle of superposition, just like for single homogeneous equations.

Theorem 3.3.1 (Superposition). *Let $\vec{x}' = P\vec{x}$ be a linear homogeneous system of ODEs. Suppose that $\vec{x}_1, \ldots, \vec{x}_n$ are n solutions of the equation, then*

$$\vec{x} = c_1 \vec{x}_1 + c_2 \vec{x}_2 + \cdots + c_n \vec{x}_n, \tag{3.1}$$

is also a solution. Furthermore, if this is a system of n equations (P is $n \times n$), and $\vec{x}_1, \vec{x}_2, \ldots, \vec{x}_n$ are linearly independent, then every solution $\vec{x}$ can be written as (3.1).

Linear independence for vector valued functions is the same idea as for normal functions. The vector valued functions $\vec{x}_1, \vec{x}_2, \ldots, \vec{x}_n$ are linearly independent when

$$c_1 \vec{x}_1 + c_2 \vec{x}_2 + \cdots + c_n \vec{x}_n = \vec{0}$$

has only the solution $c_1 = c_2 = \cdots = c_n = 0$, where the equation must hold for all t.

Example 3.3.1: $\vec{x}_1 = \begin{bmatrix} t^2 \\ t \end{bmatrix}$, $\vec{x}_2 = \begin{bmatrix} 0 \\ 1+t \end{bmatrix}$, $\vec{x}_3 = \begin{bmatrix} -t^2 \\ 1 \end{bmatrix}$ are linearly depdendent because $\vec{x}_1 + \vec{x}_3 = \vec{x}_2$, and this holds for all t. So $c_1 = 1$, $c_2 = -1$, and $c_3 = 1$ above will work.

On the other hand if we change the example just slightly $\vec{x}_1 = \begin{bmatrix} t^2 \\ t \end{bmatrix}$, $\vec{x}_2 = \begin{bmatrix} 0 \\ t \end{bmatrix}$, $\vec{x}_3 = \begin{bmatrix} -t^2 \\ 1 \end{bmatrix}$, then the functions are linearly independent. First write $c_1 \vec{x}_1 + c_2 \vec{x}_2 + c_3 \vec{x}_3 = \vec{0}$ and note that it has to hold for all t. We get that

$$c_1 \vec{x}_1 + c_2 \vec{x}_2 + c_3 \vec{x}_3 = \begin{bmatrix} c_1 t^2 - c_3 t^3 \\ c_1 t + c_2 t + c_3 \end{bmatrix} = \begin{bmatrix} 0 \\ 0 \end{bmatrix}.$$

In other words $c_1 t^2 - c_3 t^3 = 0$ and $c_1 t + c_2 t + c_3 = 0$. If we set $t = 0$, then the second equation becomes $c_3 = 0$. But then the first equation becomes $c_1 t^2 = 0$ for all t and so $c_1 = 0$. Thus the second equation is just $c_2 t = 0$, which means $c_2 = 0$. So $c_1 = c_2 = c_3 = 0$ is the only solution and $\vec{x}_1$, $\vec{x}_2$, and $\vec{x}_3$ are linearly independent.

The linear combination $c_1 \vec{x}_1 + c_2 \vec{x}_2 + \cdots + c_n \vec{x}_n$ could always be written as

$$X(t) \vec{c},$$

where $X(t)$ is the matrix with columns $\vec{x}_1, \vec{x}_2, \ldots, \vec{x}_n$, and $\vec{c}$ is the column vector with entries $c_1, c_2, \ldots, c_n$. The matrix valued function $X(t)$ is called the *fundamental matrix*, or the *fundamental matrix solution*.

To solve nonhomogeneous first order linear systems, we use the same technique as we applied to solve single linear nonhomogeneous equations.

Theorem 3.3.2. *Let $\vec{x}' = P\vec{x} + \vec{f}$ be a linear system of ODEs. Suppose $\vec{x}_p$ is one particular solution. Then every solution can be written as*

$$\vec{x} = \vec{x}_c + \vec{x}_p,$$

where $\vec{x}_c$ is a solution to the associated homogeneous equation ($\vec{x}' = P\vec{x}$).

So the procedure for systems is the same as for single equations. We find a particular solution to the nonhomogeneous equation, then we find the general solution to the associated homogeneous equation, and finally we add the two together.

Alright, suppose you have found the general solution of $\vec{x}' = P\vec{x} + \vec{f}$. Next suppose you are given an initial condition of the form $\vec{x}(t_0) = \vec{b}$ for some constant vector $\vec{b}$. Let $X(t)$ be the fundamental matrix solution of the associated homogeneous equation (i.e. columns of $X(t)$ are solutions). The general solution can be written as

$$\vec{x}(t) = X(t)\vec{c} + \vec{x}_p(t).$$

We are seeking a vector $\vec{c}$ such that

$$\vec{b} = \vec{x}(t_0) = X(t_0)\vec{c} + \vec{x}_p(t_0).$$

In other words, we are solving for $\vec{c}$ the nonhomogeneous system of linear equations

$$X(t_0)\vec{c} = \vec{b} - \vec{x}_p(t_0).$$

Example 3.3.2: In § 3.1 we solved the system

$$\begin{aligned} x_1' &= x_1, \\ x_2' &= x_1 - x_2, \end{aligned}$$

with initial conditions $x_1(0) = 1$, $x_2(0) = 2$. Let us consider this problem in the language of this section.

The system is homogeneous, so $\vec{f}(t) = \vec{0}$. We write the system and the initial conditions as

$$\vec{x}' = \begin{bmatrix} 1 & 0 \\ 1 & -1 \end{bmatrix} \vec{x}, \qquad \vec{x}(0) = \begin{bmatrix} 1 \\ 2 \end{bmatrix}.$$

We found the general solution was $x_1 = c_1 e^t$ and $x_2 = \frac{c_1}{2}e^t + c_2 e^{-t}$. Letting $c_1 = 1$ and $c_2 = 0$, we obtain the solution $\begin{bmatrix} e^t \\ (1/2)e^t \end{bmatrix}$. Letting $c_1 = 0$ and $c_2 = 1$, we obtain $\begin{bmatrix} 0 \\ e^{-t} \end{bmatrix}$. These two solutions are linearly independent, as can be seen by setting $t = 0$, and noting that the resulting constant vectors are linearly independent. In matrix notation, the fundamental matrix solution is, therefore,

$$X(t) = \begin{bmatrix} e^t & 0 \\ \frac{1}{2}e^t & e^{-t} \end{bmatrix}.$$

To solve the initial value problem we solve for $\vec{c}$ the equation

$$X(0)\,\vec{c} = \vec{b},$$

or in other words,

$$\begin{bmatrix} 1 & 0 \\ \frac{1}{2} & 1 \end{bmatrix} \vec{c} = \begin{bmatrix} 1 \\ 2 \end{bmatrix}.$$

A single elementary row operation shows $\vec{c} = \begin{bmatrix} 1 \\ 3/2 \end{bmatrix}$. Our solution is

$$\vec{x}(t) = X(t)\,\vec{c} = \begin{bmatrix} e^t & 0 \\ \frac{1}{2}e^t & e^{-t} \end{bmatrix} \begin{bmatrix} 1 \\ \frac{3}{2} \end{bmatrix} = \begin{bmatrix} e^t \\ \frac{1}{2}e^t + \frac{3}{2}e^{-t} \end{bmatrix}.$$

This new solution agrees with our previous solution from § 3.1.

3.3.1 Exercises

Exercise 3.3.1: *Write the system $x_1' = 2x_1 - 3tx_2 + \sin t$, $x_2' = e^t x_1 + 3x_2 + \cos t$ in the form $\vec{x}' = P(t)\vec{x} + \vec{f}(t)$.*

Exercise 3.3.2: *a) Verify that the system $\vec{x}' = \begin{bmatrix} 1 & 3 \\ 3 & 1 \end{bmatrix} \vec{x}$ has the two solutions $\begin{bmatrix} 1 \\ 1 \end{bmatrix} e^{4t}$ and $\begin{bmatrix} 1 \\ -1 \end{bmatrix} e^{-2t}$. b) Write down the general solution. c) Write down the general solution in the form $x_1 = ?$, $x_2 = ?$ (i.e. write down a formula for each element of the solution).*

Exercise 3.3.3: *Verify that $\begin{bmatrix} 1 \\ 1 \end{bmatrix} e^t$ and $\begin{bmatrix} 1 \\ -1 \end{bmatrix} e^t$ are linearly independent. Hint: Just plug in $t = 0$.*

Exercise 3.3.4: *Verify that $\begin{bmatrix} 1 \\ 1 \\ 0 \end{bmatrix} e^t$ and $\begin{bmatrix} 1 \\ -1 \\ 1 \end{bmatrix} e^t$ and $\begin{bmatrix} 1 \\ -1 \\ 1 \end{bmatrix} e^{2t}$ are linearly independent. Hint: You must be a bit more tricky than in the previous exercise.*

Exercise 3.3.5: *Verify that $\begin{bmatrix} t \\ t^2 \end{bmatrix}$ and $\begin{bmatrix} t^3 \\ t^4 \end{bmatrix}$ are linearly independent.*

Exercise 3.3.6: *Take the system $x_1' + x_2' = x_1$, $x_1' - x_2' = x_2$. a) Write it in the form $A\vec{x}' = B\vec{x}$ for matrices A and B. b) Compute A^{-1} and use that to write the system in the form $\vec{x}' = P\vec{x}$.*

Exercise 3.3.101: *Are $\begin{bmatrix} e^{2t} \\ e^t \end{bmatrix}$ and $\begin{bmatrix} e^t \\ e^{2t} \end{bmatrix}$ linearly independent? Justify.*

Exercise 3.3.102: *Are $\begin{bmatrix} \cosh(t) \\ 1 \end{bmatrix}$, $\begin{bmatrix} e^t \\ 1 \end{bmatrix}$, and $\begin{bmatrix} e^{-t} \\ 1 \end{bmatrix}$ linearly independent? Justify.*

Exercise 3.3.103: *Write $x' = 3x - y + e^t$, $y' = tx$ in matrix notation.*

Exercise 3.3.104: *a) Write $x_1' = 2tx_2$, $x_2' = 2tx_2$ in matrix notation. b) Solve and write the solution in matrix notation.*

3.4 Eigenvalue method

Note: 2 lectures, §5.2 in [EP], part of §7.3, §7.5, and §7.6 in [BD]

In this section we will learn how to solve linear homogeneous constant coefficient systems of ODEs by the eigenvalue method. Suppose we have such a system

$$\vec{x}' = P\vec{x},$$

where P is a constant square matrix. We wish to adapt the method for the single constant coefficient equation by trying the function $e^{\lambda t}$. However, $\vec{x}$ is a vector. So we try $\vec{x} = \vec{v}e^{\lambda t}$, where $\vec{v}$ is an arbitrary constant vector. We plug this $\vec{x}$ into the equation to get

$$\lambda \vec{v}e^{\lambda t} = P\vec{v}e^{\lambda t}.$$

We divide by $e^{\lambda t}$ and notice that we are looking for a scalar λ and a vector $\vec{v}$ that satisfy the equation

$$\lambda \vec{v} = P\vec{v}.$$

To solve this equation we need a little bit more linear algebra, which we now review.

3.4.1 Eigenvalues and eigenvectors of a matrix

Let A be a constant square matrix. Suppose there is a scalar λ and a nonzero vector $\vec{v}$ such that

$$A\vec{v} = \lambda \vec{v}.$$

We then call λ an *eigenvalue* of A and $\vec{v}$ is said to be a corresponding *eigenvector*.

Example 3.4.1: The matrix $\begin{bmatrix} 2 & 1 \\ 0 & 1 \end{bmatrix}$ has an eigenvalue of $\lambda = 2$ with a corresponding eigenvector $\begin{bmatrix} 1 \\ 0 \end{bmatrix}$ because

$$\begin{bmatrix} 2 & 1 \\ 0 & 1 \end{bmatrix}\begin{bmatrix} 1 \\ 0 \end{bmatrix} = \begin{bmatrix} 2 \\ 0 \end{bmatrix} = 2\begin{bmatrix} 1 \\ 0 \end{bmatrix}.$$

Let us see how to compute the eigenvalues for any matrix. We rewrite the equation for an eigenvalue as

$$(A - \lambda I)\vec{v} = \vec{0}.$$

We notice that this equation has a nonzero solution $\vec{v}$ only if $A - \lambda I$ is not invertible. Were it invertible, we could write $(A - \lambda I)^{-1}(A - \lambda I)\vec{v} = (A - \lambda I)^{-1}\vec{0}$, which implies $\vec{v} = \vec{0}$. Therefore, A has the eigenvalue λ if and only if λ solves the equation

$$\det(A - \lambda I) = 0.$$

Consequently, we will be able to find an eigenvalue of A without finding a corresponding eigenvector. An eigenvector will have to be found later, once λ is known.

Example 3.4.2: Find all eigenvalues of $\begin{bmatrix} 2 & 1 & 1 \\ 1 & 2 & 0 \\ 0 & 0 & 2 \end{bmatrix}$.

We write

$$
\det\left(\begin{bmatrix} 2 & 1 & 1 \\ 1 & 2 & 0 \\ 0 & 0 & 2 \end{bmatrix} - \lambda \begin{bmatrix} 1 & 0 & 0 \\ 0 & 1 & 0 \\ 0 & 0 & 1 \end{bmatrix}\right) = \det\left(\begin{bmatrix} 2-\lambda & 1 & 1 \\ 1 & 2-\lambda & 0 \\ 0 & 0 & 2-\lambda \end{bmatrix}\right) =
$$

$$
= (2-\lambda)((2-\lambda)^2 - 1) = -(\lambda - 1)(\lambda - 2)(\lambda - 3).
$$

So the eigenvalues are $\lambda = 1$, $\lambda = 2$, and $\lambda = 3$.

For an $n \times n$ matrix, the polynomial we get by computing $\det(A - \lambda I)$ is of degree n, and hence in general, we have n eigenvalues. Some may be repeated, some may be complex.

To find an eigenvector corresponding to an eigenvalue λ, we write

$$
(A - \lambda I)\vec{v} = \vec{0},
$$

and solve for a nontrivial (nonzero) vector $\vec{v}$. If λ is an eigenvalue, this is always possible.

Example 3.4.3: Find an eigenvector of $\begin{bmatrix} 2 & 1 & 1 \\ 1 & 2 & 0 \\ 0 & 0 & 2 \end{bmatrix}$ corresponding to the eigenvalue $\lambda = 3$.

We write

$$
(A - \lambda I)\vec{v} = \left(\begin{bmatrix} 2 & 1 & 1 \\ 1 & 2 & 0 \\ 0 & 0 & 2 \end{bmatrix} - 3 \begin{bmatrix} 1 & 0 & 0 \\ 0 & 1 & 0 \\ 0 & 0 & 1 \end{bmatrix}\right)\begin{bmatrix} v_1 \\ v_2 \\ v_3 \end{bmatrix} = \begin{bmatrix} -1 & 1 & 1 \\ 1 & -1 & 0 \\ 0 & 0 & -1 \end{bmatrix}\begin{bmatrix} v_1 \\ v_2 \\ v_3 \end{bmatrix} = \vec{0}.
$$

It is easy to solve this system of linear equations. We write down the augmented matrix

$$
\begin{bmatrix} -1 & 1 & 1 & | & 0 \\ 1 & -1 & 0 & | & 0 \\ 0 & 0 & -1 & | & 0 \end{bmatrix},
$$

and perform row operations (exercise: which ones?) until we get:

$$
\begin{bmatrix} 1 & -1 & 0 & | & 0 \\ 0 & 0 & 1 & | & 0 \\ 0 & 0 & 0 & | & 0 \end{bmatrix}.
$$

The entries of $\vec{v}$ have to satisfy the equations $v_1 - v_2 = 0$, $v_3 = 0$, and v_2 is a free variable. We can pick v_2 to be arbitrary (but nonzero), let $v_1 = v_2$, and of course $v_3 = 0$. For example, if we pick $v_2 = 1$, then $\vec{v} = \begin{bmatrix} 1 \\ 1 \\ 0 \end{bmatrix}$. Let us verify that $\vec{v}$ really is an eigenvector corresponding to $\lambda = 3$:

$$
\begin{bmatrix} 2 & 1 & 1 \\ 1 & 2 & 0 \\ 0 & 0 & 2 \end{bmatrix}\begin{bmatrix} 1 \\ 1 \\ 0 \end{bmatrix} = \begin{bmatrix} 3 \\ 3 \\ 0 \end{bmatrix} = 3\begin{bmatrix} 1 \\ 1 \\ 0 \end{bmatrix}.
$$

Yay! It worked.

***Exercise* 3.4.1** (easy)*: Are eigenvectors unique? Can you find a different eigenvector for $\lambda = 3$ in the example above? How are the two eigenvectors related?*

Exercise* 3.4.2: When the matrix is 2×2 you do not need to write down the augmented matrix and do row operations when computing eigenvectors (if you have computed the eigenvalues correctly). Can you see why? Explain. Try it for the matrix $\left[\begin{smallmatrix} 2 & 1 \\ 1 & 2 \end{smallmatrix}\right]$.*

3.4.2 The eigenvalue method with distinct real eigenvalues

OK. We have the system of equations

$$\vec{x}' = P\vec{x}.$$

We find the eigenvalues $\lambda_1, \lambda_2, \ldots, \lambda_n$ of the matrix P, and corresponding eigenvectors $\vec{v}_1, \vec{v}_2, \ldots, \vec{v}_n$. Now we notice that the functions $\vec{v}_1 e^{\lambda_1 t}, \vec{v}_2 e^{\lambda_2 t}, \ldots, \vec{v}_n e^{\lambda_n t}$ are solutions of the system of equations and hence $\vec{x} = c_1 \vec{v}_1 e^{\lambda_1 t} + c_2 \vec{v}_2 e^{\lambda_2 t} + \cdots + c_n \vec{v}_n e^{\lambda_n t}$ is a solution.

Theorem 3.4.1. *Take $\vec{x}' = P\vec{x}$. If P is an $n \times n$ constant matrix that has n distinct real eigenvalues $\lambda_1, \lambda_2, \ldots, \lambda_n$, then there exist n linearly independent corresponding eigenvectors $\vec{v}_1, \vec{v}_2, \ldots, \vec{v}_n$, and the general solution to $\vec{x}' = P\vec{x}$ can be written as*

$$\boxed{\vec{x} = c_1 \vec{v}_1 e^{\lambda_1 t} + c_2 \vec{v}_2 e^{\lambda_2 t} + \cdots + c_n \vec{v}_n e^{\lambda_n t}.}$$

The corresponding fundamental matrix solution is $X(t) = [\, \vec{v}_1 e^{\lambda_1 t} \quad \vec{v}_2 e^{\lambda_2 t} \quad \cdots \quad \vec{v}_n e^{\lambda_n t}\,]$. That is, $X(t)$ is the matrix whose j^{th} column is $\vec{v}_j e^{\lambda_j t}$.

Example 3.4.4: Consider the system

$$\vec{x}' = \begin{bmatrix} 2 & 1 & 1 \\ 1 & 2 & 0 \\ 0 & 0 & 2 \end{bmatrix} \vec{x}.$$

Find the general solution.

Earlier, we found the eigenvalues are $1, 2, 3$. We found the eigenvector $\begin{bmatrix} 1 \\ 1 \\ 0 \end{bmatrix}$ for the eigenvalue 3. Similarly we find the eigenvector $\begin{bmatrix} 1 \\ -1 \\ 0 \end{bmatrix}$ for the eigenvalue 1, and $\begin{bmatrix} 0 \\ 1 \\ -1 \end{bmatrix}$ for the eigenvalue 2 (exercise: check). Hence our general solution is

$$\vec{x} = c_1 \begin{bmatrix} 1 \\ -1 \\ 0 \end{bmatrix} e^t + c_2 \begin{bmatrix} 0 \\ 1 \\ -1 \end{bmatrix} e^{2t} + c_3 \begin{bmatrix} 1 \\ 1 \\ 0 \end{bmatrix} e^{3t} = \begin{bmatrix} c_1 e^t + c_3 e^{3t} \\ -c_1 e^t + c_2 e^{2t} + c_3 e^{3t} \\ -c_2 e^{2t} \end{bmatrix}.$$

In terms of a fundamental matrix solution

$$\vec{x} = X(t)\vec{c} = \begin{bmatrix} e^t & 0 & e^{3t} \\ -e^t & e^{2t} & e^{3t} \\ 0 & -e^{2t} & 0 \end{bmatrix} \begin{bmatrix} c_1 \\ c_2 \\ c_3 \end{bmatrix}.$$

***Exercise* 3.4.3:** *Check that this $\vec{x}$ really solves the system.*

Note: If we write a homogeneous linear constant coefficient n^{th} order equation as a first order system (as we did in § 3.1), then the eigenvalue equation

$$\det(P - \lambda I) = 0$$

is essentially the same as the characteristic equation we got in § 2.2 and § 2.3.

3.4.3   Complex eigenvalues

A matrix might very well have complex eigenvalues even if all the entries are real. For example, suppose that we have the system

$$\vec{x}' = \begin{bmatrix} 1 & 1 \\ -1 & 1 \end{bmatrix} \vec{x}.$$

Let us compute the eigenvalues of the matrix $P = \begin{bmatrix} 1 & 1 \\ -1 & 1 \end{bmatrix}$.

$$\det(P - \lambda I) = \det\left(\begin{bmatrix} 1 - \lambda & 1 \\ -1 & 1 - \lambda \end{bmatrix}\right) = (1 - \lambda)^2 + 1 = \lambda^2 - 2\lambda + 2 = 0.$$

Thus $\lambda = 1 \pm i$. The corresponding eigenvectors are also complex. First take $\lambda = 1 - i$,

$$(P - (1 - i)I)\vec{v} = \vec{0},$$

$$\begin{bmatrix} i & 1 \\ -1 & i \end{bmatrix} \vec{v} = \vec{0}.$$

The equations $iv_1 + v_2 = 0$ and $-v_1 + iv_2 = 0$ are multiples of each other. So we only need to consider one of them. After picking $v_2 = 1$, for example, we have an eigenvector $\vec{v} = \begin{bmatrix} i \\ 1 \end{bmatrix}$. In similar fashion we find that $\begin{bmatrix} -i \\ 1 \end{bmatrix}$ is an eigenvector corresponding to the eigenvalue $1 + i$.

We could write the solution as

$$\vec{x} = c_1 \begin{bmatrix} i \\ 1 \end{bmatrix} e^{(1-i)t} + c_2 \begin{bmatrix} -i \\ 1 \end{bmatrix} e^{(1+i)t} = \begin{bmatrix} c_1 i e^{(1-i)t} - c_2 i e^{(1+i)t} \\ c_1 e^{(1-i)t} + c_2 e^{(1+i)t} \end{bmatrix}.$$

We would then need to look for complex values c_1 and c_2 to solve any initial conditions. It is perhaps not completely clear that we get a real solution. We could use Euler's formula and do the whole song and dance we did before, but we will not. We will do something a bit smarter first.

We claim that we did not have to look for a second eigenvector (nor for the second eigenvalue). All complex eigenvalues come in pairs (because the matrix P is real).

First a small side note. The real part of a complex number z can be computed as $\frac{z + \bar{z}}{2}$, where the bar above z means $\overline{a + ib} = a - ib$. This operation is called the *complex conjugate*. If a is a real number, then $\bar{a} = a$. Similarly we can bar whole vectors or matrices. If a matrix P is real, then $\overline{P} = P$. We note that $\overline{P\vec{x}} = \overline{P}\,\overline{x} = P\overline{x}$. Therefore,

$$\overline{(P - \lambda I)\vec{v}} = (P - \bar{\lambda}I)\overline{v}.$$

So if $\vec{v}$ is an eigenvector corresponding to the eigenvalue $\lambda = a + ib$, then $\overline{v}$ is an eigenvector corresponding to the eigenvalue $\bar{\lambda} = a - ib$.

Suppose $a + ib$ is a complex eigenvalue of P, and $\vec{v}$ is a corresponding eigenvector. Then

$$\vec{x}_1 = \vec{v}e^{(a+ib)t}$$

is a solution (complex valued) of $\vec{x}' = P\vec{x}$. Euler's formula shows that $\overline{e^{a+ib}} = e^{a-ib}$, and so

$$\vec{x}_2 = \overline{\vec{x}_1} = \overline{v}e^{(a-ib)t}$$

is also a solution. The function

$$\vec{x}_3 = \operatorname{Re}\vec{x}_1 = \operatorname{Re}\vec{v}e^{(a+ib)t} = \frac{\vec{x}_1 + \overline{\vec{x}_1}}{2} = \frac{\vec{x}_1 + \vec{x}_2}{2}$$

is also a solution. And $\vec{x}_3$ is real-valued! Similarly as $\operatorname{Im} z = \frac{z-\bar{z}}{2i}$ is the imaginary part, we find that

$$\vec{x}_4 = \operatorname{Im}\vec{x}_1 = \frac{\vec{x}_1 - \overline{\vec{x}_1}}{2i} = \frac{\vec{x}_1 - \vec{x}_2}{2i}.$$

is also a real-valued solution. It turns out that $\vec{x}_3$ and $\vec{x}_4$ are linearly independent. We will use Euler's formula to separate out the real and imaginary part.

Returning to our problem,

$$\vec{x}_1 = \begin{bmatrix} i \\ 1 \end{bmatrix} e^{(1-i)t} = \begin{bmatrix} i \\ 1 \end{bmatrix} (e^t \cos t - ie^t \sin t) = \begin{bmatrix} ie^t \cos t + e^t \sin t \\ e^t \cos t - ie^t \sin t \end{bmatrix}.$$

Then

$$\operatorname{Re}\vec{x}_1 = \begin{bmatrix} e^t \sin t \\ e^t \cos t \end{bmatrix},$$

$$\operatorname{Im}\vec{x}_1 = \begin{bmatrix} e^t \cos t \\ -e^t \sin t \end{bmatrix},$$

are the two real-valued linearly independent solutions we seek.

***Exercise* 3.4.4:** *Check that these really are solutions.*

The general solution is

$$\vec{x} = c_1 \begin{bmatrix} e^t \sin t \\ e^t \cos t \end{bmatrix} + c_2 \begin{bmatrix} e^t \cos t \\ -e^t \sin t \end{bmatrix} = \begin{bmatrix} c_1 e^t \sin t + c_2 e^t \cos t \\ c_1 e^t \cos t - c_2 e^t \sin t \end{bmatrix}.$$

This solution is real-valued for real c_1 and c_2. Now we can solve for any initial conditions that we may have.

Let us summarize as a theorem.

Theorem 3.4.2. *Let P be a real-valued constant matrix. If P has a complex eigenvalue $a + ib$ and a corresponding eigenvector $\vec{v}$, then P also has a complex eigenvalue $a - ib$ with a corresponding eigenvector $\vec{\bar{v}}$. Furthermore, $\vec{x}' = P\vec{x}$ has two linearly independent real-valued solutions*

$$\vec{x_1} = \operatorname{Re} \vec{v} e^{(a+ib)t}, \qquad and \qquad \vec{x_2} = \operatorname{Im} \vec{v} e^{(a+ib)t}.$$

For each pair of complex eigenvalues $a+ib$ and $a-ib$, we get two real-valued linearly independent solutions. We then go on to the next eigenvalue, which is either a real eigenvalue or another complex eigenvalue pair. If we have n distinct eigenvalues (real or complex), then we end up with n linearly independent solutions.

We can now find a real-valued general solution to any homogeneous system where the matrix has distinct eigenvalues. When we have repeated eigenvalues, matters get a bit more complicated and we will look at that situation in § 3.7.

3.4.4 Exercises

Exercise 3.4.5 (easy): *Let A be a 3×3 matrix with an eigenvalue of 3 and a corresponding eigenvector $\vec{v} = \begin{bmatrix} 1 \\ -1 \\ 3 \end{bmatrix}$. Find $A\vec{v}$.*

Exercise 3.4.6: *a) Find the general solution of $x_1' = 2x_1$, $x_2' = 3x_2$ using the eigenvalue method (first write the system in the form $\vec{x}' = A\vec{x}$). b) Solve the system by solving each equation separately and verify you get the same general solution.*

Exercise 3.4.7: *Find the general solution of $x_1' = 3x_1 + x_2$, $x_2' = 2x_1 + 4x_2$ using the eigenvalue method.*

Exercise 3.4.8: *Find the general solution of $x_1' = x_1 - 2x_2$, $x_2' = 2x_1 + x_2$ using the eigenvalue method. Do not use complex exponentials in your solution.*

Exercise 3.4.9: *a) Compute eigenvalues and eigenvectors of $A = \begin{bmatrix} 9 & -2 & -6 \\ -8 & 3 & 6 \\ 10 & -2 & -6 \end{bmatrix}$. b) Find the general solution of $\vec{x}' = A\vec{x}$.*

Exercise 3.4.10: *Compute eigenvalues and eigenvectors of $\begin{bmatrix} -2 & -1 & -1 \\ 3 & 2 & 1 \\ -3 & -1 & 0 \end{bmatrix}$.*

***Exercise* 3.4.11:** *Let a, b, c, d, e, f be numbers. Find the eigenvalues of $\begin{bmatrix} a & b & c \\ 0 & d & e \\ 0 & 0 & f \end{bmatrix}$.*

***Exercise* 3.4.101:** *a) Compute eigenvalues and eigenvectors of $A = \begin{bmatrix} 1 & 0 & 3 \\ -1 & 0 & 1 \\ 2 & 0 & 2 \end{bmatrix}$. b) Solve the system $\vec{x}' = A\vec{x}$.*

***Exercise* 3.4.102:** *a) Compute eigenvalues and eigenvectors of $A = \begin{bmatrix} 1 & 1 \\ -1 & 0 \end{bmatrix}$. b) Solve the system $\vec{x}' = A\vec{x}$.*

***Exercise* 3.4.103:** *Solve $x_1' = x_2$, $x_2' = x_1$ using the eigenvalue method.*

***Exercise* 3.4.104:** *Solve $x_1' = x_2$, $x_2' = -x_1$ using the eigenvalue method.*

3.5   Two dimensional systems and their vector fields

Note: 1 lecture, part of §6.2 in [EP], parts of §7.5 and §7.6 in [BD]

Let us take a moment to talk about constant coefficient linear homogeneous systems in the plane. Much intuition can be obtained by studying this simple case. Suppose we have a 2×2 matrix P and the system

$$\begin{bmatrix} x \\ y \end{bmatrix}' = P \begin{bmatrix} x \\ y \end{bmatrix}. \tag{3.2}$$

The system is autonomous (compare this section to § 1.6) and so we can draw a vector field (see end of § 3.1). We will be able to visually tell what the vector field looks like and how the solutions behave, once we find the eigenvalues and eigenvectors of the matrix P. For this section, we assume that P has two eigenvalues and two corresponding eigenvectors.

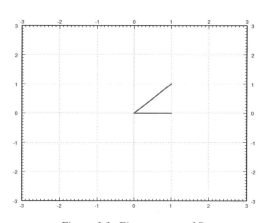

Figure 3.3: Eigenvectors of P.

Case 1. Suppose that the eigenvalues of P are real and positive. We find two corresponding eigenvectors and plot them in the plane. For example, take the matrix $\left[\begin{smallmatrix} 1 & 1 \\ 0 & 2 \end{smallmatrix}\right]$. The eigenvalues are 1 and 2 and corresponding eigenvectors are $\left[\begin{smallmatrix} 1 \\ 0 \end{smallmatrix}\right]$ and $\left[\begin{smallmatrix} 1 \\ 1 \end{smallmatrix}\right]$. See Figure 3.3.

Now suppose that x and y are on the line determined by an eigenvector $\vec{v}$ for an eigenvalue λ. That is, $\left[\begin{smallmatrix} x \\ y \end{smallmatrix}\right] = a\vec{v}$ for some scalar a. Then

$$\begin{bmatrix} x \\ y \end{bmatrix}' = P \begin{bmatrix} x \\ y \end{bmatrix} = P(a\vec{v}) = a(P\vec{v}) = a\lambda\vec{v}.$$

The derivative is a multiple of $\vec{v}$ and hence points along the line determined by $\vec{v}$. As $\lambda > 0$, the derivative points in the direction of $\vec{v}$ when a is positive and in the opposite direction when a is negative. Let us draw the lines determined by the eigenvectors, and let us draw arrows on the lines to indicate the directions. See Figure 3.4 on the facing page.

We fill in the rest of the arrows for the vector field and we also draw a few solutions. See Figure 3.5 on the next page. The picture looks like a source with arrows coming out from the origin. Hence we call this type of picture a *source* or sometimes an *unstable node*.

Case 2. Suppose both eigenvalues are negative. For example, take the negation of the matrix in case 1, $\left[\begin{smallmatrix} -1 & -1 \\ 0 & -2 \end{smallmatrix}\right]$. The eigenvalues are -1 and -2 and corresponding eigenvectors are the same, $\left[\begin{smallmatrix} 1 \\ 0 \end{smallmatrix}\right]$ and $\left[\begin{smallmatrix} 1 \\ 1 \end{smallmatrix}\right]$. The calculation and the picture are almost the same. The only difference is that the eigenvalues are negative and hence all arrows are reversed. We get the picture in Figure 3.6 on the facing page. We call this kind of picture a *sink* or sometimes a *stable node*.

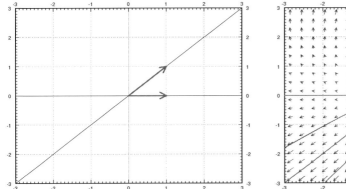

Figure 3.4: Eigenvectors of P with directions.

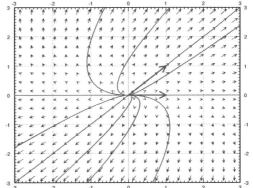

Figure 3.5: Example source vector field with eigenvectors and solutions.

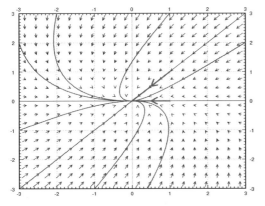

Figure 3.6: Example sink vector field with eigenvectors and solutions.

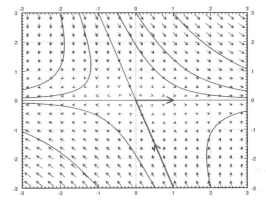

Figure 3.7: Example saddle vector field with eigenvectors and solutions.

Case 3. Suppose one eigenvalue is positive and one is negative. For example the matrix $\left[\begin{smallmatrix} 1 & 1 \\ 0 & -2 \end{smallmatrix}\right]$. The eigenvalues are 1 and -2 and corresponding eigenvectors are $\left[\begin{smallmatrix} 1 \\ 0 \end{smallmatrix}\right]$ and $\left[\begin{smallmatrix} 1 \\ -3 \end{smallmatrix}\right]$. We reverse the arrows on one line (corresponding to the negative eigenvalue) and we obtain the picture in Figure 3.7. We call this picture a *saddle point*.

For the next three cases we will assume the eigenvalues are complex. In this case the eigenvectors are also complex and we cannot just plot them in the plane.

Case 4. Suppose the eigenvalues are purely imaginary. That is, suppose the eigenvalues are $\pm ib$. For example, let $P = \left[\begin{smallmatrix} 0 & 1 \\ -4 & 0 \end{smallmatrix}\right]$. The eigenvalues turn out to be $\pm 2i$ and eigenvectors are $\left[\begin{smallmatrix} 1 \\ 2i \end{smallmatrix}\right]$ and $\left[\begin{smallmatrix} 1 \\ -2i \end{smallmatrix}\right]$.

Consider the eigenvalue $2i$ and its eigenvector $\begin{bmatrix} 1 \\ 2i \end{bmatrix}$. The real and imaginary parts of $\vec{v}e^{i2t}$ are

$$\mathrm{Re} \begin{bmatrix} 1 \\ 2i \end{bmatrix} e^{i2t} = \begin{bmatrix} \cos(2t) \\ -2\sin(2t) \end{bmatrix},$$

$$\mathrm{Im} \begin{bmatrix} 1 \\ 2i \end{bmatrix} e^{i2t} = \begin{bmatrix} \sin(2t) \\ 2\cos(2t) \end{bmatrix}.$$

We can take any linear combination of them to get other solutions, which one we take depends on the initial conditions. Now note that the real part is a parametric equation for an ellipse. Same with the imaginary part and in fact any linear combination of the two. This is what happens in general when the eigenvalues are purely imaginary. So when the eigenvalues are purely imaginary, we get *ellipses* for the solutions. This type of picture is sometimes called a *center*. See Figure 3.8.

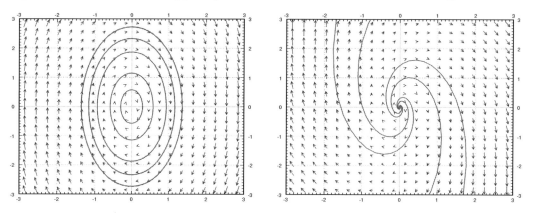

Figure 3.8: Example center vector field.   Figure 3.9: Example spiral source vector field.

Case 5. Now suppose the complex eigenvalues have a positive real part. That is, suppose the eigenvalues are $a \pm ib$ for some $a > 0$. For example, let $P = \begin{bmatrix} 1 & 1 \\ -4 & 1 \end{bmatrix}$. The eigenvalues turn out to be $1 \pm 2i$ and eigenvectors are $\begin{bmatrix} 1 \\ 2i \end{bmatrix}$ and $\begin{bmatrix} 1 \\ -2i \end{bmatrix}$. We take $1 + 2i$ and its eigenvector $\begin{bmatrix} 1 \\ 2i \end{bmatrix}$ and find the real and imaginary parts of $\vec{v}e^{(1+2i)t}$ are

$$\mathrm{Re} \begin{bmatrix} 1 \\ 2i \end{bmatrix} e^{(1+2i)t} = e^{t} \begin{bmatrix} \cos(2t) \\ -2\sin(2t) \end{bmatrix},$$

$$\mathrm{Im} \begin{bmatrix} 1 \\ 2i \end{bmatrix} e^{(1+2i)t} = e^{t} \begin{bmatrix} \sin(2t) \\ 2\cos(2t) \end{bmatrix}.$$

Note the e^{t} in front of the solutions. This means that the solutions grow in magnitude while spinning around the origin. Hence we get a *spiral source*. See Figure 3.9.

Case 6. Finally suppose the complex eigenvalues have a negative real part. That is, suppose the eigenvalues are $-a \pm ib$ for some $a > 0$. For example, let $P = \begin{bmatrix} -1 & -1 \\ 4 & -1 \end{bmatrix}$. The eigenvalues turn out to be $-1 \pm 2i$ and eigenvectors are $\left[\begin{smallmatrix} 1 \\ -2i \end{smallmatrix}\right]$ and $\left[\begin{smallmatrix} 1 \\ 2i \end{smallmatrix}\right]$. We take $-1 - 2i$ and its eigenvector $\left[\begin{smallmatrix} 1 \\ 2i \end{smallmatrix}\right]$ and find the real and imaginary parts of $\vec{v}e^{(-1-2i)t}$ are

$$\operatorname{Re}\begin{bmatrix} 1 \\ 2i \end{bmatrix} e^{(-1-2i)t} = e^{-t}\begin{bmatrix} \cos(2t) \\ 2\sin(2t) \end{bmatrix},$$

$$\operatorname{Im}\begin{bmatrix} 1 \\ 2i \end{bmatrix} e^{(-1-2i)t} = e^{-t}\begin{bmatrix} -\sin(2t) \\ 2\cos(2t) \end{bmatrix}.$$

Note the e^{-t} in front of the solutions. This means that the solutions shrink in magnitude while spinning around the origin. Hence we get a *spiral sink*. See Figure 3.10.

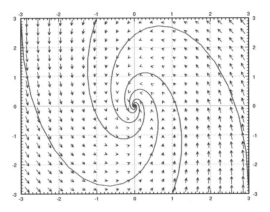

Figure 3.10: Example spiral sink vector field.

We summarize the behavior of linear homogeneous two dimensional systems in Table 3.1.

Eigenvalues	Behavior
real and both positive	source / unstable node
real and both negative	sink / stable node
real and opposite signs	saddle
purely imaginary	center point / ellipses
complex with positive real part	spiral source
complex with negative real part	spiral sink

Table 3.1: Summary of behavior of linear homogeneous two dimensional systems.

3.5.1   Exercises

Exercise 3.5.1: *Take the equation $mx'' + cx' + kx = 0$, with $m > 0$, $c \geq 0$, $k > 0$ for the mass-spring system. a) Convert this to a system of first order equations. b) Classify for what m, c, k do you get which behavior. c) Can you explain from physical intuition why you do not get all the different kinds of behavior here?*

Exercise 3.5.2: *What happens in the case when $P = \left[\begin{smallmatrix} 1 & 1 \\ 0 & 1 \end{smallmatrix}\right]$? In this case the eigenvalue is repeated and there is only one eigenvector. What picture does this look like?*

Exercise 3.5.3: *What happens in the case when $P = \left[\begin{smallmatrix} 1 & 1 \\ 1 & 1 \end{smallmatrix}\right]$? Does this look like any of the pictures we have drawn?*

Exercise 3.5.4: *Which behaviours are possible if P is diagonal, that is $P = \left[\begin{smallmatrix} a & 0 \\ 0 & b \end{smallmatrix}\right]$? You can assume that a and b are not zero.*

Exercise 3.5.101: *Describe the behavior of the following systems without solving:*
a) $x' = x + y$, $y' = x - y$.
b) $x_1' = x_1 + x_2$, $x_2' = 2x_2$.
c) $x_1' = -2x_2$, $x_2' = 2x_1$.
d) $x' = x + 3y$, $y' = -2x - 4y$.
e) $x' = x - 4y$, $y' = -4x + y$.

Exercise 3.5.102: *Suppose that $\vec{x}' = A\vec{x}$ where A is a 2 by 2 matrix with eigenvalues $2 \pm i$. Describe the behavior.*

Exercise 3.5.103: *Take $\left[\begin{smallmatrix} x \\ y \end{smallmatrix}\right]' = \left[\begin{smallmatrix} 0 & 1 \\ 0 & 0 \end{smallmatrix}\right]\left[\begin{smallmatrix} x \\ y \end{smallmatrix}\right]$. Draw the vector field and describe the behavior. Is it one of the behaviors that we have seen before?*

3.6 Second order systems and applications

Note: more than 2 lectures, §5.3 in [EP], not in [BD]

3.6.1 Undamped mass-spring systems

While we did say that we will usually only look at first order systems, it is sometimes more convenient to study the system in the way it arises naturally. For example, suppose we have 3 masses connected by springs between two walls. We could pick any higher number, and the math would be essentially the same, but for simplicity we pick 3 right now. Let us also assume no friction, that is, the system is undamped. The masses are m_1, m_2, and m_3 and the spring constants are k_1, k_2, k_3, and k_4. Let x_1 be the displacement from rest position of the first mass, and x_2 and x_3 the displacement of the second and third mass. We will make, as usual, positive values go right (as x_1 grows, the first mass is moving right). See Figure 3.11.

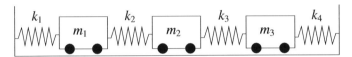

Figure 3.11: System of masses and springs.

This simple system turns up in unexpected places. For example, our world really consists of many small particles of matter interacting together. When we try the above system with many more masses, we obtain a good approximation to how an elastic material behaves. By somehow taking a limit of the number of masses going to infinity, we obtain the continuous one dimensional wave equation (that we study in § 4.7). But we digress.

Let us set up the equations for the three mass system. By Hooke's law we have that the force acting on the mass equals the spring compression times the spring constant. By Newton's second law we have that force is mass times acceleration. So if we sum the forces acting on each mass and put the right sign in front of each term, depending on the direction in which it is acting, we end up with the desired system of equations.

$$\begin{aligned}
m_1 x_1'' &= -k_1 x_1 + k_2(x_2 - x_1) &&= -(k_1 + k_2)x_1 + k_2 x_2, \\
m_2 x_2'' &= -k_2(x_2 - x_1) + k_3(x_3 - x_2) &&= k_2 x_1 - (k_2 + k_3)x_2 + k_3 x_3, \\
m_3 x_3'' &= -k_3(x_3 - x_2) - k_4 x_3 &&= k_3 x_2 - (k_3 + k_4)x_3.
\end{aligned}$$

We define the matrices

$$M = \begin{bmatrix} m_1 & 0 & 0 \\ 0 & m_2 & 0 \\ 0 & 0 & m_3 \end{bmatrix} \quad \text{and} \quad K = \begin{bmatrix} -(k_1 + k_2) & k_2 & 0 \\ k_2 & -(k_2 + k_3) & k_3 \\ 0 & k_3 & -(k_3 + k_4) \end{bmatrix}.$$

We write the equation simply as

$$M\vec{x}'' = K\vec{x}.$$

At this point we could introduce 3 new variables and write out a system of 6 first order equations. We claim this simple setup is easier to handle as a second order system. We call $\vec{x}$ the *displacement vector*, M the *mass matrix*, and K the *stiffness matrix*.

Exercise 3.6.1: *Repeat this setup for 4 masses (find the matrices M and K). Do it for 5 masses. Can you find a prescription to do it for n masses?*

As with a single equation we want to "divide by M." This means computing the inverse of M. The masses are all nonzero and M is a diagonal matrix, so comping the inverse is easy:

$$M^{-1} = \begin{bmatrix} \frac{1}{m_1} & 0 & 0 \\ 0 & \frac{1}{m_2} & 0 \\ 0 & 0 & \frac{1}{m_3} \end{bmatrix}.$$

This fact follows readily by how we multiply diagonal matrices. As an exercise, you should verify that $MM^{-1} = M^{-1}M = I$.

Let $A = M^{-1}K$. We look at the system $\vec{x}'' = M^{-1}K\vec{x}$, or

$$\vec{x}'' = A\vec{x}.$$

Many real world systems can be modeled by this equation. For simplicity, we will only talk about the given masses-and-springs problem. We try a solution of the form

$$\vec{x} = \vec{v}e^{\alpha t}.$$

We compute that for this guess, $\vec{x}'' = \alpha^2\vec{v}e^{\alpha t}$. We plug our guess into the equation and get

$$\alpha^2\vec{v}e^{\alpha t} = A\vec{v}e^{\alpha t}.$$

We divide by $e^{\alpha t}$ to arrive at $\alpha^2\vec{v} = A\vec{v}$. Hence if α^2 is an eigenvalue of A and $\vec{v}$ is a corresponding eigenvector, we have found a solution.

In our example, and in other common applications, A has only real negative eigenvalues (and possibly a zero eigenvalue). So we study only this case. When an eigenvalue λ is negative, it means that $\alpha^2 = \lambda$ is negative. Hence there is some real number ω such that $-\omega^2 = \lambda$. Then $\alpha = \pm i\omega$. The solution we guessed was

$$\vec{x} = \vec{v}\bigl(\cos(\omega t) + i\sin(\omega t)\bigr).$$

By taking the real and imaginary parts (note that $\vec{v}$ is real), we find that $\vec{v}\cos(\omega t)$ and $\vec{v}\sin(\omega t)$ are linearly independent solutions.

If an eigenvalue is zero, it turns out that both $\vec{v}$ and $\vec{v}t$ are solutions, where $\vec{v}$ is an eigenvector corresponding to the eigenvalue 0.

Exercise 3.6.2: *Show that if A has a zero eigenvalue and $\vec{v}$ is a corresponding eigenvector, then $\vec{x} = \vec{v}(a + bt)$ is a solution of $\vec{x}'' = A\vec{x}$ for arbitrary constants a and b.*

Theorem 3.6.1. *Let A be an n × n matrix with n distinct real negative eigenvalues we denote by $-\omega_1^2 > -\omega_2^2 > \cdots > -\omega_n^2$, and corresponding eigenvectors by $\vec{v}_1, \vec{v}_2, \ldots, \vec{v}_n$. If A is invertible (that is, if $\omega_1 > 0$), then*

$$\vec{x}(t) = \sum_{i=1}^{n} \vec{v}_i(a_i \cos(\omega_i t) + b_i \sin(\omega_i t)),$$

is the general solution of

$$\vec{x}'' = A\vec{x},$$

for some arbitrary constants a_i and b_i. If A has a zero eigenvalue, that is $\omega_1 = 0$, and all other eigenvalues are distinct and negative, then the general solution can be written as

$$\vec{x}(t) = \vec{v}_1(a_1 + b_1 t) + \sum_{i=2}^{n} \vec{v}_i(a_i \cos(\omega_i t) + b_i \sin(\omega_i t)).$$

We use this solution and the setup from the introduction of this section even when some of the masses and springs are missing. For example, when there are only 2 masses and only 2 springs, simply take only the equations for the two masses and set all the spring constants for the springs that are missing to zero.

3.6.2 Examples

Example 3.6.1: Suppose we have the system in Figure 3.12, with $m_1 = 2$, $m_2 = 1$, $k_1 = 4$, and $k_2 = 2$.

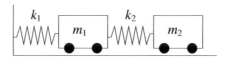

Figure 3.12: System of masses and springs.

The equations we write down are

$$\begin{bmatrix} 2 & 0 \\ 0 & 1 \end{bmatrix} \vec{x}'' = \begin{bmatrix} -(4 + 2) & 2 \\ 2 & -2 \end{bmatrix} \vec{x},$$

or

$$\vec{x}'' = \begin{bmatrix} -3 & 1 \\ 2 & -2 \end{bmatrix} \vec{x}.$$

We find the eigenvalues of A to be $\lambda = -1, -4$ (exercise). We find corresponding eigenvectors to be $\begin{bmatrix} 1 \\ 2 \end{bmatrix}$ and $\begin{bmatrix} 1 \\ -1 \end{bmatrix}$ respectively (exercise).

We check the theorem and note that $\omega_1 = 1$ and $\omega_2 = 2$. Hence the general solution is

$$\vec{x} = \begin{bmatrix} 1 \\ 2 \end{bmatrix} (a_1 \cos(t) + b_1 \sin(t)) + \begin{bmatrix} 1 \\ -1 \end{bmatrix} (a_2 \cos(2t) + b_2 \sin(2t)).$$

The two terms in the solution represent the two so-called *natural* or *normal modes of oscillation*. And the two (angular) frequencies are the *natural frequencies*. The two modes are plotted in Figure 3.13.

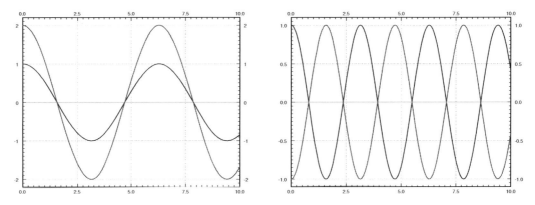

Figure 3.13: The two modes of the mass-spring system. In the left plot the masses are moving in unison and in the right plot are masses moving in the opposite direction.

Let us write the solution as

$$\vec{x} = \begin{bmatrix} 1 \\ 2 \end{bmatrix} c_1 \cos(t - \alpha_1) + \begin{bmatrix} 1 \\ -1 \end{bmatrix} c_2 \cos(2t - \alpha_2).$$

The first term,

$$\begin{bmatrix} 1 \\ 2 \end{bmatrix} c_1 \cos(t - \alpha_1) = \begin{bmatrix} c_1 \cos(t - \alpha_1) \\ 2c_1 \cos(t - \alpha_1) \end{bmatrix},$$

corresponds to the mode where the masses move synchronously in the same direction.

The second term,

$$\begin{bmatrix} 1 \\ -1 \end{bmatrix} c_2 \cos(2t - \alpha_2) = \begin{bmatrix} c_2 \cos(2t - \alpha_2) \\ -c_2 \cos(2t - \alpha_2) \end{bmatrix},$$

corresponds to the mode where the masses move synchronously but in opposite directions.

The general solution is a combination of the two modes. That is, the initial conditions determine the amplitude and phase shift of each mode.

Example 3.6.2: We have two toy rail cars. Car 1 of mass 2 kg is traveling at 3 m/s towards the second rail car of mass 1 kg. There is a bumper on the second rail car that engages at the moment the cars hit (it connects to two cars) and does not let go. The bumper acts like a spring of spring constant $k = 2$ N/m. The second car is 10 meters from a wall. See Figure 3.14.

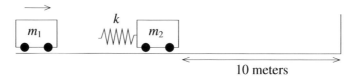

$$\text{Figure 3.14: The crash of two rail cars.}$$

We want to ask several questions. At what time after the cars link does impact with the wall happen? What is the speed of car 2 when it hits the wall?

OK, let us first set the system up. Let $t = 0$ be the time when the two cars link up. Let x_1 be the displacement of the first car from the position at $t = 0$, and let x_2 be the displacement of the second car from its original location. Then the time when $x_2(t) = 10$ is exactly the time when impact with wall occurs. For this t, $x_2'(t)$ is the speed at impact. This system acts just like the system of the previous example but without k_1. Hence the equation is

$$\begin{bmatrix} 2 & 0 \\ 0 & 1 \end{bmatrix} \vec{x}'' = \begin{bmatrix} -2 & 2 \\ 2 & -2 \end{bmatrix} \vec{x}.$$

or

$$\vec{x}'' = \begin{bmatrix} -1 & 1 \\ 2 & -2 \end{bmatrix} \vec{x}.$$

We compute the eigenvalues of A. It is not hard to see that the eigenvalues are 0 and -3 (exercise). Furthermore, eigenvectors are $\begin{bmatrix} 1 \\ 1 \end{bmatrix}$ and $\begin{bmatrix} 1 \\ -2 \end{bmatrix}$ respectively (exercise). Then $\omega_2 = \sqrt{3}$ and by the second part of the theorem we find our general solution to be

$$\vec{x} = \begin{bmatrix} 1 \\ 1 \end{bmatrix} (a_1 + b_1 t) + \begin{bmatrix} 1 \\ -2 \end{bmatrix} \left(a_2 \cos(\sqrt{3} t) + b_2 \sin(\sqrt{3} t) \right)$$

$$= \begin{bmatrix} a_1 + b_1 t + a_2 \cos(\sqrt{3} t) + b_2 \sin(\sqrt{3} t) \\ a_1 + b_1 t - 2a_2 \cos(\sqrt{3} t) - 2b_2 \sin(\sqrt{3} t). \end{bmatrix}$$

We now apply the initial conditions. First the cars start at position 0 so $x_1(0) = 0$ and $x_2(0) = 0$. The first car is traveling at 3 m/s, so $x_1'(0) = 3$ and the second car starts at rest, so $x_2'(0) = 0$. The first conditions says

$$\vec{0} = \vec{x}(0) = \begin{bmatrix} a_1 + a_2 \\ a_1 - 2a_2 \end{bmatrix}.$$

It is not hard to see that $a_1 = a_2 = 0$. We set $a_1 = 0$ and $a_2 = 0$ in $\vec{x}(t)$ and differentiate to get

$$\vec{x}'(t) = \begin{bmatrix} b_1 + \sqrt{3}\, b_2 \cos(\sqrt{3}\, t) \\ b_1 - 2\sqrt{3}\, b_2 \cos(\sqrt{3}\, t) \end{bmatrix}.$$

So

$$\begin{bmatrix} 3 \\ 0 \end{bmatrix} = \vec{x}'(0) = \begin{bmatrix} b_1 + \sqrt{3}\, b_2 \\ b_1 - 2\sqrt{3}\, b_2 \end{bmatrix}.$$

Solving these two equations we find $b_1 = 2$ and $b_2 = \frac{1}{\sqrt{3}}$. Hence the position of our cars is (until the impact with the wall)

$$\vec{x} = \begin{bmatrix} 2t + \frac{1}{\sqrt{3}} \sin(\sqrt{3}\, t) \\ 2t - \frac{2}{\sqrt{3}} \sin(\sqrt{3}\, t) \end{bmatrix}.$$

Note how the presence of the zero eigenvalue resulted in a term containing t. This means that the carts will be traveling in the positive direction as time grows, which is what we expect.

What we are really interested in is the second expression, the one for x_2. We have $x_2(t) = 2t - \frac{2}{\sqrt{3}} \sin(\sqrt{3}\, t)$. See Figure 3.15 for the plot of x_2 versus time.

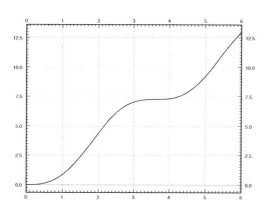

Just from the graph we can see that time of impact will be a little more than 5 seconds from time zero. For this we have to solve the equation $10 = x_2(t) = 2t - \frac{2}{\sqrt{3}} \sin(\sqrt{3}\, t)$. Using a computer (or even a graphing calculator) we find that $t_{\text{impact}} \approx 5.22$ seconds.

The speed of the second cart is $x_2' = 2 - 2\cos(\sqrt{3}\, t)$. At the time of impact (5.22 seconds from $t = 0$) we get $x_2'(t_{\text{impact}}) \approx 3.85$. The maximum speed is the maximum of $2 - 2\cos(\sqrt{3}\, t)$, which is 4. We are traveling at almost the maximum speed when we hit the wall.

Figure 3.15: Position of the second car in time (ignoring the wall).

Suppose that Bob is a tiny person sitting on car 2. Bob has a Martini in his hand and would like not to spill it. Let us suppose Bob would not spill his Martini when the first car links up with car 2, but if car 2 hits the wall at any speed greater than zero, Bob will spill his drink. Suppose Bob can move car 2 a few meters towards or away from the wall (he cannot go all the way to the wall, nor can he get out of the way of the first car). Is there a "safe" distance for him to be at? A distance such that the impact with the wall is at zero speed?

The answer is yes. Looking at Figure 3.15, we note the "plateau" between $t = 3$ and $t = 4$. There is a point where the speed is zero. To find it we need to solve $x_2'(t) = 0$. This is when

$\cos(\sqrt{3}\,t) = 1$ or in other words when $t = \frac{2\pi}{\sqrt{3}}, \frac{4\pi}{\sqrt{3}}, \ldots$ and so on. We plug in the first value to obtain $x_2\left(\frac{2\pi}{\sqrt{3}}\right) = \frac{4\pi}{\sqrt{3}} \approx 7.26$. So a "safe" distance is about 7 and a quarter meters from the wall.

Alternatively Bob could move away from the wall towards the incoming car 2 where another safe distance is $\frac{8\pi}{\sqrt{3}} \approx 14.51$ and so on, using all the different t such that $x_2'(t) = 0$. Of course $t = 0$ is always a solution here, corresponding to $x_2 = 0$, but that means standing right at the wall.

3.6.3 Forced oscillations

Finally we move to forced oscillations. Suppose that now our system is

$$\vec{x}'' = A\vec{x} + \vec{F}\cos(\omega t). \tag{3.3}$$

That is, we are adding periodic forcing to the system in the direction of the vector $\vec{F}$.

As before, this system just requires us to find one particular solution $\vec{x}_p$, add it to the general solution of the associated homogeneous system $\vec{x}_c$, and we will have the general solution to (3.3). Let us suppose that ω is not one of the natural frequencies of $\vec{x}'' = A\vec{x}$, then we can guess

$$\vec{x}_p = \vec{c}\cos(\omega t),$$

where $\vec{c}$ is an unknown constant vector. Note that we do not need to use sine since there are only second derivatives. We solve for $\vec{c}$ to find $\vec{x}_p$. This is really just the method of *undetermined coefficients* for systems. Let us differentiate $\vec{x}_p$ twice to get

$$\vec{x}_p'' = -\omega^2 \vec{c}\cos(\omega t).$$

Plug $\vec{x}_p$ and $\vec{x}_p''$ into the equation:

$$-\omega^2 \vec{c}\cos(\omega t) = A\vec{c}\cos(\omega t) + \vec{F}\cos(\omega t).$$

We cancel out the cosine and rearrange the equation to obtain

$$(A + \omega^2 I)\vec{c} = -\vec{F}.$$

So

$$\vec{c} = (A + \omega^2 I)^{-1}(-\vec{F}).$$

Of course this is possible only if $(A + \omega^2 I) = (A - (-\omega^2)I)$ is invertible. That matrix is invertible if and only if $-\omega^2$ is not an eigenvalue of A. That is true if and only if ω is not a natural frequency of the system.

Example 3.6.3: Let us take the example in Figure 3.12 on page 135 with the same parameters as before: $m_1 = 2$, $m_2 = 1$, $k_1 = 4$, and $k_2 = 2$. Now suppose that there is a force $2\cos(3t)$ acting on the second cart.

The equation is

$$\vec{x}'' = \begin{bmatrix} -3 & 1 \\ 2 & -2 \end{bmatrix} \vec{x} + \begin{bmatrix} 0 \\ 2 \end{bmatrix} \cos(3t).$$

We solved the associated homogeneous equation before and found the complementary solution to be

$$\vec{x}_c = \begin{bmatrix} 1 \\ 2 \end{bmatrix} (a_1 \cos(t) + b_1 \sin(t)) + \begin{bmatrix} 1 \\ -1 \end{bmatrix} (a_2 \cos(2t) + b_2 \sin(2t)).$$

The natural frequencies are 1 and 2. Hence as 3 is not a natural frequency, we can try $\vec{c} \cos(3t)$. We invert $(A + 3^2 I)$:

$$\left(\begin{bmatrix} -3 & 1 \\ 2 & -2 \end{bmatrix} + 3^2 I \right)^{-1} = \begin{bmatrix} 6 & 1 \\ 2 & 7 \end{bmatrix}^{-1} = \begin{bmatrix} \frac{7}{40} & \frac{-1}{40} \\ \frac{-1}{20} & \frac{3}{20} \end{bmatrix}.$$

Hence,

$$\vec{c} = (A + \omega^2 I)^{-1} (-\vec{F}) = \begin{bmatrix} \frac{7}{40} & \frac{-1}{40} \\ \frac{-1}{20} & \frac{3}{20} \end{bmatrix} \begin{bmatrix} 0 \\ -2 \end{bmatrix} = \begin{bmatrix} \frac{1}{20} \\ \frac{-3}{10} \end{bmatrix}.$$

Combining with what we know the general solution of the associated homogeneous problem to be, we get that the general solution to $\vec{x}'' = A\vec{x} + \vec{F} \cos(\omega t)$ is

$$\vec{x} = \vec{x}_c + \vec{x}_p = \begin{bmatrix} 1 \\ 2 \end{bmatrix} (a_1 \cos(t) + b_1 \sin(t)) + \begin{bmatrix} 1 \\ -1 \end{bmatrix} (a_2 \cos(2t) + b_2 \sin(2t)) + \begin{bmatrix} \frac{1}{20} \\ \frac{-3}{10} \end{bmatrix} \cos(3t).$$

The constants a_1, a_2, b_1, and b_2 must then be solved for given any initial conditions.

If ω is a natural frequency of the system *resonance* occurs because we will have to try a particular solution of the form

$$\vec{x}_p = \vec{c}\, t \sin(\omega t) + \vec{d} \cos(\omega t).$$

That is assuming that the eigenvalues of the coefficient matrix are distinct. Next, note that the amplitude of this solution grows without bound as t grows.

3.6.4 Exercises

Exercise 3.6.3: *Find a particular solution to*

$$\vec{x}'' = \begin{bmatrix} -3 & 1 \\ 2 & -2 \end{bmatrix} \vec{x} + \begin{bmatrix} 0 \\ 2 \end{bmatrix} \cos(2t).$$

Exercise 3.6.4 (challenging): *Let us take the example in Figure 3.12 on page 135 with the same parameters as before: $m_1 = 2$, $k_1 = 4$, and $k_2 = 2$, except for m_2, which is unknown. Suppose that there is a force $\cos(5t)$ acting on the first mass. Find an m_2 such that there exists a particular solution where the first mass does not move.*

Note: This idea is called dynamic damping. *In practice there will be a small amount of damping and so any transient solution will disappear and after long enough time, the first mass will always come to a stop.*

Exercise **3.6.5:** *Let us take the Example 3.6.2 on page 137, but that at time of impact, cart 2 is moving to the left at the speed of 3 m/s. a) Find the behavior of the system after linkup. b) Will the second car hit the wall, or will it be moving away from the wall as time goes on? c) At what speed would the first car have to be traveling for the system to essentially stay in place after linkup?*

Exercise **3.6.6:** *Let us take the example in Figure 3.12 on page 135 with parameters $m_1 = m_2 = 1$, $k_1 = k_2 = 1$. Does there exist a set of initial conditions for which the first cart moves but the second cart does not? If so, find those conditions. If not, argue why not.*

Exercise **3.6.101:** *Find the general solution to $\begin{bmatrix} 1 & 0 & 0 \\ 0 & 2 & 0 \\ 0 & 0 & 3 \end{bmatrix} \vec{x}'' = \begin{bmatrix} -3 & 0 & 0 \\ 2 & -4 & 0 \\ 0 & 6 & -3 \end{bmatrix} \vec{x} + \begin{bmatrix} \cos(2t) \\ 0 \\ 0 \end{bmatrix}$.*

Exercise **3.6.102:** *Suppose there are three carts of equal mass m and connected by two springs of constant k (and no connections to walls). Set up the system and find its general solution.*

Exercise **3.6.103:** *Suppose a cart of mass 2 kg is attached by a spring of constant $k = 1$ to a cart of mass 3 kg, which is attached to the wall by a spring also of constant $k = 1$. Suppose that the initial position of the first cart is 1 meter in the positive direction from the rest position, and the second mass starts at the rest position. The masses are not moving and are let go. Find the position of the second mass as a function of time.*

3.7   Multiple eigenvalues

Note: 1 or 1.5 lectures, §5.4 in [EP], §7.8 in [BD]

It may happen that a matrix A has some "repeated" eigenvalues. That is, the characteristic equation $\det(A - \lambda I) = 0$ may have repeated roots. As we said before, this is actually unlikely to happen for a random matrix. If we take a small perturbation of A (we change the entries of A slightly), we get a matrix with distinct eigenvalues. As any system we want to solve in practice is an approximation to reality anyway, it is not absolutely indispensable to know how to solve these corner cases. But it happens on occasion that it is necessary to solve such a system.

3.7.1   Geometric multiplicity

Take the diagonal matrix

$$A = \begin{bmatrix} 3 & 0 \\ 0 & 3 \end{bmatrix}.$$

A has an eigenvalue 3 of multiplicity 2. We call the multiplicity of the eigenvalue in the characteristic equation the *algebraic multiplicity*. In this case, there also exist 2 linearly independent eigenvectors, $\begin{bmatrix} 1 \\ 0 \end{bmatrix}$ and $\begin{bmatrix} 0 \\ 1 \end{bmatrix}$ corresponding to the eigenvalue 3. This means that the so-called *geometric multiplicity* of this eigenvalue is also 2.

In all the theorems where we required a matrix to have n distinct eigenvalues, we only really needed to have n linearly independent eigenvectors. For example, $\vec{x}' = A\vec{x}$ has the general solution

$$\vec{x} = c_1 \begin{bmatrix} 1 \\ 0 \end{bmatrix} e^{3t} + c_2 \begin{bmatrix} 0 \\ 1 \end{bmatrix} e^{3t}.$$

Let us restate the theorem about real eigenvalues. In the following theorem we will repeat eigenvalues according to (algebraic) multiplicity. So for the above matrix A, we would say that it has eigenvalues 3 and 3.

Theorem 3.7.1. *Suppose the $n \times n$ matrix P has n real eigenvalues (not necessarily distinct), $\lambda_1, \lambda_2,$ ..., λ_n, and there are n linearly independent corresponding eigenvectors $\vec{v}_1, \vec{v}_2, \ldots, \vec{v}_n$. Then the general solution to $\vec{x}' = P\vec{x}$ can be written as*

$$\vec{x} = c_1\vec{v}_1 e^{\lambda_1 t} + c_2\vec{v}_2 e^{\lambda_2 t} + \cdots + c_n\vec{v}_n e^{\lambda_n t}.$$

The *geometric multiplicity* of an eigenvalue of algebraic multiplicity n is equal to the number of corresponding linearly independent eigenvectors. The geometric multiplicity is always less than or equal to the algebraic multiplicity. The theorem handles the case when these two multiplicities are equal for all eigenvalues. If for an eigenvalue the geometric multiplicity is equal to the algebraic multiplicity, then we say the eigenvalue is *complete*.

In other words, the hypothesis of the theorem could be stated as saying that if all the eigenvalues of P are complete, then there are n linearly independent eigenvectors and thus we have the given general solution.

If the geometric multiplicity of an eigenvalue is 2 or greater, then the set of linearly independent eigenvectors is not unique up to multiples as it was before. For example, for the diagonal matrix $A = \begin{bmatrix} 3 & 0 \\ 0 & 3 \end{bmatrix}$ we could also pick eigenvectors $\begin{bmatrix} 1 \\ 1 \end{bmatrix}$ and $\begin{bmatrix} 1 \\ -1 \end{bmatrix}$, or in fact any pair of two linearly independent vectors. The number of linearly independent eigenvectors corresponding to λ is the number of free variables we obtain when solving $A\vec{v} = \lambda\vec{v}$. We pick specific values for those free variables to obtain eigenvectors. If you pick different values, you may get different eigenvectors.

3.7.2 Defective eigenvalues

If an $n \times n$ matrix has less than n linearly independent eigenvectors, it is said to be *deficient*. Then there is at least one eigenvalue with an algebraic multiplicity that is higher than its geometric multiplicity. We call this eigenvalue *defective* and the difference between the two multiplicities we call the *defect*.

Example 3.7.1: The matrix

$$\begin{bmatrix} 3 & 1 \\ 0 & 3 \end{bmatrix}$$

has an eigenvalue 3 of algebraic multiplicity 2. Let us try to compute eigenvectors.

$$\begin{bmatrix} 0 & 1 \\ 0 & 0 \end{bmatrix}\begin{bmatrix} v_1 \\ v_2 \end{bmatrix} = \vec{0}.$$

We must have that $v_2 = 0$. Hence any eigenvector is of the form $\begin{bmatrix} v_1 \\ 0 \end{bmatrix}$. Any two such vectors are linearly dependent, and hence the geometric multiplicity of the eigenvalue is 1. Therefore, the defect is 1, and we can no longer apply the eigenvalue method directly to a system of ODEs with such a coefficient matrix.

The key observation we will use here is that if λ is an eigenvalue of A of algebraic multiplicity m, then we will be able to find m linearly independent vectors solving the equation $(A - \lambda I)^m \vec{v} = \vec{0}$. We will call these *generalized eigenvectors*.

Let us continue with the example $A = \begin{bmatrix} 3 & 1 \\ 0 & 3 \end{bmatrix}$ and the equation $\vec{x}' = A\vec{x}$. We have an eigenvalue $\lambda = 3$ of (algebraic) multiplicity 2 and defect 1. We have found one eigenvector $\vec{v}_1 = \begin{bmatrix} 1 \\ 0 \end{bmatrix}$. We have the solution

$$\vec{x}_1 = \vec{v}_1 e^{3t}.$$

In this case, let us try (in the spirit of repeated roots of the characteristic equation for a single equation) another solution of the form

$$\vec{x}_2 = (\vec{v}_2 + \vec{v}_1 t)\, e^{3t}.$$

We differentiate to get

$$\vec{x_2}' = \vec{v_1}e^{3t} + 3(\vec{v_2} + \vec{v_1}t)\,e^{3t} = (3\vec{v_2} + \vec{v_1})\,e^{3t} + 3\vec{v_1}te^{3t}.$$

As we are assuming that $\vec{x_2}$ is a solution, $\vec{x_2}'$ must equal $A\vec{x_2}$, and

$$A\vec{x_2} = A(\vec{v_2} + \vec{v_1}t)\,e^{3t} = A\vec{v_2}e^{3t} + A\vec{v_1}te^{3t}.$$

By looking at the coefficients of e^{3t} and te^{3t} we see $3\vec{v_2} + \vec{v_1} = A\vec{v_2}$ and $3\vec{v_1} = A\vec{v_1}$. This means that

$$(A - 3I)\vec{v_2} = \vec{v_1}, \qquad \text{and} \qquad (A - 3I)\vec{v_1} = \vec{0}.$$

Therefore, $\vec{x_2}$ is a solution if these two equations are satisfied. We know the second of these two equations is satisfied as $\vec{v_1}$ is an eigenvector. If we plug the first equation into the second we obtain

$$(A - 3I)(A - 3I)\vec{v_2} = \vec{0}, \qquad \text{or} \qquad (A - 3I)^2\vec{v_2} = \vec{0}.$$

If we can, therefore, find a $\vec{v_2}$ that solves $(A - 3I)^2\vec{v_2} = \vec{0}$ and such that $(A - 3I)\vec{v_2} = \vec{v_1}$, then we are done. This is just a bunch of linear equations to solve and we are by now very good at that.

We notice that in this simple case $(A - 3I)^2$ is just the zero matrix (exercise). Hence, any vector $\vec{v_2}$ solves $(A - 3I)^2\vec{v_2} = \vec{0}$. We just have to make sure that $(A - 3I)\vec{v_2} = \vec{v_1}$. Write

$$\begin{bmatrix} 0 & 1 \\ 0 & 0 \end{bmatrix}\begin{bmatrix} a \\ b \end{bmatrix} = \begin{bmatrix} 1 \\ 0 \end{bmatrix}.$$

By inspection we see that letting $a = 0$ (a could be anything in fact) and $b = 1$ does the job. Hence we can take $\vec{v_2} = \begin{bmatrix} 0 \\ 1 \end{bmatrix}$. Our general solution to $\vec{x}' = A\vec{x}$ is

$$\vec{x} = c_1\begin{bmatrix} 1 \\ 0 \end{bmatrix}e^{3t} + c_2\left(\begin{bmatrix} 0 \\ 1 \end{bmatrix} + \begin{bmatrix} 1 \\ 0 \end{bmatrix}t\right)e^{3t} = \begin{bmatrix} c_1e^{3t} + c_2te^{3t} \\ c_2e^{3t} \end{bmatrix}.$$

Let us check that we really do have the solution. First $x_1' = c_13e^{3t} + c_2e^{3t} + 3c_2te^{3t} = 3x_1 + x_2$. Good. Now $x_2' = 3c_2e^{3t} = 3x_2$. Good.

Note that the system $\vec{x}' = A\vec{x}$ has a simpler solution since A is a so-called *upper triangular matrix*, that is every entry below the diagonal is zero. In particular, the equation for x_2 does not depend on x_1. Mind you, not every defective matrix is triangular.

Exercise 3.7.1: *Solve* $\vec{x}' = \begin{bmatrix} 3 & 1 \\ 0 & 3 \end{bmatrix}\vec{x}$ *by first solving for x_2 and then for x_1 independently. Check that you got the same solution as we did above.*

Let us describe the general algorithm. Suppose that λ is an eigenvalue of multiplicity 2, defect 1. First find an eigenvector $\vec{v_1}$ of λ. Then, find a vector $\vec{v_2}$ such that

$$(A - \lambda I)^2\vec{v_2} = \vec{0},$$
$$(A - \lambda I)\vec{v_2} = \vec{v_1}.$$

This gives us two linearly independent solutions

$$\vec{x}_1 = \vec{v}_1 e^{\lambda t},$$
$$\vec{x}_2 = (\vec{v}_2 + \vec{v}_1 t) e^{\lambda t}.$$

This machinery can also be generalized to higher multiplicities and higher defects. We will not go over this method in detail, but let us just sketch the ideas. Suppose that A has an eigenvalue λ of multiplicity m. We find vectors such that

$$(A - \lambda I)^k \vec{v} = \vec{0}, \qquad \text{but} \qquad (A - \lambda I)^{k-1} \vec{v} \neq \vec{0}.$$

Such vectors are called *generalized eigenvectors*. For every eigenvector $\vec{v}_1$ we find a chain of generalized eigenvectors $\vec{v}_2$ through $\vec{v}_k$ such that:

$$(A - \lambda I)\vec{v}_1 = \vec{0},$$
$$(A - \lambda I)\vec{v}_2 = \vec{v}_1,$$
$$\vdots$$
$$(A - \lambda I)\vec{v}_k = \vec{v}_{k-1}.$$

We form the linearly independent solutions

$$\vec{x}_1 = \vec{v}_1 e^{\lambda t},$$
$$\vec{x}_2 = (\vec{v}_2 + \vec{v}_1 t) e^{\lambda t},$$
$$\vdots$$
$$\vec{x}_k = \left(\vec{v}_k + \vec{v}_{k-1} t + \vec{v}_{k-2} \frac{t^2}{2} + \cdots + \vec{v}_2 \frac{t^{k-2}}{(k-2)!} + \vec{v}_1 \frac{t^{k-1}}{(k-1)!} \right) e^{\lambda t}.$$

Recall that $k! = 1 \cdot 2 \cdot 3 \cdots (k-1) \cdot k$ is the factorial. We proceed to find chains until we form m linearly independent solutions (m is the multiplicity). You may need to find several chains for every eigenvalue.

3.7.3 Exercises

Exercise 3.7.2: *Let $A = \begin{bmatrix} 5 & -3 \\ 3 & -1 \end{bmatrix}$. Find the general solution of $\vec{x}' = A\vec{x}$.*

Exercise 3.7.3: *Let $A = \begin{bmatrix} 5 & -4 & 4 \\ 0 & 3 & 0 \\ -2 & 4 & -1 \end{bmatrix}$. a) What are the eigenvalues? b) What is/are the defect(s) of the eigenvalue(s)? c) Find the general solution of $\vec{x}' = A\vec{x}$.*

Exercise 3.7.4: *Let $A = \begin{bmatrix} 2 & 1 & 0 \\ 0 & 2 & 0 \\ 0 & 0 & 2 \end{bmatrix}$. a) What are the eigenvalues? b) What is/are the defect(s) of the eigenvalue(s)? c) Find the general solution of $\vec{x}' = A\vec{x}$ in two different ways and verify you get the same answer.*

***Exercise* 3.7.5:** *Let* $A = \begin{bmatrix} 0 & 1 & 2 \\ -1 & -2 & -2 \\ -4 & 4 & 7 \end{bmatrix}$. *a) What are the eigenvalues? b) What is/are the defect(s) of the eigenvalue(s)? c) Find the general solution of* $\vec{x}' = A\vec{x}$.

***Exercise* 3.7.6:** *Let* $A = \begin{bmatrix} 0 & 4 & -2 \\ -1 & -4 & 1 \\ 0 & 0 & -2 \end{bmatrix}$. *a) What are the eigenvalues? b) What is/are the defect(s) of the eigenvalue(s)? c) Find the general solution of* $\vec{x}' = A\vec{x}$.

***Exercise* 3.7.7:** *Let* $A = \begin{bmatrix} 2 & 1 & -1 \\ -1 & 0 & 2 \\ -1 & -2 & 4 \end{bmatrix}$. *a) What are the eigenvalues? b) What is/are the defect(s) of the eigenvalue(s)? c) Find the general solution of* $\vec{x}' = A\vec{x}$.

***Exercise* 3.7.8:** *Suppose that A is a* 2×2 *matrix with a repeated eigenvalue* λ. *Suppose that there are two linearly independent eigenvectors. Show that* $A = \lambda I$.

***Exercise* 3.7.101:** *Let* $A = \begin{bmatrix} 1 & 1 & 1 \\ 1 & 1 & 1 \\ 1 & 1 & 1 \end{bmatrix}$. *a) What are the eigenvalues? b) What is/are the defect(s) of the eigenvalue(s)? c) Find the general solution of* $\vec{x}' = A\vec{x}$.

***Exercise* 3.7.102:** *Let* $A = \begin{bmatrix} 1 & 3 & 3 \\ 1 & 1 & 0 \\ -1 & 1 & 2 \end{bmatrix}$. *a) What are the eigenvalues? b) What is/are the defect(s) of the eigenvalue(s)? c) Find the general solution of* $\vec{x}' = A\vec{x}$.

***Exercise* 3.7.103:** *Let* $A = \begin{bmatrix} 2 & 0 & 0 \\ -1 & -1 & 9 \\ 0 & -1 & 5 \end{bmatrix}$. *a) What are the eigenvalues? b) What is/are the defect(s) of the eigenvalue(s)? c) Find the general solution of* $\vec{x}' = A\vec{x}$.

***Exercise* 3.7.104:** *Let* $A = \begin{bmatrix} a & a \\ b & c \end{bmatrix}$, *where a, b, and c are unknowns. Suppose that 5 is a doubled eigenvalue of defect 1, and suppose that* $\begin{bmatrix} 1 \\ 0 \end{bmatrix}$ *is the eigenvector. Find A and show that there is only one solution.*

3.8 Matrix exponentials

Note: 2 lectures, §5.5 in [EP], §7.7 in [BD]

3.8.1 Definition

In this section we present a different way of finding the fundamental matrix solution of a system. Suppose that we have the constant coefficient equation

$$\vec{x}' = P\vec{x},$$

as usual. Now suppose that this was one equation (P is a number or a 1×1 matrix). Then the solution to this would be

$$\vec{x} = e^{Pt}.$$

The same computation works for matrices when we define e^{Pt} properly. First let us write down the Taylor series for e^{at} for some number a:

$$e^{at} = 1 + at + \frac{(at)^2}{2} + \frac{(at)^3}{6} + \frac{(at)^4}{24} + \cdots = \sum_{k=0}^{\infty} \frac{(at)^k}{k!}.$$

Recall $k! = 1 \cdot 2 \cdot 3 \cdots k$ is the factorial, and $0! = 1$. We differentiate this series term by term

$$\frac{d}{dt}\left(e^{at}\right) = a + a^2 t + \frac{a^3 t^2}{2} + \frac{a^4 t^3}{6} + \cdots = a\left(1 + at + \frac{(at)^2}{2} + \frac{(at)^3}{6} + \cdots\right) = ae^{at}.$$

Maybe we can try the same trick with matrices. Suppose that for an $n \times n$ matrix A we define the *matrix exponential* as

$$\boxed{e^A \overset{\text{def}}{=} I + A + \frac{1}{2}A^2 + \frac{1}{6}A^3 + \cdots + \frac{1}{k!}A^k + \cdots}$$

Let us not worry about convergence. The series really does always converge. We usually write Pt as tP by convention when P is a matrix. With this small change and by the exact same calculation as above we have that

$$\frac{d}{dt}\left(e^{tP}\right) = Pe^{tP}.$$

Now P and hence e^{tP} is an $n \times n$ matrix. What we are looking for is a vector. In the 1×1 case we would at this point multiply by an arbitrary constant to get the general solution. In the matrix case we multiply by a column vector $\vec{c}$.

Theorem 3.8.1. *Let P be an $n \times n$ matrix. Then the general solution to $\vec{x}' = P\vec{x}$ is*

$$\vec{x} = e^{tP}\vec{c},$$

where $\vec{c}$ is an arbitrary constant vector. In fact $\vec{x}(0) = \vec{c}$.

Let us check:

$$\frac{d}{dt}\vec{x} = \frac{d}{dt}\left(e^{tP}\vec{c}\right) = Pe^{tP}\vec{c} = P\vec{x}.$$

Hence e^{tP} is the fundamental matrix solution of the homogeneous system. If we find a way to compute the matrix exponential, we will have another method of solving constant coefficient homogeneous systems. It also makes it easy to solve for initial conditions. To solve $\vec{x}' = A\vec{x}$, $\vec{x}(0) = \vec{b}$, we take the solution

$$\vec{x} = e^{tA}\vec{b}.$$

This equation follows because $e^{0A} = I$, so $\vec{x}(0) = e^{0A}\vec{b} = \vec{b}$.

We mention a drawback of matrix exponentials. In general $e^{A+B} \neq e^A e^B$. The trouble is that matrices do not commute, that is, in general $AB \neq BA$. If you try to prove $e^{A+B} \neq e^A e^B$ using the Taylor series, you will see why the lack of commutativity becomes a problem. However, it is still true that if $AB = BA$, that is, if A and B commute, then $e^{A+B} = e^A e^B$. We will find this fact useful. Let us restate this as a theorem to make a point.

Theorem 3.8.2. *If $AB = BA$, then $e^{A+B} = e^A e^B$. Otherwise $e^{A+B} \neq e^A e^B$ in general.*

3.8.2   Simple cases

In some instances it may work to just plug into the series definition. Suppose the matrix is diagonal. For example, $D = \left[\begin{smallmatrix} a & 0 \\ 0 & b \end{smallmatrix}\right]$. Then

$$D^k = \begin{bmatrix} a^k & 0 \\ 0 & b^k \end{bmatrix},$$

and

$$e^D = I + D + \frac{1}{2}D^2 + \frac{1}{6}D^3 + \cdots = \begin{bmatrix} 1 & 0 \\ 0 & 1 \end{bmatrix} + \begin{bmatrix} a & 0 \\ 0 & b \end{bmatrix} + \frac{1}{2}\begin{bmatrix} a^2 & 0 \\ 0 & b^2 \end{bmatrix} + \frac{1}{6}\begin{bmatrix} a^3 & 0 \\ 0 & b^3 \end{bmatrix} + \cdots = \begin{bmatrix} e^a & 0 \\ 0 & e^b \end{bmatrix}.$$

So by this rationale we have that

$$e^I = \begin{bmatrix} e & 0 \\ 0 & e \end{bmatrix} \qquad \text{and} \qquad e^{aI} = \begin{bmatrix} e^a & 0 \\ 0 & e^a \end{bmatrix}.$$

This makes exponentials of certain other matrices easy to compute. Notice for example that the matrix $A = \left[\begin{smallmatrix} 5 & 4 \\ -1 & 1 \end{smallmatrix}\right]$ can be written as $3I + B$ where $B = \left[\begin{smallmatrix} 2 & 4 \\ -1 & -2 \end{smallmatrix}\right]$. Notice that $B^2 = \left[\begin{smallmatrix} 0 & 0 \\ 0 & 0 \end{smallmatrix}\right]$. So $B^k = 0$ for all $k \geq 2$. Therefore, $e^B = I + B$. Suppose we actually want to compute e^{tA}. The matrices $3tI$ and tB commute (exercise: check this) and $e^{tB} = I + tB$, since $(tB)^2 = t^2 B^2 = 0$. We write

$$e^{tA} = e^{3tI + tB} = e^{3tI}e^{tB} = \begin{bmatrix} e^{3t} & 0 \\ 0 & e^{3t} \end{bmatrix}(I + tB) =$$

$$= \begin{bmatrix} e^{3t} & 0 \\ 0 & e^{3t} \end{bmatrix}\begin{bmatrix} 1 + 2t & 4t \\ -t & 1 - 2t \end{bmatrix} = \begin{bmatrix} (1 + 2t)\,e^{3t} & 4te^{3t} \\ -te^{3t} & (1 - 2t)\,e^{3t} \end{bmatrix}.$$

So we have found the fundamental matrix solution for the system $\vec{x}' = A\vec{x}$. Note that this matrix has a repeated eigenvalue with a defect; there is only one eigenvector for the eigenvalue 3. So we have found a perhaps easier way to handle this case. In fact, if a matrix A is 2×2 and has an eigenvalue λ of multiplicity 2, then either A is diagonal, or $A = \lambda I + B$ where $B^2 = 0$. This is a good exercise.

Exercise 3.8.1: *Suppose that A is 2×2 and λ is the only eigenvalue. Show that $(A - \lambda I)^2 = 0$, and therefore that we can write $A = \lambda I + B$, where $B^2 = 0$. Hint: First write down what does it mean for the eigenvalue to be of multiplicity 2. You will get an equation for the entries. Now compute the square of B.*

Matrices B such that $B^k = 0$ for some k are called *nilpotent*. Computation of the matrix exponential for nilpotent matrices is easy by just writing down the first k terms of the Taylor series.

3.8.3 General matrices

In general, the exponential is not as easy to compute as above. We usually cannot write a matrix as a sum of commuting matrices where the exponential is simple for each one. But fear not, it is still not too difficult provided we can find enough eigenvectors. First we need the following interesting result about matrix exponentials. For two square matrices A and B, with B invertible, we have

$$e^{BAB^{-1}} = Be^A B^{-1}.$$

This can be seen by writing down the Taylor series. First

$$(BAB^{-1})^2 = BAB^{-1}BAB^{-1} = BAIAB^{-1} = BA^2 B^{-1}.$$

And by the same reasoning $(BAB^{-1})^k = BA^k B^{-1}$. Now write down the Taylor series for $e^{BAB^{-1}}$:

$$\begin{aligned} e^{BAB^{-1}} &= I + BAB^{-1} + \frac{1}{2}(BAB^{-1})^2 + \frac{1}{6}(BAB^{-1})^3 + \cdots \\ &= BB^{-1} + BAB^{-1} + \frac{1}{2}BA^2 B^{-1} + \frac{1}{6}BA^3 B^{-1} + \cdots \\ &= B(I + A + \frac{1}{2}A^2 + \frac{1}{6}A^3 + \cdots)B^{-1} \\ &= Be^A B^{-1}. \end{aligned}$$

Given a square matrix A, we can sometimes write $A = EDE^{-1}$, where D is diagonal and E invertible. This procedure is called *diagonalization*. If we can do that, the computation of the exponential becomes easy. Adding t into the mix, we can then easily compute the exponential

$$e^{tA} = Ee^{tD}E^{-1}.$$

To diagonalize A we will need n linearly independent eigenvectors of A. Otherwise this method of computing the exponential does not work and we need to be trickier, but we will not get into such details. We let E be the matrix with the eigenvectors as columns. Let $\lambda_1, \lambda_2, \ldots, \lambda_n$ be the eigenvalues and let $\vec{v}_1, \vec{v}_2, \ldots, \vec{v}_n$ be the eigenvectors, then $E = [\, \vec{v}_1 \quad \vec{v}_2 \quad \cdots \quad \vec{v}_n \,]$. Let D be the diagonal matrix with the eigenvalues on the main diagonal. That is

$$D = \begin{bmatrix} \lambda_1 & 0 & \cdots & 0 \\ 0 & \lambda_2 & \cdots & 0 \\ \vdots & \vdots & \ddots & \vdots \\ 0 & 0 & \cdots & \lambda_n \end{bmatrix}.$$

We compute

$$\begin{aligned} AE &= A[\, \vec{v}_1 \quad \vec{v}_2 \quad \cdots \quad \vec{v}_n \,] \\ &= [\, A\vec{v}_1 \quad A\vec{v}_2 \quad \cdots \quad A\vec{v}_n \,] \\ &= [\, \lambda_1\vec{v}_1 \quad \lambda_2\vec{v}_2 \quad \cdots \quad \lambda_n\vec{v}_n \,] \\ &= [\, \vec{v}_1 \quad \vec{v}_2 \quad \cdots \quad \vec{v}_n \,]D \\ &= ED. \end{aligned}$$

The columns of E are linearly independent as these are linearly independent eigenvectors of A. Hence E is invertible. Since $AE = ED$, we multiply on the right by E^{-1} and we get

$$A = EDE^{-1}.$$

This means that $e^A = Ee^D E^{-1}$. Multiplying the matrix by t we obtain

$$e^{tA} = Ee^{tD}E^{-1} = E \begin{bmatrix} e^{\lambda_1 t} & 0 & \cdots & 0 \\ 0 & e^{\lambda_2 t} & \cdots & 0 \\ \vdots & \vdots & \ddots & \vdots \\ 0 & 0 & \cdots & e^{\lambda_n t} \end{bmatrix} E^{-1}. \tag{3.4}$$

The formula (3.4), therefore, gives the formula for computing the fundamental matrix solution e^{tA} for the system $\vec{x}' = A\vec{x}$, in the case where we have n linearly independent eigenvectors.

Notice that this computation still works when the eigenvalues and eigenvectors are complex, though then you will have to compute with complex numbers. It is clear from the definition that if A is real, then e^{tA} is real. So you will only need complex numbers in the computation and you may need to apply Euler's formula to simplify the result. If simplified properly the final matrix will not have any complex numbers in it.

Example 3.8.1: Compute the fundamental matrix solution using the matrix exponentials for the system

$$\begin{bmatrix} x \\ y \end{bmatrix}' = \begin{bmatrix} 1 & 2 \\ 2 & 1 \end{bmatrix} \begin{bmatrix} x \\ y \end{bmatrix}.$$

Then compute the particular solution for the initial conditions $x(0) = 4$ and $y(0) = 2$.

Let A be the coefficient matrix $\begin{bmatrix} 1 & 2 \\ 2 & 1 \end{bmatrix}$. We first compute (exercise) that the eigenvalues are 3 and -1 and corresponding eigenvectors are $\begin{bmatrix} 1 \\ 1 \end{bmatrix}$ and $\begin{bmatrix} 1 \\ -1 \end{bmatrix}$. Hence we write

$$
\begin{aligned}
e^{tA} &= \begin{bmatrix} 1 & 1 \\ 1 & -1 \end{bmatrix} \begin{bmatrix} e^{3t} & 0 \\ 0 & e^{-t} \end{bmatrix} \begin{bmatrix} 1 & 1 \\ 1 & -1 \end{bmatrix}^{-1} \\
&= \begin{bmatrix} 1 & 1 \\ 1 & -1 \end{bmatrix} \begin{bmatrix} e^{3t} & 0 \\ 0 & e^{-t} \end{bmatrix} \frac{-1}{2} \begin{bmatrix} -1 & -1 \\ -1 & 1 \end{bmatrix} \\
&= \frac{-1}{2} \begin{bmatrix} e^{3t} & e^{-t} \\ e^{3t} & -e^{-t} \end{bmatrix} \begin{bmatrix} -1 & -1 \\ -1 & 1 \end{bmatrix} \\
&= \frac{-1}{2} \begin{bmatrix} -e^{3t} - e^{-t} & -e^{3t} + e^{-t} \\ -e^{3t} + e^{-t} & -e^{3t} - e^{-t} \end{bmatrix} = \begin{bmatrix} \frac{e^{3t}+e^{-t}}{2} & \frac{e^{3t}-e^{-t}}{2} \\ \frac{e^{3t}-e^{-t}}{2} & \frac{e^{3t}+e^{-t}}{2} \end{bmatrix}.
\end{aligned}
$$

The initial conditions are $x(0) = 4$ and $y(0) = 2$. Hence, by the property that $e^{0A} = I$ we find that the particular solution we are looking for is $e^{tA}\vec{b}$ where $\vec{b}$ is $\begin{bmatrix} 4 \\ 2 \end{bmatrix}$. Then the particular solution we are looking for is

$$
\begin{bmatrix} x \\ y \end{bmatrix} = \begin{bmatrix} \frac{e^{3t}+e^{-t}}{2} & \frac{e^{3t}-e^{-t}}{2} \\ \frac{e^{3t}-e^{-t}}{2} & \frac{e^{3t}+e^{-t}}{2} \end{bmatrix} \begin{bmatrix} 4 \\ 2 \end{bmatrix} = \begin{bmatrix} 2e^{3t} + 2e^{-t} + e^{3t} - e^{-t} \\ 2e^{3t} - 2e^{-t} + e^{3t} + e^{-t} \end{bmatrix} = \begin{bmatrix} 3e^{3t} + e^{-t} \\ 3e^{3t} - e^{-t} \end{bmatrix}.
$$

3.8.4   Fundamental matrix solutions

We note that if you can compute the fundamental matrix solution in a different way, you can use this to find the matrix exponential e^{tA}. The fundamental matrix solution of a system of ODEs is not unique. The exponential is the fundamental matrix solution with the property that for $t = 0$ we get the identity matrix. So we must find the right fundamental matrix solution. Let X be any fundamental matrix solution to $\vec{x}' = A\vec{x}$. Then we claim

$$
e^{tA} = X(t)\,[X(0)]^{-1}.
$$

Clearly, if we plug $t = 0$ into $X(t)\,[X(0)]^{-1}$ we get the identity. We can multiply a fundamental matrix solution on the right by any constant invertible matrix and we still get a fundamental matrix solution. All we are doing is changing what the arbitrary constants are in the general solution $\vec{x}(t) = X(t)\,\vec{c}$.

3.8.5   Approximations

If you think about it, the computation of any fundamental matrix solution X using the eigenvalue method is just as difficult as the computation of e^{tA}. So perhaps we did not gain much by this new tool. However, the Taylor series expansion actually gives us a way to approximate solutions, which the eigenvalue method did not.

The simplest thing we can do is to just compute the series up to a certain number of terms. There are better ways to approximate the exponential*. In many cases however, few terms of the Taylor series give a reasonable approximation for the exponential and may suffice for the application. For example, let us compute the first 4 terms of the series for the matrix $A = \left[\begin{smallmatrix} 1 & 2 \\ 2 & 1 \end{smallmatrix}\right]$.

$$e^{tA} \approx I + tA + \frac{t^2}{2}A^2 + \frac{t^3}{6}A^3 = I + t\begin{bmatrix} 1 & 2 \\ 2 & 1 \end{bmatrix} + t^2 \begin{bmatrix} \frac{5}{2} & 2 \\ 2 & \frac{5}{2} \end{bmatrix} + t^3 \begin{bmatrix} \frac{13}{6} & \frac{7}{3} \\ \frac{7}{3} & \frac{13}{6} \end{bmatrix} =$$

$$= \begin{bmatrix} 1 + t + \frac{5}{2}t^2 + \frac{13}{6}t^3 & 2t + 2t^2 + \frac{7}{3}t^3 \\ 2t + 2t^2 + \frac{7}{3}t^3 & 1 + t + \frac{5}{2}t^2 + \frac{13}{6}t^3 \end{bmatrix}.$$

Just like the scalar version of the Taylor series approximation, the approximation will be better for small t and worse for larger t. For larger t, we will generally have to compute more terms. Let us see how we stack up against the real solution with $t = 0.1$. The approximate solution is approximately (rounded to 8 decimal places)

$$e^{0.1A} \approx I + 0.1\,A + \frac{0.1^2}{2}A^2 + \frac{0.1^3}{6}A^3 = \begin{bmatrix} 1.12716667 & 0.22233333 \\ 0.22233333 & 1.12716667 \end{bmatrix}.$$

And plugging $t = 0.1$ into the real solution (rounded to 8 decimal places) we get

$$e^{0.1A} = \begin{bmatrix} 1.12734811 & 0.22251069 \\ 0.22251069 & 1.12734811 \end{bmatrix}.$$

Not bad at all! Although if we take the same approximation for $t = 1$ we get

$$I + A + \frac{1}{2}A^2 + \frac{1}{6}A^3 = \begin{bmatrix} 6.66666667 & 6.33333333 \\ 6.33333333 & 6.66666667 \end{bmatrix},$$

while the real value is (again rounded to 8 decimal places)

$$e^A = \begin{bmatrix} 10.22670818 & 9.85882874 \\ 9.85882874 & 10.22670818 \end{bmatrix}.$$

So the approximation is not very good once we get up to $t = 1$. To get a good approximation at $t = 1$ (say up to 2 decimal places) we would need to go up to the 11th power (exercise).

3.8.6 Exercises

***Exercise* 3.8.2:** *Using the matrix exponential, find a fundamental matrix solution for the system* $x' = 3x + y$, $y' = x + 3y$.

---

*C. Moler and C.F. Van Loan, *Nineteen Dubious Ways to Compute the Exponential of a Matrix, Twenty-Five Years Later*, SIAM Review 45 (1), 2003, 3–49

Exercise 3.8.3: *Find e^{tA} for the matrix $A = \begin{bmatrix} 2 & 3 \\ 0 & 2 \end{bmatrix}$.*

Exercise 3.8.4: *Find a fundamental matrix solution for the system $x_1' = 7x_1 + 4x_2 + 12x_3$, $x_2' = x_1 + 2x_2 + x_3$, $x_3' = -3x_1 - 2x_2 - 5x_3$. Then find the solution that satisfies $\vec{x}(0) = \begin{bmatrix} 0 \\ 1 \\ -2 \end{bmatrix}$.*

Exercise 3.8.5: *Compute the matrix exponential e^A for $A = \begin{bmatrix} 1 & 2 \\ 0 & 1 \end{bmatrix}$.*

Exercise 3.8.6 (challenging)*: Suppose $AB = BA$. Show that under this assumption, $e^{A+B} = e^A e^B$.*

Exercise 3.8.7: *Use Exercise 3.8.6 to show that $(e^A)^{-1} = e^{-A}$. In particular this means that e^A is invertible even if A is not.*

Exercise 3.8.8: *Suppose A is a matrix with eigenvalues -1, 1, and corresponding eigenvectors $\begin{bmatrix} 1 \\ 1 \end{bmatrix}$, $\begin{bmatrix} 0 \\ 1 \end{bmatrix}$. a) Find matrix A with these properties. b) Find the fundamental matrix solution to $\vec{x}' = A\vec{x}$. c) Solve the system in with initial conditions $\vec{x}(0) = \begin{bmatrix} 2 \\ 3 \end{bmatrix}$.*

Exercise 3.8.9: *Suppose that A is an $n \times n$ matrix with a repeated eigenvalue λ of multiplicity n. Suppose that there are n linearly independent eigenvectors. Show that the matrix is diagonal, in particular $A = \lambda I$. Hint: Use diagonalization and the fact that the identity matrix commutes with every other matrix.*

Exercise 3.8.10: *Let $A = \begin{bmatrix} -1 & -1 \\ 1 & -3 \end{bmatrix}$. a) Find e^{tA}. b) Solve $\vec{x}' = A\vec{x}$, $\vec{x}(0) = \begin{bmatrix} 1 \\ -2 \end{bmatrix}$.*

Exercise 3.8.11: *Let $A = \begin{bmatrix} 1 & 2 \\ 3 & 4 \end{bmatrix}$. Approximate e^{tA} by expanding the power series up to the third order.*

Exercise 3.8.101: *Compute e^{tA} where $A = \begin{bmatrix} 1 & -2 \\ -2 & 1 \end{bmatrix}$.*

Exercise 3.8.102: *Compute e^{tA} where $A = \begin{bmatrix} 1 & -3 & 2 \\ -2 & 1 & 2 \\ -1 & -3 & 4 \end{bmatrix}$.*

Exercise 3.8.103: *a) Compute e^{tA} where $A = \begin{bmatrix} 3 & -1 \\ 1 & 1 \end{bmatrix}$. b) Solve $\vec{x}' = A\vec{x}$ for $\vec{x}(0) = \begin{bmatrix} 1 \\ 2 \end{bmatrix}$.*

Exercise 3.8.104: *Compute the first 3 terms (up to the second degree) of the Taylor expansion of e^{tA} where $A = \begin{bmatrix} 2 & 3 \\ 2 & 2 \end{bmatrix}$ (Write as a single matrix). Then use it to approximate $e^{0.1A}$.*

3.9   Nonhomogeneous systems

Note: 3 lectures (may have to skip a little), somewhat different from §5.6 in [EP], §7.9 in [BD]

3.9.1   First order constant coefficient

Integrating factor

Let us first focus on the nonhomogeneous first order equation

$$\vec{x}'(t) = A\vec{x}(t) + \vec{f}(t),$$

where A is a constant matrix. The first method we look at is the *integrating factor method*. For simplicity we rewrite the equation as

$$\vec{x}'(t) + P\vec{x}(t) = \vec{f}(t),$$

where $P = -A$. We multiply both sides of the equation by e^{tP} (being mindful that we are dealing with matrices that may not commute) to obtain

$$e^{tP}\vec{x}'(t) + e^{tP}P\vec{x}(t) = e^{tP}\vec{f}(t).$$

We notice that $Pe^{tP} = e^{tP}P$. This fact follows by writing down the series definition of e^{tP},

$$Pe^{tP} = P\left(I + tP + \frac{1}{2}(tP)^2 + \cdots\right) = P + tP^2 + \frac{1}{2}t^2P^3 + \cdots =$$

$$= \left(I + tP + \frac{1}{2}(tP)^2 + \cdots\right)P = e^{tP}P.$$

We have already seen that $\frac{d}{dt}\left(e^{tP}\right) = Pe^{tP}$. Hence,

$$\frac{d}{dt}\left(e^{tP}\vec{x}(t)\right) = e^{tP}\vec{f}(t).$$

We can now integrate. That is, we integrate each component of the vector separately

$$e^{tP}\vec{x}(t) = \int e^{tP}\vec{f}(t)\,dt + \vec{c}.$$

Recall from Exercise 3.8.7 that $\left(e^{tP}\right)^{-1} = e^{-tP}$. Therefore, we obtain

$$\vec{x}(t) = e^{-tP}\int e^{tP}\vec{f}(t)\,dt + e^{-tP}\vec{c}.$$

Perhaps it is better understood as a definite integral. In this case it will be easy to also solve for the initial conditions as well. Suppose we have the equation with initial conditions

$$\vec{x}'(t) + P\vec{x}(t) = \vec{f}(t), \qquad \vec{x}(0) = \vec{b}.$$

The solution can then be written as

$$\vec{x}(t) = e^{-tP} \int_0^t e^{sP} \vec{f}(s)\, ds + e^{-tP}\vec{b}. \tag{3.5}$$

Again, the integration means that each component of the vector $e^{sP}\vec{f}(s)$ is integrated separately. It is not hard to see that (3.5) really does satisfy the initial condition $\vec{x}(0) = \vec{b}$.

$$\vec{x}(0) = e^{-0P} \int_0^0 e^{sP} \vec{f}(s)\, ds + e^{-0P}\vec{b} = I\vec{b} = \vec{b}.$$

Example 3.9.1: Suppose that we have the system

$$x_1' + 5x_1 - 3x_2 = e^t,$$
$$x_2' + 3x_1 - x_2 = 0,$$

with initial conditions $x_1(0) = 1, x_2(0) = 0$.

Let us write the system as

$$\vec{x}' + \begin{bmatrix} 5 & -3 \\ 3 & -1 \end{bmatrix} \vec{x} = \begin{bmatrix} e^t \\ 0 \end{bmatrix}, \qquad \vec{x}(0) = \begin{bmatrix} 1 \\ 0 \end{bmatrix}.$$

We have previously computed e^{tP} for $P = \begin{bmatrix} 5 & -3 \\ 3 & -1 \end{bmatrix}$. We immediately have e^{-tP}, simply by negating t.

$$e^{tP} = \begin{bmatrix} (1 + 3t)\, e^{2t} & -3te^{2t} \\ 3te^{2t} & (1 - 3t)\, e^{2t} \end{bmatrix}, \qquad e^{-tP} = \begin{bmatrix} (1 - 3t)\, e^{-2t} & 3te^{-2t} \\ -3te^{-2t} & (1 + 3t)\, e^{-2t} \end{bmatrix}.$$

Instead of computing the whole formula at once. Let us do it in stages. First

$$\int_0^t e^{sP} \vec{f}(s)\, ds = \int_0^t \begin{bmatrix} (1 + 3s)\, e^{2s} & -3se^{2s} \\ 3se^{2s} & (1 - 3s)\, e^{2s} \end{bmatrix} \begin{bmatrix} e^s \\ 0 \end{bmatrix} ds$$
$$= \int_0^t \begin{bmatrix} (1 + 3s)\, e^{3s} \\ 3se^{3s} \end{bmatrix} ds$$
$$= \begin{bmatrix} te^{3t} \\ \frac{(3t-1)\, e^{3t}+1}{3} \end{bmatrix}.$$

Then

$$\vec{x}(t) = e^{-tP} \int_0^t e^{sP} \vec{f}(s)\, ds + e^{-tP} \vec{b}$$

$$= \begin{bmatrix} (1-3t)\,e^{-2t} & 3te^{-2t} \\ -3te^{-2t} & (1+3t)\,e^{-2t} \end{bmatrix} \begin{bmatrix} te^{3t} \\ \frac{(3t-1)\,e^{3t}+1}{3} \end{bmatrix} + \begin{bmatrix} (1-3t)\,e^{-2t} & 3te^{-2t} \\ -3te^{-2t} & (1+3t)\,e^{-2t} \end{bmatrix} \begin{bmatrix} 1 \\ 0 \end{bmatrix}$$

$$= \begin{bmatrix} te^{-2t} \\ -\frac{e^t}{3} + \left(\frac{1}{3}+t\right)e^{-2t} \end{bmatrix} + \begin{bmatrix} (1-3t)\,e^{-2t} \\ -3te^{-2t} \end{bmatrix}$$

$$= \begin{bmatrix} (1-2t)\,e^{-2t} \\ -\frac{e^t}{3} + \left(\frac{1}{3}-2t\right)e^{-2t} \end{bmatrix}.$$

Phew!

Let us check that this really works.

$$x_1' + 5x_1 - 3x_2 = (4te^{-2t} - 4e^{-2t}) + 5(1-2t)\,e^{-2t} + e^t - (1-6t)\,e^{-2t} = e^t.$$

Similarly (exercise) $x_2' + 3x_1 - x_2 = 0$. The initial conditions are also satisfied as well (exercise).

For systems, the integrating factor method only works if P does not depend on t, that is, P is constant. The problem is that in general

$$\frac{d}{dt}\left[e^{\int P(t)\,dt} \right] \neq P(t)\,e^{\int P(t)\,dt},$$

because matrix multiplication is not commutative.

Eigenvector decomposition

For the next method, we note that eigenvectors of a matrix give the directions in which the matrix acts like a scalar. If we solve our system along these directions these solutions would be simpler as we can treat the matrix as a scalar. We can put those solutions together to get the general solution.

Take the equation

$$\vec{x}'(t) = A\vec{x}(t) + \vec{f}(t). \tag{3.6}$$

Assume that A has n linearly independent eigenvectors $\vec{v}_1, \ldots, \vec{v}_n$. Let us write

$$\vec{x}(t) = \vec{v}_1\,\xi_1(t) + \vec{v}_2\,\xi_2(t) + \cdots + \vec{v}_n\,\xi_n(t). \tag{3.7}$$

That is, we wish to write our solution as a linear combination of eigenvectors of A. If we can solve for the scalar functions ξ_1 through ξ_n we have our solution $\vec{x}$. Let us decompose $\vec{f}$ in terms of the eigenvectors as well. Write

$$\vec{f}(t) = \vec{v}_1\,g_1(t) + \vec{v}_2\,g_2(t) + \cdots + \vec{v}_n\,g_n(t). \tag{3.8}$$

That is, we wish to find g_1 through g_n that satisfy (3.8). We note that since all the eigenvectors are independent, the matrix $E = [\, \vec{v}_1 \quad \vec{v}_2 \quad \cdots \quad \vec{v}_n \,]$ is invertible. We see that (3.8) can be written as $\vec{f} = E\vec{g}$, where the components of $\vec{g}$ are the functions g_1 through g_n. Then $\vec{g} = E^{-1}\vec{f}$. Hence it is always possible to find $\vec{g}$ when there are n linearly independent eigenvectors.

We plug (3.7) into (3.6), and note that $A\vec{v}_k = \lambda_k \vec{v}_k$.

$$
\begin{aligned}
\vec{x}' &= \vec{v}_1 \xi_1' + \vec{v}_2 \xi_2' + \cdots + \vec{v}_n \xi_n' \\
&= A \left(\vec{v}_1 \xi_1 + \vec{v}_2 \xi_2 + \cdots + \vec{v}_n \xi_n \right) + \vec{v}_1 g_1 + \vec{v}_2 g_2 + \cdots + \vec{v}_n g_n \\
&= A\vec{v}_1 \xi_1 + A\vec{v}_2 \xi_2 + \cdots + A\vec{v}_n \xi_n + \vec{v}_1 g_1 + \vec{v}_2 g_2 + \cdots + \vec{v}_n g_n \\
&= \vec{v}_1 \lambda_1 \xi_1 + \vec{v}_2 \lambda_2 \xi_2 + \cdots + \vec{v}_n \lambda_n \xi_n + \vec{v}_1 g_1 + \vec{v}_2 g_2 + \cdots + \vec{v}_n g_n \\
&= \vec{v}_1 \left(\lambda_1 \xi_1 + g_1 \right) + \vec{v}_2 \left(\lambda_2 \xi_2 + g_2 \right) + \cdots + \vec{v}_n \left(\lambda_n \xi_n + g_n \right).
\end{aligned}
$$

If we identify the coefficients of the vectors $\vec{v}_1$ through $\vec{v}_n$ we get the equations

$$
\begin{aligned}
\xi_1' &= \lambda_1 \xi_1 + g_1, \\
\xi_2' &= \lambda_2 \xi_2 + g_2, \\
&\;\;\vdots \\
\xi_n' &= \lambda_n \xi_n + g_n.
\end{aligned}
$$

Each one of these equations is independent of the others. They are all linear first order equations and can easily be solved by the standard integrating factor method for single equations. That is, for example for the kth equation we write

$$
\xi_k'(t) - \lambda_k \xi_k(t) = g_k(t).
$$

We use the integrating factor $e^{-\lambda_k t}$ to find that

$$
\frac{d}{dt} \left[\xi_k(t) e^{-\lambda_k t} \right] = e^{-\lambda_k t} g_k(t).
$$

We integrate and solve for ξ_k to get

$$
\xi_k(t) = e^{\lambda_k t} \int e^{-\lambda_k t} g_k(t) \, dt + C_k e^{\lambda_k t}.
$$

If we are looking for just any particular solution, we could set C_k to be zero. If we leave these constants in, we will get the general solution. Write $\vec{x}(t) = \vec{v}_1 \xi_1(t) + \vec{v}_2 \xi_2(t) + \cdots + \vec{v}_n \xi_n(t)$, and we are done.

Again, as always, it is perhaps better to write these integrals as definite integrals. Suppose that we have an initial condition $\vec{x}(0) = \vec{b}$. We take $\vec{c} = E^{-1}\vec{b}$ and note $\vec{b} = \vec{v}_1 a_1 + \cdots + \vec{v}_n a_n$, just like before. Then if we write

$$
\boxed{\; \xi_k(t) = e^{\lambda_k t} \int_0^t e^{-\lambda_k s} g_k(s) \, dt + a_k e^{\lambda_k t}, \;}
$$

we actually get the particular solution $\vec{x}(t) = \vec{v}_1\xi_1(t) + \vec{v}_2\xi_2(t) + \cdots + \vec{v}_n\xi_n(t)$ satisfying $\vec{x}(0) = \vec{b}$, because $\xi_k(0) = a_k$.

Example 3.9.2: Let $A = \begin{bmatrix} 1 & 3 \\ 3 & 1 \end{bmatrix}$. Solve $\vec{x}' = A\vec{x} + \vec{f}$ where $\vec{f}(t) = \begin{bmatrix} 2e^t \\ 2t \end{bmatrix}$ for $\vec{x}(0) = \begin{bmatrix} 3/16 \\ -5/16 \end{bmatrix}$.

The eigenvalues of A are -2 and 4 and corresponding eigenvectors are $\begin{bmatrix} 1 \\ -1 \end{bmatrix}$ and $\begin{bmatrix} 1 \\ 1 \end{bmatrix}$ respectively. This calculation is left as an exercise. We write down the matrix E of the eigenvectors and compute its inverse (using the inverse formula for 2×2 matrices)

$$E = \begin{bmatrix} 1 & 1 \\ -1 & 1 \end{bmatrix}, \qquad E^{-1} = \frac{1}{2}\begin{bmatrix} 1 & -1 \\ 1 & 1 \end{bmatrix}.$$

We are looking for a solution of the form $\vec{x} = \begin{bmatrix} 1 \\ -1 \end{bmatrix}\xi_1 + \begin{bmatrix} 1 \\ 1 \end{bmatrix}\xi_2$. We also wish to write $\vec{f}$ in terms of the eigenvectors. That is we wish to write $\vec{f} = \begin{bmatrix} 2e^t \\ 2t \end{bmatrix} = \begin{bmatrix} 1 \\ -1 \end{bmatrix}g_1 + \begin{bmatrix} 1 \\ 1 \end{bmatrix}g_2$. Thus

$$\begin{bmatrix} g_1 \\ g_2 \end{bmatrix} = E^{-1}\begin{bmatrix} 2e^t \\ 2t \end{bmatrix} = \frac{1}{2}\begin{bmatrix} 1 & -1 \\ 1 & 1 \end{bmatrix}\begin{bmatrix} 2e^t \\ 2t \end{bmatrix} = \begin{bmatrix} e^t - t \\ e^t + t \end{bmatrix}.$$

So $g_1 = e^t - t$ and $g_2 = e^t + t$.

We further want to write $\vec{x}(0)$ in terms of the eigenvectors. That is, we wish to write $\vec{x}(0) = \begin{bmatrix} 3/16 \\ -5/16 \end{bmatrix} = \begin{bmatrix} 1 \\ -1 \end{bmatrix}a_1 + \begin{bmatrix} 1 \\ 1 \end{bmatrix}a_2$. Hence

$$\begin{bmatrix} a_1 \\ a_2 \end{bmatrix} = E^{-1}\begin{bmatrix} 3/16 \\ -5/16 \end{bmatrix} = \begin{bmatrix} 1/4 \\ -1/16 \end{bmatrix}.$$

So $a_1 = 1/4$ and $a_2 = -1/16$. We plug our $\vec{x}$ into the equation and get that

$$\begin{bmatrix} 1 \\ -1 \end{bmatrix}\xi_1' + \begin{bmatrix} 1 \\ 1 \end{bmatrix}\xi_2' = A\begin{bmatrix} 1 \\ -1 \end{bmatrix}\xi_1 + A\begin{bmatrix} 1 \\ 1 \end{bmatrix}\xi_2 + \begin{bmatrix} 1 \\ -1 \end{bmatrix}g_1 + \begin{bmatrix} 1 \\ 1 \end{bmatrix}g_2$$

$$= \begin{bmatrix} 1 \\ -1 \end{bmatrix}(-2\xi_1) + \begin{bmatrix} 1 \\ 1 \end{bmatrix}4\xi_2 + \begin{bmatrix} 1 \\ -1 \end{bmatrix}(e^t - t) + \begin{bmatrix} 1 \\ 1 \end{bmatrix}(e^t + t).$$

We get the two equations

$$\xi_1' = -2\xi_1 + e^t - t, \qquad\qquad \text{where } \xi_1(0) = a_1 = \frac{1}{4},$$

$$\xi_2' = 4\xi_2 + e^t + t, \qquad\qquad \text{where } \xi_2(0) = a_2 = \frac{-1}{16}.$$

We solve with integrating factor. Computation of the integral is left as an exercise to the student. Note that you will need integration by parts.

$$\xi_1 = e^{-2t}\int e^{2t}(e^t - t)\,dt + C_1 e^{-2t} = \frac{e^t}{3} - \frac{t}{2} + \frac{1}{4} + C_1 e^{-2t}.$$

C_1 is the constant of integration. As $\xi_1(0) = $ ¹/4, then ¹/4 $=$ ¹/3 $+$ ¹/4 $+ C_1$ and hence $C_1 = $ ⁻¹/3. Similarly

$$\xi_2 = e^{4t} \int e^{-4t} (e^t + t) \, dt + C_2 e^{4t} = -\frac{e^t}{3} - \frac{t}{4} - \frac{1}{16} + C_2 e^{4t}.$$

As $\xi_2(0) = $ ⁻¹/16 we have ⁻¹/16 $=$ ⁻¹/3 $-$ ¹/16 $+ C_2$ and hence $C_2 = $ ¹/3. The solution is

$$\vec{x}(t) = \begin{bmatrix} 1 \\ -1 \end{bmatrix} \left(\frac{e^t - e^{-2t}}{3} + \frac{1 - 2t}{4} \right) + \begin{bmatrix} 1 \\ 1 \end{bmatrix} \left(\frac{e^{4t} - e^t}{3} - \frac{4t + 1}{16} \right) = \begin{bmatrix} \frac{e^{4t} - e^{-2t}}{3} + \frac{3 - 12t}{16} \\ \frac{e^{-2t} + e^{4t} + 2e^t}{3} + \frac{4t - 5}{16} \end{bmatrix}.$$

That is, $x_1 = \frac{e^{4t} - e^{-2t}}{3} + \frac{3 - 12t}{16}$ and $x_2 = \frac{e^{-2t} + e^{4t} + 2e^t}{3} + \frac{4t - 5}{16}$.

***Exercise* 3.9.1:** *Check that x_1 and x_2 solve the problem. Check both that they satisfy the differential equation and that they satisfy the initial conditions.*

Undetermined coefficients

We also have the method of undetermined coefficients for systems. The only difference here is that we will have to take unknown vectors rather than just numbers. Same caveats apply to undetermined coefficients for systems as for single equations. This method does not always work. Furthermore if the right hand side is complicated, we will have to solve for lots of variables. Each element of an unknown vector is an unknown number. So in system of 3 equations if we have say 4 unknown vectors (this would not be uncommon), then we already have 12 unknown numbers that we need to solve for. The method can turn into a lot of tedious work. As this method is essentially the same as it is for single equations, let us just do an example.

Example 3.9.3: Let $A = \begin{bmatrix} -1 & 0 \\ -2 & 1 \end{bmatrix}$. Find a particular solution of $\vec{x}' = A\vec{x} + \vec{f}$ where $\vec{f}(t) = \begin{bmatrix} e^t \\ t \end{bmatrix}$.

Note that we can solve this system in an easier way (can you see how?), but for the purposes of the example, let us use the eigenvalue method plus undetermined coefficients.

The eigenvalues of A are -1 and 1 and corresponding eigenvectors are $\begin{bmatrix} 1 \\ 1 \end{bmatrix}$ and $\begin{bmatrix} 0 \\ 1 \end{bmatrix}$ respectively. Hence our complementary solution is

$$\vec{x}_c = \alpha_1 \begin{bmatrix} 1 \\ 1 \end{bmatrix} e^{-t} + \alpha_2 \begin{bmatrix} 0 \\ 1 \end{bmatrix} e^t,$$

for some arbitrary constants α_1 and α_2.

We would want to guess a particular solution of

$$\vec{x} = \vec{a}e^t + \vec{b}t + \vec{c}.$$

However, something of the form $\vec{a}e^t$ appears in the complementary solution. Because we do not yet know if the vector $\vec{a}$ is a multiple of $\begin{bmatrix} 0 \\ 1 \end{bmatrix}$, we do not know if a conflict arises. It is possible that there is no conflict, but to be safe we should also try $\vec{b}te^t$. Here we find the crux of the difference for systems. We try *both* terms $\vec{a}e^t$ and $\vec{b}te^t$ in the solution, not just the term $\vec{b}te^t$. Therefore, we try

$$\vec{x} = \vec{a}e^t + \vec{b}te^t + \vec{c}t + \vec{d}.$$

Thus we have 8 unknowns. We write $\vec{a} = \left[\begin{smallmatrix} a_1 \\ a_2 \end{smallmatrix}\right]$, $\vec{b} = \left[\begin{smallmatrix} b_1 \\ b_2 \end{smallmatrix}\right]$, $\vec{c} = \left[\begin{smallmatrix} c_1 \\ c_2 \end{smallmatrix}\right]$, and $\vec{d} = \left[\begin{smallmatrix} d_1 \\ d_2 \end{smallmatrix}\right]$. We plug $\vec{x}$ into the equation. First let us compute $\vec{x}'$.

$$\vec{x}' = \left(\vec{a} + \vec{b}\right)e^t + \vec{b}te^t + \vec{c} = \begin{bmatrix} a_1 + b_1 \\ a_2 + b_2 \end{bmatrix} e^t + \begin{bmatrix} b_1 \\ b_2 \end{bmatrix} te^t + \begin{bmatrix} c_1 \\ c_2 \end{bmatrix}.$$

Now $\vec{x}'$ must equal $A\vec{x} + \vec{f}$, which is

$$A\vec{x} + \vec{f} = A\vec{a}e^t + A\vec{b}te^t + A\vec{c}t + A\vec{d} + \vec{f} =$$

$$= \begin{bmatrix} -a_1 \\ -2a_1 + a_2 \end{bmatrix} e^t + \begin{bmatrix} -b_1 \\ -2b_1 + b_2 \end{bmatrix} te^t + \begin{bmatrix} -c_1 \\ -2c_1 + c_2 \end{bmatrix} t + \begin{bmatrix} -d_1 \\ -2d_1 + d_2 \end{bmatrix} + \begin{bmatrix} 1 \\ 0 \end{bmatrix} e^t + \begin{bmatrix} 0 \\ 1 \end{bmatrix} t.$$

We identify the coefficients of e^t, te^t, t and any constant vectors.

$$a_1 + b_1 = -a_1 + 1,$$
$$a_2 + b_2 = -2a_1 + a_2,$$
$$b_1 = -b_1,$$
$$b_2 = -2b_1 + b_2,$$
$$0 = -c_1,$$
$$0 = -2c_1 + c_2 + 1,$$
$$c_1 = -d_1,$$
$$c_2 = -2d_1 + d_2.$$

We could write the 8×9 augmented matrix and start row reduction, but it is easier to just solve the equations in an ad hoc manner. Immediately we see that $b_1 = 0$, $c_1 = 0$, $d_1 = 0$. Plugging these back in, we get that $c_2 = -1$ and $d_2 = -1$. The remaining equations that tell us something are

$$a_1 = -a_1 + 1,$$
$$a_2 + b_2 = -2a_1 + a_2.$$

So $a_1 = 1/2$ and $b_2 = -1$. Finally, a_2 can be arbitrary and still satisfy the equations. We are looking for just a single solution so presumably the simplest one is when $a_2 = 0$. Therefore,

$$\vec{x} = \vec{a}e^t + \vec{b}te^t + \vec{c}t + \vec{d} = \begin{bmatrix} 1/2 \\ 0 \end{bmatrix} e^t + \begin{bmatrix} 0 \\ -1 \end{bmatrix} te^t + \begin{bmatrix} 0 \\ -1 \end{bmatrix} t + \begin{bmatrix} 0 \\ -1 \end{bmatrix} = \begin{bmatrix} \frac{1}{2} e^t \\ -te^t - t - 1 \end{bmatrix}.$$

That is, $x_1 = \frac{1}{2} e^t$, $x_2 = -te^t - t - 1$. We would add this to the complementary solution to get the general solution of the problem. Notice also that both $\vec{a}e^t$ and $\vec{b}te^t$ were really needed.

Exercise 3.9.2: *Check that x_1 and x_2 solve the problem. Also try setting $a_2 = 1$ and again check these solutions. What is the difference between the two solutions we can obtain in this way?*

As you can see, other than the handling of conflicts, undetermined coefficients works exactly the same as it did for single equations. However, the computations can get out of hand pretty quickly for systems. The equation we had done was very simple.

3.9.2 First order variable coefficient

Just as for a single equation, there is the method of variation of parameters. In fact for constant coefficient systems, this is essentially the same thing as the integrating factor method we discussed earlier. However, this method will work for any linear system, even if it is not constant coefficient, provided we can somehow solve the associated homogeneous problem.

Suppose we have the equation

$$\vec{x}' = A(t)\,\vec{x} + \vec{f}(t). \tag{3.9}$$

Further, suppose we have solved the associated homogeneous equation $\vec{x}' = A(t)\,\vec{x}$ and found the fundamental matrix solution $X(t)$. The general solution to the associated homogeneous equation is $X(t)\vec{c}$ for a constant vector $\vec{c}$. Just like for variation of parameters for single equation we try the solution to the nonhomogeneous equation of the form

$$\vec{x}_p = X(t)\,\vec{u}(t),$$

where $\vec{u}(t)$ is a vector valued function instead of a constant. Now we substitute into (3.9) to obtain

$$\vec{x}_p'(t) = X'(t)\,\vec{u}(t) + X(t)\,\vec{u}'(t) = A(t)\,X(t)\,\vec{u}(t) + \vec{f}(t).$$

But $X(t)$ is the fundamental matrix solution to the homogeneous problem, so $X'(t) = A(t)X(t)$, and

$$X'(t)\,\vec{u}(t) + X(t)\,\vec{u}'(t) = X'(t)\,\vec{u}(t) + \vec{f}(t).$$

Hence $X(t)\,\vec{u}'(t) = \vec{f}(t)$. If we compute $[X(t)]^{-1}$, then $\vec{u}'(t) = [X(t)]^{-1}\,\vec{f}(t)$. We integrate to obtain $\vec{u}$ and we have the particular solution $\vec{x}_p = X(t)\,\vec{u}(t)$. Let us write this as a formula

$$\boxed{\vec{x}_p = X(t) \int [X(t)]^{-1}\,\vec{f}(t)\,dt.}$$

Note that if A is constant and we let $X(t) = e^{tA}$, then $[X(t)]^{-1} = e^{-tA}$ and hence we get a solution $\vec{x}_p = e^{tA} \int e^{-tA}\,\vec{f}(t)\,dt$, which is precisely what we got using the integrating factor method.

Example 3.9.4: Find a particular solution to

$$\vec{x}' = \frac{1}{t^2 + 1}\begin{bmatrix} t & -1 \\ 1 & t \end{bmatrix}\vec{x} + \begin{bmatrix} t \\ 1 \end{bmatrix}(t^2 + 1). \tag{3.10}$$

Here $A = \frac{1}{t^2+1}\left[\begin{smallmatrix} t & -1 \\ 1 & t \end{smallmatrix}\right]$ is most definitely not constant. Perhaps by a lucky guess, we find that $X = \left[\begin{smallmatrix} 1 & -t \\ t & 1 \end{smallmatrix}\right]$ solves $X'(t) = A(t)X(t)$. Once we know the complementary solution we can easily find a solution to (3.10). First we find

$$[X(t)]^{-1} = \frac{1}{t^2 + 1}\begin{bmatrix} 1 & t \\ -t & 1 \end{bmatrix}.$$

Next we know a particular solution to (3.10) is

$$
\begin{aligned}
\vec{x}_p &= X(t) \int [X(t)]^{-1}\,\vec{f}(t)\,dt \\
&= \begin{bmatrix} 1 & -t \\ t & 1 \end{bmatrix} \int \frac{1}{t^2 + 1}\begin{bmatrix} 1 & t \\ -t & 1 \end{bmatrix}\begin{bmatrix} t \\ 1 \end{bmatrix}(t^2 + 1)\,dt \\
&= \begin{bmatrix} 1 & -t \\ t & 1 \end{bmatrix} \int \begin{bmatrix} 2t \\ -t^2 + 1 \end{bmatrix}\,dt \\
&= \begin{bmatrix} 1 & -t \\ t & 1 \end{bmatrix}\begin{bmatrix} t^2 \\ -\frac{1}{3}t^3 + t \end{bmatrix} \\
&= \begin{bmatrix} \frac{1}{3}t^4 \\ \frac{2}{3}t^3 + t \end{bmatrix}.
\end{aligned}
$$

Adding the complementary solution we have that the general solution to (3.10).

$$\vec{x} = \begin{bmatrix} 1 & -t \\ t & 1 \end{bmatrix}\begin{bmatrix} c_1 \\ c_2 \end{bmatrix} + \begin{bmatrix} \frac{1}{3}t^4 \\ \frac{2}{3}t^3 + t \end{bmatrix} = \begin{bmatrix} c_1 - c_2 t + \frac{1}{3}t^4 \\ c_2 + (c_1 + 1)t + \frac{2}{3}t^3 \end{bmatrix}.$$

Exercise 3.9.3: *Check that $x_1 = \frac{1}{3}t^4$ and $x_2 = \frac{2}{3}t^3 + t$ really solve (3.10).*

In the variation of parameters, just like in the integrating factor method we can obtain the general solution by adding in constants of integration. That is, we will add $X(t)\vec{c}$ for a vector of arbitrary constants. But that is precisely the complementary solution.

3.9.3   Second order constant coefficients

Undetermined coefficients

We have already seen a simple example of the method of undetermined coefficients for second order systems in § 3.6. This method is essentially the same as undetermined coefficients for first order systems. There are some simplifications that we can make, as we did in § 3.6. Let the equation be

$$\vec{x}'' = A\vec{x} + \vec{F}(t),$$

where A is a constant matrix. If $\vec{F}(t)$ is of the form $\vec{F}_0 \cos(\omega t)$, then as two derivatives of cosine is again cosine we can try a solution of the form

$$\vec{x}_p = \vec{c}\cos(\omega t),$$

and we do not need to introduce sines.

If the $\vec{F}$ is a sum of cosines, note that we still have the superposition principle. If $\vec{F}(t) = \vec{F}_0 \cos(\omega_0 t) + \vec{F}_1 \cos(\omega_1 t)$, then we would try $\vec{a} \cos(\omega_0 t)$ for the problem $\vec{x}'' = A\vec{x} + \vec{F}_0 \cos(\omega_0 t)$, and we would try $\vec{b} \cos(\omega_1 t)$ for the problem $\vec{x}'' = A\vec{x} + \vec{F}_1 \cos(\omega_1 t)$. Then we sum the solutions.

However, if there is duplication with the complementary solution, or the equation is of the form $\vec{x}'' = A\vec{x}' + B\vec{x} + \vec{F}(t)$, then we need to do the same thing as we do for first order systems.

You will never go wrong with putting in more terms than needed into your guess. You will find that the extra coefficients will turn out to be zero. But it is useful to save some time and effort.

Eigenvector decomposition

If we have the system

$$\vec{x}'' = A\vec{x} + \vec{F}(t),$$

we can do *eigenvector decomposition*, just like for first order systems.

Let $\lambda_1, \ldots, \lambda_n$ be the eigenvalues and $\vec{v}_1, \ldots, \vec{v}_n$ be eigenvectors. Again form the matrix $E = [\vec{v}_1 \cdots \vec{v}_n]$. We write

$$\vec{x}(t) = \vec{v}_1 \xi_1(t) + \vec{v}_2 \xi_2(t) + \cdots + \vec{v}_n \xi_n(t).$$

We decompose $\vec{F}$ in terms of the eigenvectors

$$\vec{F}(t) = \vec{v}_1 g_1(t) + \vec{v}_2 g_2(t) + \cdots + \vec{v}_n g_n(t).$$

And again $\vec{g} = E^{-1} \vec{F}$.

Now we plug in and doing the same thing as before we obtain

$$\begin{aligned}
\vec{x}'' &= \vec{v}_1 \xi_1'' + \vec{v}_2 \xi_2'' + \cdots + \vec{v}_n \xi_n'' \\
&= A \left(\vec{v}_1 \xi_1 + \vec{v}_2 \xi_2 + \cdots + \vec{v}_n \xi_n \right) + \vec{v}_1 g_1 + \vec{v}_2 g_2 + \cdots + \vec{v}_n g_n \\
&= A\vec{v}_1 \xi_1 + A\vec{v}_2 \xi_2 + \cdots + A\vec{v}_n \xi_n + \vec{v}_1 g_1 + \vec{v}_2 g_2 + \cdots + \vec{v}_n g_n \\
&= \vec{v}_1 \lambda_1 \xi_1 + \vec{v}_2 \lambda_2 \xi_2 + \cdots + \vec{v}_n \lambda_n \xi_n + \vec{v}_1 g_1 + \vec{v}_2 g_2 + \cdots + \vec{v}_n g_n \\
&= \vec{v}_1 (\lambda_1 \xi_1 + g_1) + \vec{v}_2 (\lambda_2 \xi_2 + g_2) + \cdots + \vec{v}_n (\lambda_n \xi_n + g_n).
\end{aligned}$$

We identify the coefficients of the eigenvectors to get the equations

$$\begin{aligned}
\xi_1'' &= \lambda_1 \xi_1 + g_1, \\
\xi_2'' &= \lambda_2 \xi_2 + g_2, \\
&\vdots \\
\xi_n'' &= \lambda_n \xi_n + g_n.
\end{aligned}$$

Each one of these equations is independent of the others. We solve each equation using the methods of chapter 2. We write $\vec{x}(t) = \vec{v}_1 \xi_1(t) + \cdots + \vec{v}_n \xi_n(t)$, and we are done; we have a particular solution.

If we have found the general solution for ξ_1 through ξ_n, then again $\vec{x}(t) = \vec{v}_1 \xi_1(t) + \cdots + \vec{v}_n \xi_n(t)$ is the general solution (and not just a particular solution).

Example 3.9.5: Let us do the example from § 3.6 using this method. The equation is

$$\vec{x}'' = \begin{bmatrix} -3 & 1 \\ 2 & -2 \end{bmatrix} \vec{x} + \begin{bmatrix} 0 \\ 2 \end{bmatrix} \cos(3t).$$

The eigenvalues were -1 and -4, with eigenvectors $\begin{bmatrix} 1 \\ 2 \end{bmatrix}$ and $\begin{bmatrix} 1 \\ -1 \end{bmatrix}$. Therefore $E = \begin{bmatrix} 1 & 1 \\ 2 & -1 \end{bmatrix}$ and $E^{-1} = \frac{1}{3} \begin{bmatrix} 1 & 1 \\ 2 & -1 \end{bmatrix}$. Therefore,

$$\begin{bmatrix} g_1 \\ g_2 \end{bmatrix} = E^{-1} \vec{F}(t) = \frac{1}{3} \begin{bmatrix} 1 & 1 \\ 2 & -1 \end{bmatrix} \begin{bmatrix} 0 \\ 2\cos(3t) \end{bmatrix} = \begin{bmatrix} \frac{2}{3}\cos(3t) \\ \frac{-2}{3}\cos(3t) \end{bmatrix}.$$

So after the whole song and dance of plugging in, the equations we get are

$$\xi_1'' = -\xi_1 + \frac{2}{3}\cos(3t),$$

$$\xi_2'' = -4\xi_2 - \frac{2}{3}\cos(3t).$$

For each equation we use the method of undetermined coefficients. We try $C_1 \cos(3t)$ for the first equation and $C_2 \cos(3t)$ for the second equation. We plug in to get

$$-9C_1 \cos(3t) = -C_1 \cos(3t) + \frac{2}{3}\cos(3t),$$

$$-9C_2 \cos(3t) = -4C_2 \cos(3t) - \frac{2}{3}\cos(3t).$$

We solve each of these equations separately. We get $-9C_1 = -C_1 + 2/3$ and $-9C_2 = -4C_2 - 2/3$. And hence $C_1 = -1/12$ and $C_2 = 2/15$. So our particular solution is

$$\vec{x} = \begin{bmatrix} 1 \\ 2 \end{bmatrix} \left(\frac{-1}{12}\cos(3t) \right) + \begin{bmatrix} 1 \\ -1 \end{bmatrix} \left(\frac{2}{15}\cos(3t) \right) = \begin{bmatrix} 1/20 \\ -3/10 \end{bmatrix} \cos(3t).$$

This solution matches what we got previously in § 3.6.

3.9.4   Exercises

Exercise 3.9.4: *Find a particular solution to $x' = x + 2y + 2t$, $y' = 3x + 2y - 4$, a) using integrating factor method, b) using eigenvector decomposition, c) using undetermined coefficients.*

Exercise 3.9.5: *Find the general solution to $x' = 4x + y - 1$, $y' = x + 4y - e^t$, a) using integrating factor method, b) using eigenvector decomposition, c) using undetermined coefficients.*

Exercise 3.9.6: *Find the general solution to* $x_1'' = -6x_1 + 3x_2 + \cos(t)$, $x_2'' = 2x_1 - 7x_2 + 3\cos(t)$, *a) using eigenvector decomposition, b) using undetermined coefficients.*

Exercise 3.9.7: *Find the general solution to* $x_1'' = -6x_1 + 3x_2 + \cos(2t)$, $x_2'' = 2x_1 - 7x_2 + 3\cos(2t)$, *a) using eigenvector decomposition, b) using undetermined coefficients.*

Exercise 3.9.8: *Take the equation*

$$\vec{x}' = \begin{bmatrix} \frac{1}{t} & -1 \\ 1 & \frac{1}{t} \end{bmatrix} \vec{x} + \begin{bmatrix} t^2 \\ -t \end{bmatrix}.$$

a) Check that

$$\vec{x}_c = c_1 \begin{bmatrix} t \sin t \\ -t \cos t \end{bmatrix} + c_2 \begin{bmatrix} t \cos t \\ t \sin t \end{bmatrix}$$

is the complementary solution. b) Use variation of parameters to find a particular solution.

Exercise 3.9.101: *Find a particular solution to* $x' = 5x + 4y + t$, $y' = x + 8y - t$, *a) using integrating factor method, b) using eigenvector decomposition, c) using undetermined coefficients.*

Exercise 3.9.102: *Find a particular solution to* $x' = y + e^t$, $y' = x + e^t$, *a) using integrating factor method, b) using eigenvector decomposition, c) using undetermined coefficients.*

Exercise 3.9.103: *Solve* $x_1' = x_2 + t$, $x_2' = x_1 + t$ *with initial conditions* $x_1(0) = 1$, $x_2(0) = 2$, *using eigenvector decomposition.*

Exercise 3.9.104: *Solve* $x_1'' = -3x_1 + x_2 + t$, $x_2'' = 9x_1 + 5x_2 + \cos(t)$ *with initial conditions* $x_1(0) = 0$, $x_2(0) = 0$, $x_1'(0) = 0$, $x_2'(0) = 0$, *using eigenvector decomposition.*

Chapter 4

Fourier series and PDEs

4.1 Boundary value problems

Note: 2 lectures, similar to §3.8 in [EP], §10.1 and §11.1 in [BD]

4.1.1 Boundary value problems

Before we tackle the Fourier series, we need to study the so-called *boundary value problems* (or *endpoint problems*). For example, suppose we have

$$x'' + \lambda x = 0, \quad x(a) = 0, \quad x(b) = 0,$$

for some constant λ, where $x(t)$ is defined for t in the interval $[a, b]$. Unlike before, when we specified the value of the solution and its derivative at a single point, we now specify the value of the solution at two different points. Note that $x = 0$ is a solution to this equation, so existence of solutions is not an issue here. Uniqueness of solutions is another issue. The general solution to $x'' + \lambda x = 0$ has two arbitrary constants present. It is, therefore, natural (but wrong) to believe that requiring two conditions guarantees a unique solution.

Example 4.1.1: Take $\lambda = 1$, $a = 0$, $b = \pi$. That is,

$$x'' + x = 0, \quad x(0) = 0, \quad x(\pi) = 0.$$

Then $x = \sin t$ is another solution (besides $x = 0$) satisfying both boundary conditions. There are more. Write down the general solution of the differential equation, which is $x = A \cos t + B \sin t$. The condition $x(0) = 0$ forces $A = 0$. Letting $x(\pi) = 0$ does not give us any more information as $x = B \sin t$ already satisfies both boundary conditions. Hence, there are infinitely many solutions of the form $x = B \sin t$, where B is an arbitrary constant.

Example 4.1.2: On the other hand, change to $\lambda = 2$.

$$x'' + 2x = 0, \quad x(0) = 0, \quad x(\pi) = 0.$$

Then the general solution is $x = A\cos(\sqrt{2}\,t) + B\sin(\sqrt{2}\,t)$. Letting $x(0) = 0$ still forces $A = 0$. We apply the second condition to find $0 = x(\pi) = B\sin(\sqrt{2}\,\pi)$. As $\sin(\sqrt{2}\,\pi) \neq 0$ we obtain $B = 0$. Therefore $x = 0$ is the unique solution to this problem.

What is going on? We will be interested in finding which constants λ allow a nonzero solution, and we will be interested in finding those solutions. This problem is an analogue of finding eigenvalues and eigenvectors of matrices.

4.1.2   Eigenvalue problems

For basic Fourier series theory we will need the following three eigenvalue problems. We will consider more general equations, but we will postpone this until chapter 5.

$$x'' + \lambda x = 0, \quad x(a) = 0, \quad x(b) = 0, \tag{4.1}$$

$$x'' + \lambda x = 0, \quad x'(a) = 0, \quad x'(b) = 0, \tag{4.2}$$

and

$$x'' + \lambda x = 0, \quad x(a) = x(b), \quad x'(a) = x'(b), \tag{4.3}$$

A number λ is called an *eigenvalue* of (4.1) (resp. (4.2) or (4.3)) if and only if there exists a nonzero (not identically zero) solution to (4.1) (resp. (4.2) or (4.3)) given that specific λ. The nonzero solution we found is called the corresponding *eigenfunction*.

Note the similarity to eigenvalues and eigenvectors of matrices. The similarity is not just coincidental. If we think of the equations as differential operators, then we are doing the same exact thing. For example, let $L = -\frac{d^2}{dt^2}$. We are looking for nonzero functions f satisfying certain endpoint conditions that solve $(L - \lambda)f = 0$. A lot of the formalism from linear algebra can still apply here, though we will not pursue this line of reasoning too far.

Example 4.1.3: Let us find the eigenvalues and eigenfunctions of

$$x'' + \lambda x = 0, \quad x(0) = 0, \quad x(\pi) = 0.$$

For reasons that will be clear from the computations, we will have to handle the cases $\lambda > 0$, $\lambda = 0$, $\lambda < 0$ separately. First suppose that $\lambda > 0$, then the general solution to $x'' + \lambda x = 0$ is

$$x = A\cos(\sqrt{\lambda}\,t) + B\sin(\sqrt{\lambda}\,t).$$

The condition $x(0) = 0$ implies immediately $A = 0$. Next

$$0 = x(\pi) = B\sin(\sqrt{\lambda}\,\pi).$$

If B is zero, then x is not a nonzero solution. So to get a nonzero solution we must have that $\sin(\sqrt{\lambda}\,\pi) = 0$. Hence, $\sqrt{\lambda}\,\pi$ must be an integer multiple of π. In other words, $\sqrt{\lambda} = k$ for a positive integer k. Hence the positive eigenvalues are k^2 for all integers $k \geq 1$. The corresponding eigenfunctions can be taken as $x = \sin(kt)$. Just like for eigenvectors, we get all the multiples of an eigenfunction, so we only need to pick one.

Now suppose that $\lambda = 0$. In this case the equation is $x'' = 0$ and the general solution is $x = At + B$. The condition $x(0) = 0$ implies that $B = 0$, and $x(\pi) = 0$ implies that $A = 0$. This means that $\lambda = 0$ is *not* an eigenvalue.

Finally, suppose that $\lambda < 0$. In this case we have the general solution

$$x = A \cosh(\sqrt{-\lambda}\, t) + B \sinh(\sqrt{-\lambda}\, t).$$

Letting $x(0) = 0$ implies that $A = 0$ (recall $\cosh 0 = 1$ and $\sinh 0 = 0$). So our solution must be $x = B \sinh(\sqrt{-\lambda}\, t)$ and satisfy $x(\pi) = 0$. This is only possible if B is zero. Why? Because $\sinh \xi$ is only zero when $\xi = 0$. You should plot sinh to see this fact. We can also see this from the definition of sinh. We get $0 = \sinh t = \frac{e^t - e^{-t}}{2}$. Hence $e^t = e^{-t}$, which implies $t = -t$ and that is only true if $t = 0$. So there are no negative eigenvalues.

In summary, the eigenvalues and corresponding eigenfunctions are

$$\lambda_k = k^2 \qquad \text{with an eigenfunction} \qquad x_k = \sin(kt) \qquad \text{for all integers } k \geq 1.$$

Example 4.1.4: Let us compute the eigenvalues and eigenfunctions of

$$x'' + \lambda x = 0, \quad x'(0) = 0, \quad x'(\pi) = 0.$$

Again we will have to handle the cases $\lambda > 0$, $\lambda = 0$, $\lambda < 0$ separately. First suppose that $\lambda > 0$. The general solution to $x'' + \lambda x = 0$ is $x = A \cos(\sqrt{\lambda}\, t) + B \sin(\sqrt{\lambda}\, t)$. So

$$x' = -A \sqrt{\lambda}\, \sin(\sqrt{\lambda}\, t) + B \sqrt{\lambda}\, \cos(\sqrt{\lambda}\, t).$$

The condition $x'(0) = 0$ implies immediately $B = 0$. Next

$$0 = x'(\pi) = -A \sqrt{\lambda}\, \sin(\sqrt{\lambda}\, \pi).$$

Again A cannot be zero if λ is to be an eigenvalue, and $\sin(\sqrt{\lambda}\, \pi)$ is only zero if $\sqrt{\lambda} = k$ for a positive integer k. Hence the positive eigenvalues are again k^2 for all integers $k \geq 1$. And the corresponding eigenfunctions can be taken as $x = \cos(kt)$.

Now suppose that $\lambda = 0$. In this case the equation is $x'' = 0$ and the general solution is $x = At + B$ so $x' = A$. The condition $x'(0) = 0$ implies that $A = 0$. Now $x'(\pi) = 0$ also simply implies $A = 0$. This means that B could be anything (let us take it to be 1). So $\lambda = 0$ is an eigenvalue and $x = 1$ is a corresponding eigenfunction.

Finally, let $\lambda < 0$. In this case we have the general solution $x = A \cosh(\sqrt{-\lambda}\, t) + B \sinh(\sqrt{-\lambda}\, t)$ and hence

$$x' = A \sqrt{-\lambda}\, \sinh(\sqrt{-\lambda}\, t) + B \sqrt{-\lambda}\, \cosh(\sqrt{-\lambda}\, t).$$

We have already seen (with roles of A and B switched) that for this to be zero at $t = 0$ and $t = \pi$ it implies that $A = B = 0$. Hence there are no negative eigenvalues.

In summary, the eigenvalues and corresponding eigenfunctions are

$$\lambda_k = k^2 \qquad \text{with an eigenfunction} \qquad x_k = \cos(kt) \qquad \text{for all integers } k \geq 1,$$

and there is another eigenvalue

$$\lambda_0 = 0 \qquad \text{with an eigenfunction} \qquad x_0 = 1.$$

The following problem is the one that leads to the general Fourier series.

Example 4.1.5: Let us compute the eigenvalues and eigenfunctions of

$$x'' + \lambda x = 0, \quad x(-\pi) = x(\pi), \quad x'(-\pi) = x'(\pi).$$

Notice that we have not specified the values or the derivatives at the endpoints, but rather that they are the same at the beginning and at the end of the interval.

Let us skip $\lambda < 0$. The computations are the same as before, and again we find that there are no negative eigenvalues.

For $\lambda = 0$, the general solution is $x = At + B$. The condition $x(-\pi) = x(\pi)$ implies that $A = 0$ ($A\pi + B = -A\pi + B$ implies $A = 0$). The second condition $x'(-\pi) = x'(\pi)$ says nothing about B and hence $\lambda = 0$ is an eigenvalue with a corresponding eigenfunction $x = 1$.

For $\lambda > 0$ we get that $x = A \cos(\sqrt{\lambda}\, t) + B \sin(\sqrt{\lambda}\, t)$. Now

$$A \cos(-\sqrt{\lambda}\, \pi) + B \sin(-\sqrt{\lambda}\, \pi) = A \cos(\sqrt{\lambda}\, \pi) + B \sin(\sqrt{\lambda}\, \pi).$$

We remember that $\cos(-\theta) = \cos(\theta)$ and $\sin(-\theta) = -\sin(\theta)$. Therefore,

$$A \cos(\sqrt{\lambda}\, \pi) - B \sin(\sqrt{\lambda}\, \pi) = A \cos(\sqrt{\lambda}\, \pi) + B \sin(\sqrt{\lambda}\, \pi).$$

Hence either $B = 0$ or $\sin(\sqrt{\lambda}\, \pi) = 0$. Similarly (exercise) if we differentiate x and plug in the second condition we find that $A = 0$ or $\sin(\sqrt{\lambda}\, \pi) = 0$. Therefore, unless we want A and B to both be zero (which we do not) we must have $\sin(\sqrt{\lambda}\, \pi) = 0$. Hence, $\sqrt{\lambda}$ is an integer and the eigenvalues are yet again $\lambda = k^2$ for an integer $k \geq 1$. In this case, however, $x = A \cos(kt) + B \sin(kt)$ is an eigenfunction for any A and any B. So we have two linearly independent eigenfunctions $\sin(kt)$ and $\cos(kt)$. Remember that for a matrix we could also have had two eigenvectors corresponding to a single eigenvalue if the eigenvalue was repeated.

In summary, the eigenvalues and corresponding eigenfunctions are

$$\lambda_k = k^2 \qquad \text{with the eigenfunctions} \qquad \cos(kt) \quad \text{and} \quad \sin(kt) \qquad \text{for all integers } k \geq 1,$$
$$\lambda_0 = 0 \qquad \text{with an eigenfunction} \qquad x_0 = 1.$$

4.1.3 Orthogonality of eigenfunctions

Something that will be very useful in the next section is the *orthogonality* property of the eigen-functions. This is an analogue of the following fact about eigenvectors of a matrix. A matrix is called *symmetric* if $A = A^T$. *Eigenvectors for two distinct eigenvalues of a symmetric matrix are orthogonal.* That symmetry is required. We will not prove this fact here. The differential operators we are dealing with act much like a symmetric matrix. We, therefore, get the following theorem.

Theorem 4.1.1. *Suppose that $x_1(t)$ and $x_2(t)$ are two eigenfunctions of the problem* (4.1), (4.2) *or* (4.3) *for two different eigenvalues λ_1 and λ_2. Then they are* orthogonal *in the sense that*

$$\int_a^b x_1(t)x_2(t)\, dt = 0.$$

Note that the terminology comes from the fact that the integral is a type of inner product. We will expand on this in the next section. The theorem has a very short, elegant, and illuminating proof so let us give it here. First note that we have the following two equations.

$$x_1'' + \lambda_1 x_1 = 0 \qquad \text{and} \qquad x_2'' + \lambda_2 x_2 = 0.$$

Multiply the first by x_2 and the second by x_1 and subtract to get

$$(\lambda_1 - \lambda_2)x_1 x_2 = x_2'' x_1 - x_2 x_1''.$$

Now integrate both sides of the equation.

$$
\begin{aligned}
(\lambda_1 - \lambda_2)\int_a^b x_1 x_2\, dt &= \int_a^b x_2'' x_1 - x_2 x_1''\, dt \\
&= \int_a^b \frac{d}{dt}\left(x_2' x_1 - x_2 x_1'\right)\, dt \\
&= \left[x_2' x_1 - x_2 x_1'\right]_{t=a}^{b} = 0.
\end{aligned}
$$

The last equality holds because of the boundary conditions. For example, if we consider (4.1) we have $x_1(a) = x_1(b) = x_2(a) = x_2(b) = 0$ and so $x_2' x_1 - x_2 x_1'$ is zero at both a and b. As $\lambda_1 \neq \lambda_2$, the theorem follows.

Exercise 4.1.1 (easy)*: Finish the theorem (check the last equality in the proof) for the cases* (4.2) *and* (4.3).

We have seen previously that $\sin(nt)$ was an eigenfunction for the problem $x'' + \lambda x = 0$, $x(0) = 0$, $x(\pi) = 0$. Hence we have the integral

$$\int_0^\pi \sin(mt)\sin(nt)\, dt = 0, \qquad \text{when } m \neq n.$$

Similarly

$$\int_0^\pi \cos(mt)\cos(nt)\,dt = 0, \qquad \text{when } m \neq n.$$

And finally we also get

$$\int_{-\pi}^\pi \sin(mt)\sin(nt)\,dt = 0, \qquad \text{when } m \neq n,$$

$$\int_{-\pi}^\pi \cos(mt)\cos(nt)\,dt = 0, \qquad \text{when } m \neq n,$$

and

$$\int_{-\pi}^\pi \cos(mt)\sin(nt)\,dt = 0.$$

4.1.4   Fredholm alternative

We now touch on a very useful theorem in the theory of differential equations. The theorem holds in a more general setting than we are going to state it, but for our purposes the following statement is sufficient. We will give a slightly more general version in chapter 5.

Theorem 4.1.2 (Fredholm alternative*). *Exactly one of the following statements holds. Either*

$$x'' + \lambda x = 0, \quad x(a) = 0, \quad x(b) = 0 \tag{4.4}$$

has a nonzero solution, or

$$x'' + \lambda x = f(t), \quad x(a) = 0, \quad x(b) = 0 \tag{4.5}$$

has a unique solution for every function f continuous on $[a, b]$.

The theorem is also true for the other types of boundary conditions we considered. The theorem means that if λ is not an eigenvalue, the nonhomogeneous equation (4.5) has a unique solution for every right hand side. On the other hand if λ is an eigenvalue, then (4.5) need not have a solution for every f, and furthermore, even if it happens to have a solution, the solution is not unique.

We also want to reinforce the idea here that linear differential operators have much in common with matrices. So it is no surprise that there is a finite dimensional version of Fredholm alternative for matrices as well. Let A be an $n \times n$ matrix. The Fredholm alternative then states that either $(A - \lambda I)\vec{x} = \vec{0}$ has a nontrivial solution, or $(A - \lambda I)\vec{x} = \vec{b}$ has a solution for every $\vec{b}$.

A lot of intuition from linear algebra can be applied to linear differential operators, but one must be careful of course. For example, one difference we have already seen is that in general a differential operator will have infinitely many eigenvalues, while a matrix has only finitely many.

---

*Named after the Swedish mathematician Erik Ivar Fredholm (1866–1927).

4.1.5 Application

Let us consider a physical application of an endpoint problem. Suppose we have a tightly stretched quickly spinning elastic string or rope of uniform linear density ρ. Let us put this problem into the xy-plane. The x axis represents the position on the string. The string rotates at angular velocity ω, so we will assume that the whole xy-plane rotates at angular velocity ω. We will assume that the string stays in this xy-plane and y will measure its deflection from the equilibrium position, $y = 0$, on the x axis. Hence, we will find a graph giving the shape of the string. We will idealize the string to have no volume to just be a mathematical curve. If we take a small segment and we look at the tension at the endpoints, we see that this force is tangential and we will assume that the magnitude is the same at both end points. Hence the magnitude is constant everywhere and we will call its magnitude T. If we assume that the deflection is small, then we can use Newton's second law to get an equation

$$Ty'' + \rho\omega^2 y = 0.$$

Let L be the length of the string and the string is fixed at the beginning and end points. Hence, $y(0) = 0$ and $y(L) = 0$. See Figure 4.1.

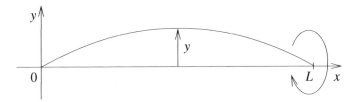

Figure 4.1: Whirling string.

We rewrite the equation as $y'' + \frac{\rho\omega^2}{T}y = 0$. The setup is similar to Example 4.1.3 on page 168, except for the interval length being L instead of π. We are looking for eigenvalues of $y'' + \lambda y = 0, y(0) = 0, y(L) = 0$ where $\lambda = \frac{\rho\omega^2}{T}$. As before there are no nonpositive eigenvalues. With $\lambda > 0$, the general solution to the equation is $y = A\cos(\sqrt{\lambda}\,x) + B\sin(\sqrt{\lambda}\,x)$. The condition $y(0) = 0$ implies that $A = 0$ as before. The condition $y(L) = 0$ implies that $\sin(\sqrt{\lambda}\,L) = 0$ and hence $\sqrt{\lambda}\,L = k\pi$ for some integer $k > 0$, so

$$\frac{\rho\omega^2}{T} = \lambda = \frac{k^2\pi^2}{L^2}.$$

What does this say about the shape of the string? It says that for all parameters ρ, ω, T not satisfying the above equation, the string is in the equilibrium position, $y = 0$. When $\frac{\rho\omega^2}{T} = \frac{k^2\pi^2}{L^2}$, then the string will "pop out" some distance B at the midpoint. We cannot compute B with the information we have.

Let us assume that ρ and T are fixed and we are changing ω. For most values of ω the string is in the equilibrium state. When the angular velocity ω hits a value $\omega = \frac{k\pi\sqrt{T}}{L\sqrt{\rho}}$, then the string will pop out and will have the shape of a sin wave crossing the x axis k times. When ω changes again, the string returns to the equilibrium position. You can see that the higher the angular velocity the more times it crosses the x axis when it is popped out.

4.1.6   Exercises

Hint for the following exercises: Note that when $\lambda > 0$, then $\cos(\sqrt{\lambda}\,(t - a))$ and $\sin(\sqrt{\lambda}\,(t - a))$ are also solutions of the homogeneous equation.

***Exercise* 4.1.2:** *Compute all eigenvalues and eigenfunctions of* $x'' + \lambda x = 0$, $x(a) = 0$, $x(b) = 0$ *(assume $a < b$).*

***Exercise* 4.1.3:** *Compute all eigenvalues and eigenfunctions of* $x'' + \lambda x = 0$, $x'(a) = 0$, $x'(b) = 0$ *(assume $a < b$).*

***Exercise* 4.1.4:** *Compute all eigenvalues and eigenfunctions of* $x'' + \lambda x = 0$, $x'(a) = 0$, $x(b) = 0$ *(assume $a < b$).*

***Exercise* 4.1.5:** *Compute all eigenvalues and eigenfunctions of* $x'' + \lambda x = 0$, $x(a) = x(b)$, $x'(a) = x'(b)$ *(assume $a < b$).*

***Exercise* 4.1.6:** *We have skipped the case of $\lambda < 0$ for the boundary value problem $x'' + \lambda x = 0$, $x(-\pi) = x(\pi)$, $x'(-\pi) = x'(\pi)$. Finish the calculation and show that there are no negative eigenvalues.*

***Exercise* 4.1.101:** *Consider a spinning string of length 2 and linear density 0.1 and tension 3. Find smallest angular velocity when the string pops out.*

***Exercise* 4.1.102:** *Suppose $x'' + \lambda x = 0$ and $x(0) = 1$, $x(1) = 1$. Find all λ for which there is more than one solution. Also find the corresponding solutions (only for the eigenvalues).*

***Exercise* 4.1.103:** *Suppose $x'' + x = 0$ and $x(0) = 0$, $x'(\pi) = 1$. Find all the solution(s) if any exist.*

***Exercise* 4.1.104:** *Consider $x' + \lambda x = 0$ and $x(0) = 0$, $x(1) = 0$. Why does it not have any eigenvalues? Why does any first order equation with two endpoint conditions such as above have no eigenvalues?*

***Exercise* 4.1.105** (challenging): *Suppose $x''' + \lambda x = 0$ and $x(0) = 0$, $x'(0) = 0$, $x(1) = 0$. Suppose that $\lambda > 0$. Find an equation that all such eigenvalues must satisfy. Hint: Note that $-\sqrt[3]{\lambda}$ is a root of $r^3 + \lambda = 0$.*

4.2   The trigonometric series

Note: 2 lectures, §9.1 in [EP], §10.2 in [BD]

4.2.1   Periodic functions and motivation

As motivation for studying Fourier series, suppose we have the problem

$$x'' + \omega_0^2 x = f(t), \tag{4.6}$$

for some periodic function $f(t)$. We have already solved

$$x'' + \omega_0^2 x = F_0 \cos(\omega t). \tag{4.7}$$

One way to solve (4.6) is to decompose $f(t)$ as a sum of cosines (and sines) and then solve many problems of the form (4.7). We then use the principle of superposition, to sum up all the solutions we got to get a solution to (4.6).

Before we proceed, let us talk a little bit more in detail about periodic functions. A function is said to be *periodic* with period P if $f(t) = f(t + P)$ for all t. For brevity we will say $f(t)$ is P-periodic. Note that a P-periodic function is also $2P$-periodic, $3P$-periodic and so on. For example, $\cos(t)$ and $\sin(t)$ are 2π-periodic. So are $\cos(kt)$ and $\sin(kt)$ for all integers k. The constant functions are an extreme example. They are periodic for any period (exercise).

Normally we will start with a function $f(t)$ defined on some interval $[-L, L]$ and we will want to *extend periodically* to make it a $2L$-periodic function. We do this extension by defining a new function $F(t)$ such that for t in $[-L, L]$, $F(t) = f(t)$. For t in $[L, 3L]$, we define $F(t) = f(t - 2L)$, for t in $[-3L, -L]$, $F(t) = f(t + 2L)$, and so on. We assumed that $f(-L) = f(L)$. We could have also started with f defined only on the half-open interval $(-L, L]$ and then define $f(-L) = f(L)$.

Example 4.2.1: Define $f(t) = 1 - t^2$ on $[-1, 1]$. Now extend periodically to a 2-periodic function. See Figure 4.2 on the following page.

You should be careful to distinguish between $f(t)$ and its extension. A common mistake is to assume that a formula for $f(t)$ holds for its extension. It can be confusing when the formula for $f(t)$ is periodic, but with perhaps a different period.

Exercise 4.2.1: Define $f(t) = \cos t$ on $[-\pi/2, \pi/2]$. Take the π-periodic extension and sketch its graph. How does it compare to the graph of $\cos t$?

4.2.2   Inner product and eigenvector decomposition

Suppose we have a *symmetric matrix*, that is $A^T = A$. We have said before that the eigenvectors of A are then orthogonal. Here the word *orthogonal* means that if $\vec{v}$ and $\vec{w}$ are two distinct (and not

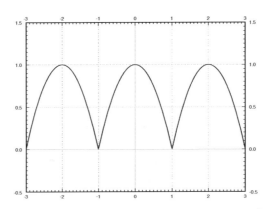

Figure 4.2: Periodic extension of the function $1 - t^2$.

multiples of each other) eigenvectors of A, then $\langle \vec{v}, \vec{w} \rangle = 0$. In this case the inner product $\langle \vec{v}, \vec{w} \rangle$ is the *dot product*, which can be computed as $\vec{v}^T \vec{w}$.

To decompose a vector $\vec{v}$ in terms of mutually orthogonal vectors $\vec{w}_1$ and $\vec{w}_2$ we write

$$\vec{v} = a_1 \vec{w}_1 + a_2 \vec{w}_2.$$

Let us find the formula for a_1 and a_2. First let us compute

$$\langle \vec{v}, \vec{w}_1 \rangle = \langle a_1 \vec{w}_1 + a_2 \vec{w}_2, \vec{w}_1 \rangle = a_1 \langle \vec{w}_1, \vec{w}_1 \rangle + a_2 \langle \vec{w}_2, \vec{w}_1 \rangle = a_1 \langle \vec{w}_1, \vec{w}_1 \rangle.$$

Therefore,

$$a_1 = \frac{\langle \vec{v}, \vec{w}_1 \rangle}{\langle \vec{w}_1, \vec{w}_1 \rangle}.$$

Similarly

$$a_2 = \frac{\langle \vec{v}, \vec{w}_2 \rangle}{\langle \vec{w}_2, \vec{w}_2 \rangle}.$$

You probably remember this formula from vector calculus.

Example 4.2.2: Write $\vec{v} = \left[\begin{smallmatrix} 2 \\ 3 \end{smallmatrix}\right]$ as a linear combination of $\vec{w}_1 = \left[\begin{smallmatrix} 1 \\ -1 \end{smallmatrix}\right]$ and $\vec{w}_2 = \left[\begin{smallmatrix} 1 \\ 1 \end{smallmatrix}\right]$.

First note that $\vec{w}_1$ and $\vec{w}_2$ are orthogonal as $\langle \vec{w}_1, \vec{w}_2 \rangle = 1(1) + (-1)1 = 0$. Then

$$a_1 = \frac{\langle \vec{v}, \vec{w}_1 \rangle}{\langle \vec{w}_1, \vec{w}_1 \rangle} = \frac{2(1) + 3(-1)}{1(1) + (-1)(-1)} = \frac{-1}{2},$$
$$a_2 = \frac{\langle \vec{v}, \vec{w}_2 \rangle}{\langle \vec{w}_2, \vec{w}_2 \rangle} = \frac{2+3}{1+1} = \frac{5}{2}.$$

Hence

$$\begin{bmatrix} 2 \\ 3 \end{bmatrix} = \frac{-1}{2} \begin{bmatrix} 1 \\ -1 \end{bmatrix} + \frac{5}{2} \begin{bmatrix} 1 \\ 1 \end{bmatrix}.$$

4.2.3 The trigonometric series

Instead of decomposing a vector in terms of eigenvectors of a matrix, we will decompose a function in terms of eigenfunctions of a certain eigenvalue problem. The eigenvalue problem we will use for the Fourier series is

$$x'' + \lambda x = 0, \quad x(-\pi) = x(\pi), \quad x'(-\pi) = x'(\pi).$$

We have previously computed that the eigenfunctions are 1, $\cos(kt)$, $\sin(kt)$. That is, we will want to find a representation of a 2π-periodic function $f(t)$ as

$$f(t) = \frac{a_0}{2} + \sum_{n=1}^{\infty} a_n \cos(nt) + b_n \sin(nt).$$

This series is called the *Fourier series* or the *trigonometric series* for $f(t)$. We write the coefficient of the eigenfunction 1 as $\frac{a_0}{2}$ for convenience. We could also think of $1 = \cos(0t)$, so that we only need to look at $\cos(kt)$ and $\sin(kt)$.

As for matrices we want to find a *projection* of $f(t)$ onto the subspace generated by the eigenfunctions. So we will want to define an *inner product of functions*. For example, to find a_n we want to compute $\langle f(t), \cos(nt) \rangle$. We define the inner product as

$$\langle f(t), g(t) \rangle \overset{\text{def}}{=} \int_{-\pi}^{\pi} f(t) g(t) \, dt.$$

With this definition of the inner product, we have seen in the previous section that the eigenfunctions $\cos(kt)$ (including the constant eigenfunction), and $\sin(kt)$ are *orthogonal* in the sense that

$$
\begin{aligned}
\langle \cos(mt), \cos(nt) \rangle &= 0 \qquad \text{for } m \neq n, \\
\langle \sin(mt), \sin(nt) \rangle &= 0 \qquad \text{for } m \neq n, \\
\langle \sin(mt), \cos(nt) \rangle &= 0 \qquad \text{for all } m \text{ and } n.
\end{aligned}
$$

By elementary calculus for $n = 1, 2, 3, \ldots$ we have $\langle \cos(nt), \cos(nt) \rangle = \pi$ and $\langle \sin(nt), \sin(nt) \rangle = \pi$. For the constant we get that $\langle 1, 1 \rangle = 2\pi$. The coefficients are given by

$$
\begin{aligned}
a_n &= \frac{\langle f(t), \cos(nt) \rangle}{\langle \cos(nt), \cos(nt) \rangle} = \frac{1}{\pi} \int_{-\pi}^{\pi} f(t) \cos(nt) \, dt, \\
b_n &= \frac{\langle f(t), \sin(nt) \rangle}{\langle \sin(nt), \sin(nt) \rangle} = \frac{1}{\pi} \int_{-\pi}^{\pi} f(t) \sin(nt) \, dt.
\end{aligned}
$$

Compare these expressions with the finite-dimensional example. For a_0 we get a similar formula

$$a_0 = 2\frac{\langle f(t), 1 \rangle}{\langle 1, 1 \rangle} = \frac{1}{\pi} \int_{-\pi}^{\pi} f(t) \, dt.$$

---

*Named after the French mathematician Jean Baptiste Joseph Fourier (1768–1830).

Let us check the formulas using the orthogonality properties. Suppose for a moment that

$$f(t) = \frac{a_0}{2} + \sum_{n=1}^{\infty} a_n \cos(nt) + b_n \sin(nt).$$

Then for $m \geq 1$ we have

$$\langle f(t), \cos(mt) \rangle = \left\langle \frac{a_0}{2} + \sum_{n=1}^{\infty} a_n \cos(nt) + b_n \sin(nt), \cos(mt) \right\rangle$$

$$= \frac{a_0}{2} \langle 1, \cos(mt) \rangle + \sum_{n=1}^{\infty} a_n \langle \cos(nt), \cos(mt) \rangle + b_n \langle \sin(nt), \cos(mt) \rangle$$

$$= a_m \langle \cos(mt), \cos(mt) \rangle.$$

And hence $a_m = \frac{\langle f(t), \cos(mt) \rangle}{\langle \cos(mt), \cos(mt) \rangle}$.

Exercise 4.2.2: *Carry out the calculation for a_0 and b_m.*

Example 4.2.3: Take the function

$$f(t) = t$$

for t in $(-\pi, \pi]$. Extend $f(t)$ periodically and write it as a Fourier series. This function is called the *sawtooth*.

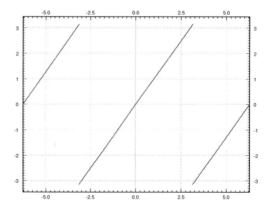

Figure 4.3: The graph of the sawtooth function.

The plot of the extended periodic function is given in Figure 4.3. Let us compute the coefficients. We start with a_0,

$$a_0 = \frac{1}{\pi} \int_{-\pi}^{\pi} t \, dt = 0.$$

We will often use the result from calculus that says that the integral of an odd function over a symmetric interval is zero. Recall that an *odd function* is a function $\varphi(t)$ such that $\varphi(-t) = -\varphi(t)$. For example the functions t, $\sin t$, or (importantly for us) $t\cos(nt)$ are all odd functions. Thus

$$a_n = \frac{1}{\pi} \int_{-\pi}^{\pi} t\cos(nt)\, dt = 0.$$

Let us move to b_n. Another useful fact from calculus is that the integral of an even function over a symmetric interval is twice the integral of the same function over half the interval. Recall an *even function* is a function $\varphi(t)$ such that $\varphi(-t) = \varphi(t)$. For example $t\sin(nt)$ is even.

$$
\begin{aligned}
b_n &= \frac{1}{\pi} \int_{-\pi}^{\pi} t\sin(nt)\, dt \\
&= \frac{2}{\pi} \int_{0}^{\pi} t\sin(nt)\, dt \\
&= \frac{2}{\pi} \left(\left[\frac{-t\cos(nt)}{n} \right]_{t=0}^{\pi} + \frac{1}{n} \int_{0}^{\pi} \cos(nt)\, dt \right) \\
&= \frac{2}{\pi} \left(\frac{-\pi\cos(n\pi)}{n} + 0 \right) \\
&= \frac{-2\cos(n\pi)}{n} = \frac{2(-1)^{n+1}}{n}.
\end{aligned}
$$

We have used the fact that

$$\cos(n\pi) = (-1)^n = \begin{cases} 1 & \text{if } n \text{ even,} \\ -1 & \text{if } n \text{ odd.} \end{cases}$$

The series, therefore, is

$$\sum_{n=1}^{\infty} \frac{2(-1)^{n+1}}{n} \sin(nt).$$

Let us write out the first 3 harmonics of the series for $f(t)$.

$$2\sin(t) - \sin(2t) + \frac{2}{3}\sin(3t) + \cdots$$

The plot of these first three terms of the series, along with a plot of the first 20 terms is given in Figure 4.4 on the following page.

Example 4.2.4: Take the function

$$f(t) = \begin{cases} 0 & \text{if } -\pi < t \leq 0, \\ \pi & \text{if } 0 < t \leq \pi. \end{cases}$$

Extend $f(t)$ periodically and write it as a Fourier series. This function or its variants appear often in applications and the function is called the *square wave*.

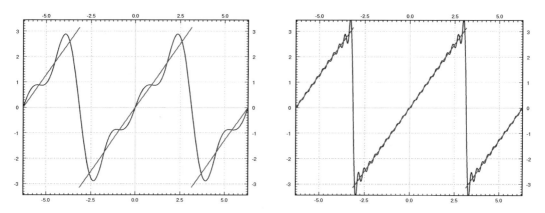

Figure 4.4: First 3 (left graph) and 20 (right graph) harmonics of the sawtooth function.

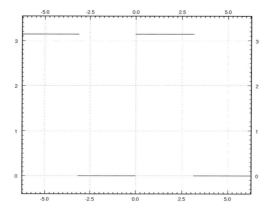

Figure 4.5: The graph of the square wave function.

The plot of the extended periodic function is given in Figure 4.5. Now we compute the coefficients. Let us start with a_0

$$a_0 = \frac{1}{\pi} \int_{-\pi}^{\pi} f(t) \, dt = \frac{1}{\pi} \int_{0}^{\pi} \pi \, dt = \pi.$$

Next,

$$a_n = \frac{1}{\pi} \int_{-\pi}^{\pi} f(t) \cos(nt) \, dt = \frac{1}{\pi} \int_{0}^{\pi} \pi \cos(nt) \, dt = 0.$$

And finally

$$b_n = \frac{1}{\pi} \int_{-\pi}^{\pi} f(t) \sin(nt)\, dt$$

$$= \frac{1}{\pi} \int_{0}^{\pi} \pi \sin(nt)\, dt$$

$$= \left[\frac{-\cos(nt)}{n} \right]_{t=0}^{\pi}$$

$$= \frac{1 - \cos(\pi n)}{n} = \frac{1 - (-1)^n}{n} = \begin{cases} \frac{2}{n} & \text{if } n \text{ is odd,} \\ 0 & \text{if } n \text{ is even.} \end{cases}$$

The Fourier series is

$$\frac{\pi}{2} + \sum_{\substack{n=1 \\ n \text{ odd}}}^{\infty} \frac{2}{n} \sin(nt) = \frac{\pi}{2} + \sum_{k=1}^{\infty} \frac{2}{2k-1} \sin((2k-1)\,t).$$

Let us write out the first 3 harmonics of the series for $f(t)$.

$$\frac{\pi}{2} + 2\,\sin(t) + \frac{2}{3}\,\sin(3t) + \cdots$$

The plot of these first three and also of the first 20 terms of the series is given in Figure 4.6.

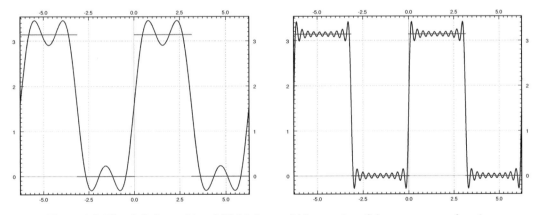

Figure 4.6: First 3 (left graph) and 20 (right graph) harmonics of the square wave function.

We have so far skirted the issue of convergence. For example, if $f(t)$ is the square wave function, the equation

$$f(t) = \frac{\pi}{2} + \sum_{k=1}^{\infty} \frac{2}{2k-1} \sin((2k-1)\,t).$$

is only an equality for such t where $f(t)$ is continuous. That is, we do not get an equality for $t = -\pi, 0, \pi$ and all the other discontinuities of $f(t)$. It is not hard to see that when t is an integer multiple of π (which includes all the discontinuities), then

$$\frac{\pi}{2} + \sum_{k=1}^{\infty} \frac{2}{2k-1} \sin((2k-1)t) = \frac{\pi}{2}.$$

We redefine $f(t)$ on $[-\pi, \pi]$ as

$$f(t) = \begin{cases} 0 & \text{if } -\pi < t < 0, \\ \pi & \text{if } \quad 0 < t < \pi, \\ \pi/2 & \text{if } \quad t = -\pi, t = 0, \text{ or } t = \pi, \end{cases}$$

and extend periodically. The series equals this extended $f(t)$ everywhere, including the discontinuities. We will generally not worry about changing the function values at several (finitely many) points.

We will say more about convergence in the next section. Let us however mention briefly an effect of the discontinuity. Let us zoom in near the discontinuity in the square wave. Further, let us plot the first 100 harmonics, see Figure 4.7. You will notice that while the series is a very good approximation away from the discontinuities, the error (the overshoot) near the discontinuity at $t = \pi$ does not seem to be getting any smaller. This behavior is known as the *Gibbs phenomenon*. The region where the error is large does get smaller, however, the more terms in the series we take.

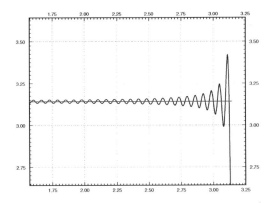

Figure 4.7: Gibbs phenomenon in action.

We can think of a periodic function as a "signal" being a superposition of many signals of pure frequency. For example, we could think of the square wave as a tone of certain base frequency. It will be, in fact, a superposition of many different pure tones of frequencies that are multiples of the base frequency. On the other hand a simple sine wave is only the pure tone. The simplest way to make sound using a computer is the square wave, and the sound is very different from nice pure tones. If you have played video games from the 1980s or so, then you have heard what square waves sound like.

4.2.4 Exercises

Exercise 4.2.3: *Suppose $f(t)$ is defined on $[-\pi, \pi]$ as $\sin(5t) + \cos(3t)$. Extend periodically and compute the Fourier series of $f(t)$.*

Exercise 4.2.4: *Suppose $f(t)$ is defined on $[-\pi, \pi]$ as $|t|$. Extend periodically and compute the Fourier series of $f(t)$.*

Exercise 4.2.5: *Suppose $f(t)$ is defined on $[-\pi, \pi]$ as $|t|^3$. Extend periodically and compute the Fourier series of $f(t)$.*

Exercise 4.2.6: *Suppose $f(t)$ is defined on $(-\pi, \pi]$ as*

$$f(t) = \begin{cases} -1 & \text{if } -\pi < t \leq 0, \\ 1 & \text{if } 0 < t \leq \pi. \end{cases}$$

Extend periodically and compute the Fourier series of $f(t)$.

Exercise 4.2.7: *Suppose $f(t)$ is defined on $(-\pi, \pi]$ as t^3. Extend periodically and compute the Fourier series of $f(t)$.*

Exercise 4.2.8: *Suppose $f(t)$ is defined on $[-\pi, \pi]$ as t^2. Extend periodically and compute the Fourier series of $f(t)$.*

There is another form of the Fourier series using complex exponentials that is sometimes easier to work with.

Exercise 4.2.9: *Let*

$$f(t) = \frac{a_0}{2} + \sum_{n=1}^{\infty} a_n \cos(nt) + b_n \sin(nt).$$

Use Euler's formula $e^{i\theta} = \cos(\theta) + i\sin(\theta)$ to show that there exist complex numbers c_m such that

$$f(t) = \sum_{m=-\infty}^{\infty} c_m e^{imt}.$$

Note that the sum now ranges over all the integers including negative ones. Do not worry about convergence in this calculation. Hint: It may be better to start from the complex exponential form and write the series as

$$c_0 + \sum_{m=1}^{\infty} c_m e^{imt} + c_{-m} e^{-imt}.$$

Exercise **4.2.101***: Suppose $f(t)$ is defined on $[-\pi, \pi]$ as $f(t) = \sin(t)$. Extend periodically and compute the Fourier series.*

Exercise **4.2.102***: Suppose $f(t)$ is defined on $(-\pi, \pi]$ as $f(t) = \sin(\pi t)$. Extend periodically and compute the Fourier series.*

Exercise **4.2.103***: Suppose $f(t)$ is defined on $(-\pi, \pi]$ as $f(t) = \sin^2(t)$. Extend periodically and compute the Fourier series.*

Exercise **4.2.104***: Suppose $f(t)$ is defined on $(-\pi, \pi]$ as $f(t) = t^4$. Extend periodically and compute the Fourier series.*

4.3 More on the Fourier series

Note: 2 lectures, §9.2–§9.3 in [EP], §10.3 in [BD]

4.3.1 2L-periodic functions

We have computed the Fourier series for a 2π-periodic function, but what about functions of different periods. Well, fear not, the computation is a simple case of change of variables. We can just rescale the independent axis. Suppose we have a $2L$-periodic function $f(t)$. Then L is called the *half period*. Let $s = \frac{\pi}{L}t$. Then the function

$$g(s) = f\left(\frac{L}{\pi}s\right)$$

is 2π-periodic. We must also rescale all our sines and cosines. In the series we use $\frac{\pi}{L}t$ as the variable. That is, we want to write

$$\boxed{f(t) = \frac{a_0}{2} + \sum_{n=1}^{\infty} a_n \cos\left(\frac{n\pi}{L}t\right) + b_n \sin\left(\frac{n\pi}{L}t\right).}$$

If we change variables to s we see that

$$g(s) = \frac{a_0}{2} + \sum_{n=1}^{\infty} a_n \cos(ns) + b_n \sin(ns).$$

We compute a_n and b_n as before. After we write down the integrals, we change variables from s back to t, noting also that $ds = \frac{\pi}{L}dt$.

$$\boxed{\begin{aligned}
a_0 &= \frac{1}{\pi}\int_{-\pi}^{\pi} g(s)\,ds = \frac{1}{L}\int_{-L}^{L} f(t)\,dt, \\
a_n &= \frac{1}{\pi}\int_{-\pi}^{\pi} g(s)\cos(ns)\,ds = \frac{1}{L}\int_{-L}^{L} f(t)\cos\left(\frac{n\pi}{L}t\right)\,dt, \\
b_n &= \frac{1}{\pi}\int_{-\pi}^{\pi} g(s)\sin(ns)\,ds = \frac{1}{L}\int_{-L}^{L} f(t)\sin\left(\frac{n\pi}{L}t\right)\,dt.
\end{aligned}}$$

The two most common half periods that show up in examples are π and 1 because of the simplicity of the formulas. We should stress that we have done no new mathematics, we have only changed variables. If you understand the Fourier series for 2π-periodic functions, you understand it for $2L$-periodic functions. You can think of it as just using different units for time. All that we are doing is moving some constants around, but all the mathematics is the same.

Example 4.3.1: Let

$$f(t) = |t| \qquad \text{for } -1 < t \le 1,$$

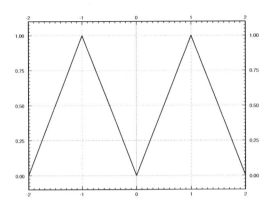

Figure 4.8: Periodic extension of the function $f(t)$.

extended periodically. The plot of the periodic extension is given in Figure 4.8. Compute the Fourier series of $f(t)$.

We want to write $f(t) = \frac{a_0}{2} + \sum_{n=1}^{\infty} a_n \cos(n\pi t) + b_n \sin(n\pi t)$. For $n \geq 1$ we note that $|t| \cos(n\pi t)$ is even and hence

$$
\begin{aligned}
a_n &= \int_{-1}^{1} f(t) \cos(n\pi t)\, dt \\
&= 2 \int_{0}^{1} t \cos(n\pi t)\, dt \\
&= 2 \left[\frac{t}{n\pi} \sin(n\pi t) \right]_{t=0}^{1} - 2 \int_{0}^{1} \frac{1}{n\pi} \sin(n\pi t)\, dt \\
&= 0 + \frac{1}{n^2\pi^2} \Big[\cos(n\pi t) \Big]_{t=0}^{1} = \frac{2((-1)^n - 1)}{n^2\pi^2} = \begin{cases} 0 & \text{if } n \text{ is even,} \\ \frac{-4}{n^2\pi^2} & \text{if } n \text{ is odd.} \end{cases}
\end{aligned}
$$

Next we find a_0

$$
a_0 = \int_{-1}^{1} |t|\, dt = 1.
$$

You should be able to find this integral by thinking about the integral as the area under the graph without doing any computation at all. Finally we can find b_n. Here, we notice that $|t| \sin(n\pi t)$ is odd and, therefore,

$$
b_n = \int_{-1}^{1} f(t) \sin(n\pi t)\, dt = 0.
$$

Hence, the series is

$$
\frac{1}{2} + \sum_{\substack{n=1 \\ n \text{ odd}}}^{\infty} \frac{-4}{n^2\pi^2} \cos(n\pi t).
$$

Let us explicitly write down the first few terms of the series up to the 3rd harmonic.

$$\frac{1}{2} - \frac{4}{\pi^2}\cos(\pi t) - \frac{4}{9\pi^2}\cos(3\pi t) - \cdots$$

The plot of these few terms and also a plot up to the 20th harmonic is given in Figure 4.9. You should notice how close the graph is to the real function. You should also notice that there is no "Gibbs phenomenon" present as there are no discontinuities.

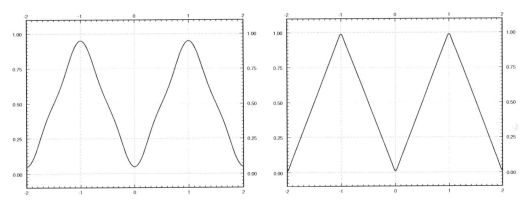

Figure 4.9: *Fourier series of $f(t)$ up to the 3rd harmonic (left graph) and up to the 20th harmonic (right graph).*

4.3.2 Convergence

We will need the one sided limits of functions. We will use the following notation

$$f(c-) = \lim_{t \uparrow c} f(t), \qquad \text{and} \qquad f(c+) = \lim_{t \downarrow c} f(t).$$

If you are unfamiliar with this notation, $\lim_{t \uparrow c} f(t)$ means we are taking a limit of $f(t)$ as t approaches c from below (i.e. $t < c$) and $\lim_{t \downarrow c} f(t)$ means we are taking a limit of $f(t)$ as t approaches c from above (i.e. $t > c$). For example, for the square wave function

$$f(t) = \begin{cases} 0 & \text{if } -\pi < t \leq 0, \\ \pi & \text{if } \quad 0 < t \leq \pi, \end{cases} \tag{4.8}$$

we have $f(0-) = 0$ and $f(0+) = \pi$.

Let $f(t)$ be a function defined on an interval $[a, b]$. Suppose that we find finitely many points $a = t_0, t_1, t_2, \ldots, t_k = b$ in the interval, such that $f(t)$ is continuous on the intervals (t_0, t_1), (t_1, t_2),

..., (t_{k-1}, t_k). Also suppose that all the one sided limits exist, that is, all of $f(t_0+)$, $f(t_1-)$, $f(t_1+)$, $f(t_2-)$, $f(t_2+)$, ..., $f(t_k-)$ exist and are finite. Then we say $f(t)$ is *piecewise continuous*.

If moreover, $f(t)$ is differentiable at all but finitely many points, and $f'(t)$ is piecewise continuous, then $f(t)$ is said to be *piecewise smooth*.

Example 4.3.2: The square wave function (4.8) is piecewise smooth on $[-\pi, \pi]$ or any other interval. In such a case we simply say that the function is piecewise smooth.

Example 4.3.3: The function $f(t) = |t|$ is piecewise smooth.

Example 4.3.4: The function $f(t) = \frac{1}{t}$ is not piecewise smooth on $[-1, 1]$ (or any other interval containing zero). In fact, it is not even piecewise continuous.

Example 4.3.5: The function $f(t) = \sqrt[3]{t}$ is not piecewise smooth on $[-1, 1]$ (or any other interval containing zero). $f(t)$ is continuous, but the derivative of $f(t)$ is unbounded near zero and hence not piecewise continuous.

Piecewise smooth functions have an easy answer on the convergence of the Fourier series.

Theorem 4.3.1. *Suppose $f(t)$ is a 2L-periodic piecewise smooth function. Let*

$$\frac{a_0}{2} + \sum_{n=1}^{\infty} a_n \cos\left(\frac{n\pi}{L}t\right) + b_n \sin\left(\frac{n\pi}{L}t\right)$$

be the Fourier series for $f(t)$. Then the series converges for all t. If $f(t)$ is continuous near t, then

$$f(t) = \frac{a_0}{2} + \sum_{n=1}^{\infty} a_n \cos\left(\frac{n\pi}{L}t\right) + b_n \sin\left(\frac{n\pi}{L}t\right).$$

Otherwise

$$\frac{f(t-) + f(t+)}{2} = \frac{a_0}{2} + \sum_{n=1}^{\infty} a_n \cos\left(\frac{n\pi}{L}t\right) + b_n \sin\left(\frac{n\pi}{L}t\right).$$

If we happen to have that $f(t) = \frac{f(t-)+f(t+)}{2}$ at all the discontinuities, the Fourier series converges to $f(t)$ everywhere. We can always just redefine $f(t)$ by changing the value at each discontinuity appropriately. Then we can write an equals sign between $f(t)$ and the series without any worry. We mentioned this fact briefly at the end last section.

Note that the theorem does not say how fast the series converges. Think back to the discussion of the Gibbs phenomenon in the last section. The closer you get to the discontinuity, the more terms you need to take to get an accurate approximation to the function.

4.3.3 Differentiation and integration of Fourier series

Not only does Fourier series converge nicely, but it is easy to differentiate and integrate the series. We can do this just by differentiating or integrating term by term.

Theorem 4.3.2. *Suppose*

$$f(t) = \frac{a_0}{2} + \sum_{n=1}^{\infty} a_n \cos\left(\frac{n\pi}{L} t\right) + b_n \sin\left(\frac{n\pi}{L} t\right)$$

is a piecewise smooth continuous function and the derivative $f'(t)$ is piecewise smooth. Then the derivative can be obtained by differentiating term by term,

$$f'(t) = \sum_{n=1}^{\infty} \frac{-a_n n\pi}{L} \sin\left(\frac{n\pi}{L} t\right) + \frac{b_n n\pi}{L} \cos\left(\frac{n\pi}{L} t\right).$$

It is important that the function is continuous. It can have corners, but no jumps. Otherwise the differentiated series will fail to converge. For an exercise, take the series obtained for the square wave and try to differentiate the series. Similarly, we can also integrate a Fourier series.

Theorem 4.3.3. *Suppose*

$$f(t) = \frac{a_0}{2} + \sum_{n=1}^{\infty} a_n \cos\left(\frac{n\pi}{L} t\right) + b_n \sin\left(\frac{n\pi}{L} t\right)$$

is a piecewise smooth function. Then the antiderivative is obtained by antidifferentiating term by term and so

$$F(t) = \frac{a_0 t}{2} + C + \sum_{n=1}^{\infty} \frac{a_n L}{n\pi} \sin\left(\frac{n\pi}{L} t\right) + \frac{-b_n L}{n\pi} \cos\left(\frac{n\pi}{L} t\right),$$

where $F'(t) = f(t)$ and C is an arbitrary constant.

Note that the series for $F(t)$ is no longer a Fourier series as it contains the $\frac{a_0 t}{2}$ term. The antiderivative of a periodic function need no longer be periodic and so we should not expect a Fourier series.

4.3.4 Rates of convergence and smoothness

Let us do an example of a periodic function with one derivative everywhere.

Example 4.3.6: Take the function

$$f(t) = \begin{cases} (t + 1)t & \text{if } -1 < t \leq 0, \\ (1 - t)t & \text{if } 0 < t \leq 1, \end{cases}$$

and extend to a 2-periodic function. The plot is given in Figure 4.10 on the following page.

Note that this function has one derivative everywhere, but it does not have a second derivative whenever t is an integer.

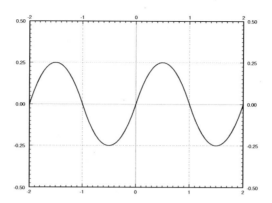

Figure 4.10: Smooth 2-periodic function.

Exercise 4.3.1: *Compute $f''(0+)$ and $f''(0-)$.*

Let us compute the Fourier series coefficients. The actual computation involves several integration by parts and is left to student.

$$a_0 = \int_{-1}^{1} f(t)\, dt = \int_{-1}^{0} (t+1)t\, dt + \int_{0}^{1} (1-t)t\, dt = 0,$$

$$a_n = \int_{-1}^{1} f(t)\cos(n\pi t)\, dt = \int_{-1}^{0} (t+1)t\,\cos(n\pi t)\, dt + \int_{0}^{1} (1-t)t\,\cos(n\pi t)\, dt = 0$$

$$b_n = \int_{-1}^{1} f(t)\sin(n\pi t)\, dt = \int_{-1}^{0} (t+1)t\,\sin(n\pi t)\, dt + \int_{0}^{1} (1-t)t\,\sin(n\pi t)\, dt$$

$$= \frac{4(1-(-1)^n)}{\pi^3 n^3} = \begin{cases} \frac{8}{\pi^3 n^3} & \text{if } n \text{ is odd,} \\ 0 & \text{if } n \text{ is even.} \end{cases}$$

That is, the series is

$$\sum_{\substack{n=1 \\ n \text{ odd}}}^{\infty} \frac{8}{\pi^3 n^3}\, \sin(n\pi t).$$

This series converges very fast. If you plot up to the third harmonic, that is the function

$$\frac{8}{\pi^3}\, \sin(\pi t) + \frac{8}{27\pi^3}\, \sin(3\pi t),$$

it is almost indistinguishable from the plot of $f(t)$ in Figure 4.10. In fact, the coefficient $\frac{8}{27\pi^3}$ is already just 0.0096 (approximately). The reason for this behavior is the n^3 term in the denominator. The coefficients b_n in this case go to zero as fast as $1/n^3$ goes to zero.

For functions constructed piecewise from polynomials as above, it is generally true that if you have one derivative, the Fourier coefficients will go to zero approximately like $1/n^3$. If you have only a continuous function, then the Fourier coefficients will go to zero as $1/n^2$. If you have discontinuities, then the Fourier coefficients will go to zero approximately as $1/n$. For more general functions the story is somewhat more complicated but the same idea holds, the more derivatives you have, the faster the coefficients go to zero. Similar reasoning works in reverse. If the coefficients go to zero like $1/n^2$ you always obtain a continuous function. If they go to zero like $1/n^3$ you obtain an everywhere differentiable function.

To justify this behavior, take for example the function defined by the Fourier series

$$f(t) = \sum_{n=1}^{\infty} \frac{1}{n^3} \sin(nt).$$

When we differentiate term by term we notice

$$f'(t) = \sum_{n=1}^{\infty} \frac{1}{n^2} \cos(nt).$$

Therefore, the coefficients now go down like $1/n^2$, which means that we have a continuous function. The derivative of $f'(t)$ is defined at most points, but there are points where $f'(t)$ is not differentiable. It has corners, but no jumps. If we differentiate again (where we can) we find that the function $f''(t)$, now fails to be continuous (has jumps)

$$f''(t) = \sum_{n=1}^{\infty} \frac{-1}{n} \sin(nt).$$

This function is similar to the sawtooth. If we tried to differentiate the series again we would obtain

$$\sum_{n=1}^{\infty} -\cos(nt),$$

which does not converge!

Exercise 4.3.2: Use a computer to plot the series we obtained for $f(t)$, $f'(t)$ and $f''(t)$. That is, plot say the first 5 harmonics of the functions. At what points does $f''(t)$ have the discontinuities?

4.3.5   Exercises

Exercise 4.3.3: Let

$$f(t) = \begin{cases} 0 & \text{if } -1 < t \le 0, \\ t & \text{if } 0 < t \le 1, \end{cases}$$

extended periodically. a) Compute the Fourier series for $f(t)$. b) Write out the series explicitly up to the 3rd harmonic.

Exercise **4.3.4:** *Let*

$$f(t) = \begin{cases} -t & \text{if } -1 < t \le 0, \\ t^2 & \text{if } \ \ 0 < t \le 1, \end{cases}$$

extended periodically. a) Compute the Fourier series for $f(t)$. b) Write out the series explicitly up to the 3rd harmonic.

Exercise **4.3.5:** *Let*

$$f(t) = \begin{cases} \frac{-t}{10} & \text{if } -10 < t \le 0, \\ \frac{t}{10} & \text{if } \ \ \ 0 < t \le 10, \end{cases}$$

extended periodically (period is 20). a) Compute the Fourier series for $f(t)$. b) Write out the series explicitly up to the 3rd harmonic.

Exercise **4.3.6:** *Let $f(t) = \sum_{n=1}^{\infty} \frac{1}{n^3} \cos(nt)$. Is $f(t)$ continuous and differentiable everywhere? Find the derivative (if it exists everywhere) or justify why $f(t)$ is not differentiable everywhere.*

Exercise **4.3.7:** *Let $f(t) = \sum_{n=1}^{\infty} \frac{(-1)^n}{n} \sin(nt)$. Is $f(t)$ differentiable everywhere? Find the derivative (if it exists everywhere) or justify why $f(t)$ is not differentiable everywhere.*

Exercise **4.3.8:** *Let*

$$f(t) = \begin{cases} 0 & \text{if } -2 < t \le 0, \\ t & \text{if } \ \ 0 < t \le 1, \\ -t + 2 & \text{if } \ \ 1 < t \le 2, \end{cases}$$

extended periodically. a) Compute the Fourier series for $f(t)$. b) Write out the series explicitly up to the 3rd harmonic.

Exercise **4.3.9:** *Let*

$$f(t) = e^t \qquad \text{for } -1 < t \le 1$$

extended periodically. a) Compute the Fourier series for $f(t)$. b) Write out the series explicitly up to the 3rd harmonic. c) What does the series converge to at $t = 1$.

Exercise **4.3.10:** *Let*

$$f(t) = t^2 \qquad \text{for } -1 < t \le 1$$

extended periodically. a) Compute the Fourier series for $f(t)$. b) By plugging in $t = 0$, evaluate
$$\sum_{n=1}^{\infty} \frac{(-1)^n}{n^2} = 1 - \frac{1}{4} + \frac{1}{9} - \cdots. \text{ c) Now evaluate } \sum_{n=1}^{\infty} \frac{1}{n^2} = 1 + \frac{1}{4} + \frac{1}{9} + \cdots.$$

Exercise **4.3.101:** *Let*

$$f(t) = t^2 \qquad \text{for } -2 < t \le 2$$

extended periodically. a) Compute the Fourier series for $f(t)$. b) Write out the series explicitly up to the 3rd harmonic.

Exercise 4.3.102: *Let*

$$f(t) = t \qquad for \ -\lambda < t \leq \lambda \ (for \ some \ \lambda > 0)$$

extended periodically. a) Compute the Fourier series for $f(t)$*. b) Write out the series explicitly up to the 3rd harmonic.*

Exercise 4.3.103: *Let*

$$f(t) = \frac{1}{2} + \sum_{n=1}^{\infty} \frac{1}{n(n^2 + 1)} \sin(n\pi t).$$

Compute $f'(t)$*.*

Exercise 4.3.104: *Let*

$$f(t) = \frac{1}{2} + \sum_{n=1}^{\infty} \frac{1}{n^3} \cos(nt).$$

a) Find the antiderivative. b) Is the antiderivative periodic?

Exercise 4.3.105: *Let*

$$f(t) = 1/2 \qquad for \ -\pi < t < \pi$$

extended periodically. a) Compute the Fourier series for $f(t)$*. b) Plug in* $t = \pi/2$ *to find a series representation for* $\pi/4$*. c) Using the first 4 terms of the result from part b) approximate* $\pi/4$*.*

4.4   Sine and cosine series

Note: 2 lectures, §9.3 in [EP], §10.4 in [BD]

4.4.1   Odd and even periodic functions

You may have noticed by now that an odd function has no cosine terms in the Fourier series and an even function has no sine terms in the Fourier series. This observation is not a coincidence. Let us look at even and odd periodic function in more detail.

Recall that a function $f(t)$ is *odd* if $f(-t) = -f(t)$. A function $f(t)$ is *even* if $f(-t) = f(t)$. For example, $\cos(nt)$ is even and $\sin(nt)$ is odd. Similarly the function t^k is even if k is even and odd when k is odd.

***Exercise* 4.4.1:** *Take two functions $f(t)$ and $g(t)$ and define their product $h(t) = f(t)g(t)$. a) Suppose both are odd, is $h(t)$ odd or even? b) Suppose one is even and one is odd, is $h(t)$ odd or even? c) Suppose both are even, is $h(t)$ odd or even?*

If $f(t)$ and $g(t)$ are both odd, then $f(t) + g(t)$ is odd. Similarly for even functions. On the other hand, if $f(t)$ is odd and $g(t)$ even, then we cannot say anything about the sum $f(t) + g(t)$. In fact, the Fourier series of any function is a sum of an odd (the sine terms) and an even (the cosine terms) function.

In this section we consider odd and even periodic functions. We have previously defined the $2L$-periodic extension of a function defined on the interval $[-L, L]$. Sometimes we are only interested in the function on the range $[0, L]$ and it would be convenient to have an odd (resp. even) function. If the function is odd (resp. even), all the cosine (resp. sine) terms will disappear. What we will do is take the odd (resp. even) extension of the function to $[-L, L]$ and then extend periodically to a $2L$-periodic function.

Take a function $f(t)$ defined on $[0, L]$. On $(-L, L]$ define the functions

$$F_{\text{odd}}(t) \overset{\text{def}}{=} \begin{cases} f(t) & \text{if} \quad 0 \leq t \leq L, \\ -f(-t) & \text{if} \quad -L < t < 0, \end{cases}$$

$$F_{\text{even}}(t) \overset{\text{def}}{=} \begin{cases} f(t) & \text{if} \quad 0 \leq t \leq L, \\ f(-t) & \text{if} \quad -L < t < 0. \end{cases}$$

Extend $F_{\text{odd}}(t)$ and $F_{\text{even}}(t)$ to be $2L$-periodic. Then $F_{\text{odd}}(t)$ is called the *odd periodic extension* of $f(t)$, and $F_{\text{even}}(t)$ is called the *even periodic extension* of $f(t)$.

***Exercise* 4.4.2:** *Check that $F_{\text{odd}}(t)$ is odd and that $F_{\text{even}}(t)$ is even.*

Example 4.4.1: Take the function $f(t) = t(1 - t)$ defined on $[0, 1]$. Figure 4.11 on the facing page shows the plots of the odd and even extensions of $f(t)$.

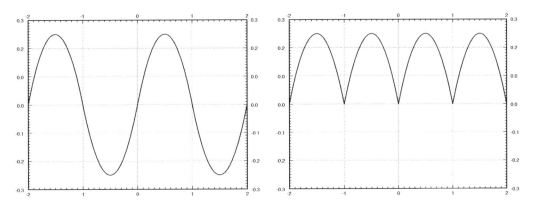

Figure 4.11: Odd and even 2-periodic extension of $f(t) = t(1 - t)$, $0 \le t \le 1$.

4.4.2   Sine and cosine series

Let $f(t)$ be an odd $2L$-periodic function. We write the Fourier series for $f(t)$. First, we compute the coefficients a_n (including $n = 0$) and get

$$a_n = \frac{1}{L} \int_{-L}^{L} f(t) \cos\left(\frac{n\pi}{L} t\right) dt = 0.$$

That is, there are no cosine terms in the Fourier series of an odd function. The integral is zero because $f(t) \cos(n\pi L t)$ is an odd function (product of an odd and an even function is odd) and the integral of an odd function over a symmetric interval is always zero. The integral of an even function over a symmetric interval $[-L, L]$ is twice the integral of the function over the interval $[0, L]$. The function $f(t) \sin\left(\frac{n\pi}{L} t\right)$ is the product of two odd functions and hence is even.

$$b_n = \frac{1}{L} \int_{-L}^{L} f(t) \sin\left(\frac{n\pi}{L} t\right) dt = \frac{2}{L} \int_{0}^{L} f(t) \sin\left(\frac{n\pi}{L} t\right) dt.$$

We now write the Fourier series of $f(t)$ as

$$\sum_{n=1}^{\infty} b_n \sin\left(\frac{n\pi}{L} t\right).$$

Similarly, if $f(t)$ is an even $2L$-periodic function. For the same exact reasons as above, we find that $b_n = 0$ and

$$a_n = \frac{2}{L} \int_{0}^{L} f(t) \cos\left(\frac{n\pi}{L} t\right) dt.$$

The formula still works for $n = 0$, in which case it becomes

$$a_0 = \frac{2}{L} \int_{0}^{L} f(t) dt.$$

The Fourier series is then

$$\frac{a_0}{2} + \sum_{n=1}^{\infty} a_n \cos\left(\frac{n\pi}{L} t\right).$$

An interesting consequence is that the coefficients of the Fourier series of an odd (or even) function can be computed by just integrating over the half interval $[0, L]$. Therefore, we can compute the Fourier series of the odd (or even) extension of a function by computing certain integrals over the interval where the original function is defined.

Theorem 4.4.1. *Let $f(t)$ be a piecewise smooth function defined on $[0, L]$. Then the odd extension of $f(t)$ has the Fourier series*

$$F_{odd}(t) = \sum_{n=1}^{\infty} b_n \sin\left(\frac{n\pi}{L} t\right),$$

where

$$b_n = \frac{2}{L} \int_0^L f(t) \sin\left(\frac{n\pi}{L} t\right) dt.$$

The even extension of $f(t)$ has the Fourier series

$$F_{even}(t) = \frac{a_0}{2} + \sum_{n=1}^{\infty} a_n \cos\left(\frac{n\pi}{L} t\right),$$

where

$$a_n = \frac{2}{L} \int_0^L f(t) \cos\left(\frac{n\pi}{L} t\right) dt.$$

The series $\sum_{n=1}^{\infty} b_n \sin\left(\frac{n\pi}{L} t\right)$ is called the *sine series* of $f(t)$ and the series $\frac{a_0}{2} + \sum_{n=1}^{\infty} a_n \cos\left(\frac{n\pi}{L} t\right)$ is called the *cosine series* of $f(t)$. We often do not actually care what happens outside of $[0, L]$. In this case, we pick whichever series fits our problem better.

It is not necessary to start with the full Fourier series to obtain the sine and cosine series. The sine series is really the eigenfunction expansion of $f(t)$ using the eigenfunctions of the eigenvalue problem $x'' + \lambda x = 0$, $x(0) = 0$, $x(L) = L$. The cosine series is the eigenfunction expansion of $f(t)$ using the eigenfunctions of the eigenvalue problem $x'' + \lambda x = 0$, $x'(0) = 0$, $x'(L) = L$. We could have, therefore, gotten the same formulas by defining the inner product

$$\langle f(t), g(t) \rangle = \int_0^L f(t)g(t) \, dt,$$

and following the procedure of § 4.2. This point of view is useful, as we commonly use a specific series that arose because our underlying question led to a certain eigenvalue problem. If the eigenvalue problem is not one of the three we covered so far, you can still do an eigenfunction expansion, generalizing the results of this chapter. We will deal with such a generalization in chapter 5.

Example 4.4.2: Find the Fourier series of the even periodic extension of the function $f(t) = t^2$ for $0 \le t \le \pi$.

We want to write

$$f(t) = \frac{a_0}{2} + \sum_{n=1}^{\infty} a_n \cos(nt),$$

where

$$a_0 = \frac{2}{\pi} \int_0^{\pi} t^2 \, dt = \frac{2\pi^2}{3},$$

and

$$a_n = \frac{2}{\pi} \int_0^{\pi} t^2 \cos(nt) \, dt = \frac{2}{\pi} \left[t^2 \frac{1}{n} \sin(nt) \right]_0^{\pi} - \frac{4}{n\pi} \int_0^{\pi} t \sin(nt) \, dt$$

$$= \frac{4}{n^2 \pi} \left[t \cos(nt) \right]_0^{\pi} + \frac{4}{n^2 \pi} \int_0^{\pi} \cos(nt) \, dt = \frac{4(-1)^n}{n^2}.$$

Note that we have "detected" the continuity of the extension since the coefficients decay as $\frac{1}{n^2}$. That is, the even extension of t^2 has no jump discontinuities. It does have corners, since the derivative, which is an odd function and a sine series, has jumps; it has a Fourier series whose coefficients decay only as $\frac{1}{n}$.

Explicitly, the first few terms of the series are

$$\frac{\pi^2}{3} - 4\cos(t) + \cos(2t) - \frac{4}{9}\cos(3t) + \cdots$$

Exercise 4.4.3: a) *Compute the derivative of the even extension of* $f(t)$ *above and verify it has jump discontinuities. Use the actual definition of* $f(t)$*, not its cosine series! b) Why is it that the derivative of the even extension of* $f(t)$ *is the odd extension of* $f'(t)$*?*

4.4.3 Application

Fourier series ties in to the boundary value problems we studied earlier. Let us see this connection in more detail.

Suppose we have the boundary value problem for $0 < t < L$,

$$x''(t) + \lambda x(t) = f(t),$$

for the *Dirichlet boundary conditions* $x(0) = 0$, $x(L) = 0$. By using the Fredholm alternative (Theorem 4.1.2 on page 172) we note that as long as λ is not an eigenvalue of the underlying homogeneous problem, there exists a unique solution. Note that the eigenfunctions of this eigenvalue problem are the functions $\sin\left(\frac{n\pi}{L}t\right)$. Therefore, to find the solution, we first find the Fourier sine series for $f(t)$. We write x also as a sine series, but with unknown coefficients. We substitute the series for x into the equation and solve for the unknown coefficients. If we have the *Neumann boundary conditions* $x'(0) = 0$, $x'(L) = 0$, we do the same procedure using the cosine series.

Let us see how this method works on examples.

Example 4.4.3: Take the boundary value problem for $0 < t < 1$,

$$x''(t) + 2x(t) = f(t),$$

where $f(t) = t$ on $0 < t < 1$, and satisfying the Dirichlet boundary conditions $x(0) = 0$, $x(1) = 0$. We write $f(t)$ as a sine series

$$f(t) = \sum_{n=1}^{\infty} c_n \sin(n\pi t),$$

where

$$c_n = 2 \int_0^1 t \sin(n\pi t)\, dt = \frac{2(-1)^{n+1}}{n\pi}.$$

We write $x(t)$ as

$$x(t) = \sum_{n=1}^{\infty} b_n \sin(n\pi t).$$

We plug in to obtain

$$x''(t) + 2x(t) = \sum_{n=1}^{\infty} -b_n n^2 \pi^2 \sin(n\pi t) + 2 \sum_{n=1}^{\infty} b_n \sin(n\pi t)$$

$$= \sum_{n=1}^{\infty} b_n (2 - n^2 \pi^2) \sin(n\pi t)$$

$$= f(t) = \sum_{n=1}^{\infty} \frac{2(-1)^{n+1}}{n\pi} \sin(n\pi t).$$

Therefore,

$$b_n (2 - n^2 \pi^2) = \frac{2(-1)^{n+1}}{n\pi}$$

or

$$b_n = \frac{2(-1)^{n+1}}{n\pi(2 - n^2 \pi^2)}.$$

We have thus obtained a Fourier series for the solution

$$x(t) = \sum_{n=1}^{\infty} \frac{2(-1)^{n+1}}{n\pi\,(2 - n^2 \pi^2)} \sin(n\pi t).$$

Example 4.4.4: Similarly we handle the Neumann conditions. Take the boundary value problem for $0 < t < 1$,

$$x''(t) + 2x(t) = f(t),$$

where again $f(t) = t$ on $0 < t < 1$, but now satisfying the Neumann boundary conditions $x'(0) = 0$, $x'(1) = 0$. We write $f(t)$ as a cosine series

$$f(t) = \frac{c_0}{2} + \sum_{n=1}^{\infty} c_n \cos(n\pi t),$$

where

$$c_0 = 2 \int_0^1 t \, dt = 1,$$

and

$$c_n = 2 \int_0^1 t \cos(n\pi t) \, dt = \frac{2((-1)^n - 1)}{\pi^2 n^2} = \begin{cases} \frac{-4}{\pi^2 n^2} & \text{if } n \text{ odd,} \\ 0 & \text{if } n \text{ even.} \end{cases}$$

We write $x(t)$ as a cosine series

$$x(t) = \frac{a_0}{2} + \sum_{n=1}^{\infty} a_n \cos(n\pi t).$$

We plug in to obtain

$$x''(t) + 2x(t) = \sum_{n=1}^{\infty} \left[-a_n n^2 \pi^2 \cos(n\pi t) \right] + a_0 + 2 \sum_{n=1}^{\infty} \left[a_n \cos(n\pi t) \right]$$

$$= a_0 + \sum_{n=1}^{\infty} a_n (2 - n^2 \pi^2) \cos(n\pi t)$$

$$= f(t) = \frac{1}{2} + \sum_{\substack{n=1 \\ n \text{ odd}}}^{\infty} \frac{-4}{\pi^2 n^2} \cos(n\pi t).$$

Therefore, $a_0 = \frac{1}{2}$, $a_n = 0$ for n even ($n \geq 2$) and for n odd we have

$$a_n(2 - n^2 \pi^2) = \frac{-4}{\pi^2 n^2},$$

or

$$a_n = \frac{-4}{n^2 \pi^2 (2 - n^2 \pi^2)}.$$

The Fourier series for the solution $x(t)$ is

$$x(t) = \frac{1}{4} + \sum_{\substack{n=1 \\ n \text{ odd}}}^{\infty} \frac{-4}{n^2 \pi^2 (2 - n^2 \pi^2)} \cos(n\pi t).$$

4.4.4   Exercises

Exercise 4.4.4: *Take* $f(t) = (t-1)^2$ *defined on* $0 \le t \le 1$. *a) Sketch the plot of the even periodic extension of* f. *b) Sketch the plot of the odd periodic extension of* f.

Exercise 4.4.5: *Find the Fourier series of both the odd and even periodic extension of the function* $f(t) = (t-1)^2$ *for* $0 \le t \le 1$. *Can you tell which extension is continuous from the Fourier series coefficients?*

Exercise 4.4.6: *Find the Fourier series of both the odd and even periodic extension of the function* $f(t) = t$ *for* $0 \le t \le \pi$.

Exercise 4.4.7: *Find the Fourier series of the even periodic extension of the function* $f(t) = \sin t$ *for* $0 \le t \le \pi$.

Exercise 4.4.8: *Consider*
$$x''(t) + 4x(t) = f(t),$$
where $f(t) = 1$ *on* $0 < t < 1$. *a) Solve for the Dirichlet conditions* $x(0) = 0, x(1) = 0$. *b) Solve for the Neumann conditions* $x'(0) = 0, x'(1) = 0$.

Exercise 4.4.9: *Consider*
$$x''(t) + 9x(t) = f(t),$$
for $f(t) = \sin(2\pi t)$ *on* $0 < t < 1$. *a) Solve for the Dirichlet conditions* $x(0) = 0, x(1) = 0$. *b) Solve for the Neumann conditions* $x'(0) = 0, x'(1) = 0$.

Exercise 4.4.10: *Consider*
$$x''(t) + 3x(t) = f(t), \quad x(0) = 0, \quad x(1) = 0,$$
where $f(t) = \sum_{n=1}^{\infty} b_n \sin(n\pi t)$. *Write the solution* $x(t)$ *as a Fourier series, where the coefficients are given in terms of* b_n.

Exercise 4.4.11: *Let* $f(t) = t^2(2-t)$ *for* $0 \le t \le 2$. *Let* $F(t)$ *be the odd periodic extension. Compute* $F(1)$, $F(2)$, $F(3)$, $F(-1)$, $F(\textit{9/2})$, $F(101)$, $F(103)$. *Note: Do **not** compute using the sine series.*

Exercise 4.4.101: *Let* $f(t) = \textit{1/3}$ *on* $0 \le t < 3$. *a) Find the Fourier series of the even periodic extension. b) Find the Fourier series of the odd periodic extension.*

Exercise 4.4.102: *Let* $f(t) = \cos(2t)$ *on* $0 \le t < \pi$. *a) Find the Fourier series of the even periodic extension. b) Find the Fourier series of the odd periodic extension.*

Exercise 4.4.103: *Let* $f(t)$ *be defined on* $0 \le t < 1$. *Now take the average of the two extensions* $g(t) = \frac{F_{odd}(t) + F_{even}(t)}{2}$. *a) What is* $g(t)$ *if* $0 \le t < 1$ *(Justify!) b) What is* $g(t)$ *if* $-1 < t < 0$ *(Justify!)*

Exercise 4.4.104: *Let* $f(t) = \sum_{n=1}^{\infty} \frac{1}{n^2} \sin(nt)$. *Solve* $x'' - x = f(t)$ *for the Dirichlet conditions* $x(0) = 0$ *and* $x(\pi) = 0$.

Exercise 4.4.105 (challenging)*: Let* $f(t) = t + \sum_{n=1}^{\infty} \frac{1}{2^n} \sin(nt)$. *Solve* $x'' + \pi x = f(t)$ *for the Dirichlet conditions* $x(0) = 0$ *and* $x(\pi) = 1$. *Hint: Note that* $\frac{t}{\pi}$ *satisfies the given Dirichlet conditions.*

4.5 Applications of Fourier series

Note: 2 lectures, §9.4 in [EP], not in [BD]

4.5.1 Periodically forced oscillation

Let us return to the forced oscillations. Consider a mass-spring system as before, where we have a mass m on a spring with spring constant k, with damping c, and a force $F(t)$ applied to the mass. Suppose the forcing function $F(t)$ is $2L$-periodic for some $L > 0$. We have already seen this problem in chapter 2 with a simple $F(t)$. The equation that governs this particular setup is

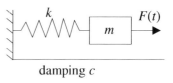

$$mx''(t) + cx'(t) + kx(t) = F(t). \tag{4.9}$$

The general solution consists of the complementary solution x_c, which solves the associated homogeneous equation $mx'' + cx' + kx = 0$, and a particular solution of (4.9) we call x_p. For $c > 0$, the complementary solution x_c will decay as time goes by. Therefore, we are mostly interested in a particular solution x_p that does not decay and is periodic with the same period as $F(t)$. We call this particular solution the *steady periodic solution* and we write it as x_{sp} as before. What will be new in this section is that we consider an arbitrary forcing function $F(t)$ instead of a simple cosine.

For simplicity, let us suppose that $c = 0$. The problem with $c > 0$ is very similar. The equation

$$mx'' + kx = 0$$

has the general solution

$$x(t) = A \cos(\omega_0 t) + B \sin(\omega_0 t),$$

where $\omega_0 = \sqrt{\frac{k}{m}}$. Any solution to $mx''(t) + kx(t) = F(t)$ is of the form $A \cos(\omega_0 t) + B \sin(\omega_0 t) + x_{sp}$. The steady periodic solution x_{sp} has the same period as $F(t)$.

In the spirit of the last section and the idea of undetermined coefficients we first write

$$F(t) = \frac{c_0}{2} + \sum_{n=1}^{\infty} c_n \cos\left(\frac{n\pi}{L} t\right) + d_n \sin\left(\frac{n\pi}{L} t\right).$$

Then we write a proposed steady periodic solution x as

$$x(t) = \frac{a_0}{2} + \sum_{n=1}^{\infty} a_n \cos\left(\frac{n\pi}{L} t\right) + b_n \sin\left(\frac{n\pi}{L} t\right),$$

where a_n and b_n are unknowns. We plug x into the differential equation and solve for a_n and b_n in terms of c_n and d_n. This process is perhaps best understood by example.

Example 4.5.1: Suppose that $k = 2$, and $m = 1$. The units are again the mks units (meters-kilograms-seconds). There is a jetpack strapped to the mass, which fires with a force of 1 newton for 1 second and then is off for 1 second, and so on. We want to find the steady periodic solution.

The equation is, therefore,

$$x'' + 2x = F(t),$$

where $F(t)$ is the step function

$$F(t) = \begin{cases} 0 & \text{if } -1 < t < 0, \\ 1 & \text{if } 0 < t < 1, \end{cases}$$

extended periodically. We write

$$F(t) = \frac{c_0}{2} + \sum_{n=1}^{\infty} c_n \cos(n\pi t) + d_n \sin(n\pi t).$$

We compute

$$c_n = \int_{-1}^{1} F(t) \cos(n\pi t)\, dt = \int_{0}^{1} \cos(n\pi t)\, dt = 0 \qquad \text{for } n \geq 1,$$

$$c_0 = \int_{-1}^{1} F(t)\, dt = \int_{0}^{1} dt = 1,$$

$$d_n = \int_{-1}^{1} F(t) \sin(n\pi t)\, dt$$

$$= \int_{0}^{1} \sin(n\pi t)\, dt$$

$$= \left[\frac{-\cos(n\pi t)}{n\pi} \right]_{t=0}^{1}$$

$$= \frac{1 - (-1)^n}{\pi n} = \begin{cases} \frac{2}{\pi n} & \text{if } n \text{ odd}, \\ 0 & \text{if } n \text{ even}. \end{cases}$$

So

$$F(t) = \frac{1}{2} + \sum_{\substack{n=1 \\ n \text{ odd}}}^{\infty} \frac{2}{\pi n} \sin(n\pi t).$$

We want to try

$$x(t) = \frac{a_0}{2} + \sum_{n=1}^{\infty} a_n \cos(n\pi t) + b_n \sin(n\pi t).$$

Once we plug x into the differential equation $x'' + 2x = F(t)$, it is clear that $a_n = 0$ for $n \geq 1$ as there are no corresponding terms in the series for $F(t)$. Similarly $b_n = 0$ for n even. Hence we try

$$x(t) = \frac{a_0}{2} + \sum_{\substack{n=1 \\ n \text{ odd}}}^{\infty} b_n \sin(n\pi t).$$

We plug into the differential equation and obtain

$$x'' + 2x = \sum_{\substack{n=1 \\ n \text{ odd}}}^{\infty} \left[-b_n n^2 \pi^2 \sin(n\pi t) \right] + a_0 + 2 \sum_{\substack{n=1 \\ n \text{ odd}}}^{\infty} \left[b_n \sin(n\pi t) \right]$$

$$= a_0 + \sum_{\substack{n=1 \\ n \text{ odd}}}^{\infty} b_n (2 - n^2 \pi^2) \sin(n\pi t)$$

$$= F(t) = \frac{1}{2} + \sum_{\substack{n=1 \\ n \text{ odd}}}^{\infty} \frac{2}{\pi n} \sin(n\pi t).$$

So $a_0 = \frac{1}{2}$, $b_n = 0$ for even n, and for odd n we get

$$b_n = \frac{2}{\pi n (2 - n^2 \pi^2)}.$$

The steady periodic solution has the Fourier series

$$x_{sp}(t) = \frac{1}{4} + \sum_{\substack{n=1 \\ n \text{ odd}}}^{\infty} \frac{2}{\pi n (2 - n^2 \pi^2)} \sin(n\pi t).$$

We know this is the steady periodic solution as it contains no terms of the complementary solution and it is periodic with the same period as $F(t)$ itself. See Figure 4.12 for the plot of this solution.

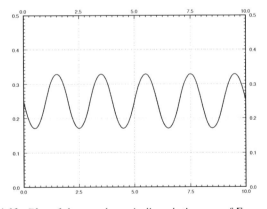

Figure 4.12: Plot of the steady periodic solution x_{sp} of Example 4.5.1.

4.5.2   Resonance

Just like when the forcing function was a simple cosine, resonance could still happen. Let us assume $c = 0$ and we will discuss only pure resonance. Again, take the equation

$$mx''(t) + kx(t) = F(t).$$

When we expand $F(t)$ and find that some of its terms coincide with the complementary solution to $mx'' + kx = 0$, we cannot use those terms in the guess. Just like before, they will disappear when we plug into the left hand side and we will get a contradictory equation (such as $0 = 1$). That is, suppose

$$x_c = A \cos(\omega_0 t) + B \sin(\omega_0 t),$$

where $\omega_0 = \frac{N\pi}{L}$ for some positive integer N. In this case we have to modify our guess and try

$$x(t) = \frac{a_0}{2} + t\left(a_N \cos\left(\frac{N\pi}{L}t\right) + b_N \sin\left(\frac{N\pi}{L}t\right)\right) + \sum_{\substack{n=1 \\ n \neq N}}^{\infty} a_n \cos\left(\frac{n\pi}{L}t\right) + b_n \sin\left(\frac{n\pi}{L}t\right).$$

In other words, we multiply the offending term by t. From then on, we proceed as before.

Of course, the solution will not be a Fourier series (it will not even be periodic) since it contains these terms multiplied by t. Further, the terms $t\left(a_N \cos\left(\frac{N\pi}{L}t\right) + b_N \sin\left(\frac{N\pi}{L}t\right)\right)$ will eventually dominate and lead to wild oscillations. As before, this behavior is called *pure resonance* or just *resonance*.

Note that there now may be infinitely many resonance frequencies to hit. That is, as we change the frequency of F (we change L), different terms from the Fourier series of F may interfere with the complementary solution and will cause resonance. However, we should note that since everything is an approximation and in particular c is never actually zero but something very close to zero, only the first few resonance frequencies will matter.

Example 4.5.2: Find the steady periodic solution to the equation

$$2x'' + 18\pi^2 x = F(t),$$

where

$$F(t) = \begin{cases} -1 & \text{if } -1 < t < 0, \\ 1 & \text{if } \quad 0 < t < 1, \end{cases}$$

extended periodically. We note that

$$F(t) = \sum_{\substack{n=1 \\ n \text{ odd}}}^{\infty} \frac{4}{\pi n} \sin(n\pi t).$$

Exercise 4.5.1: Compute the Fourier series of F to verify the above equation.

The solution must look like

$$x(t) = c_1 \cos(3\pi t) + c_2 \sin(3\pi t) + x_p(t)$$

for some particular solution x_p.

We note that if we just tried a Fourier series with $\sin(n\pi t)$ as usual, we would get duplication when $n = 3$. Therefore, we pull out that term and multiply by t. We also have to add a cosine term to get everything right. That is, we must try

$$x_p(t) = a_3 t \cos(3\pi t) + b_3 t \sin(3\pi t) + \sum_{\substack{n=1 \\ n \text{ odd} \\ n \neq 3}}^{\infty} b_n \sin(n\pi t).$$

Let us compute the second derivative.

$$x_p''(t) = -6a_3\pi \sin(3\pi t) - 9\pi^2 a_3\, t \cos(3\pi t) + 6b_3\pi \cos(3\pi t) - 9\pi^2 b_3\, t \sin(3\pi t)+$$

$$+ \sum_{\substack{n=1 \\ n \text{ odd} \\ n \neq 3}}^{\infty} (-n^2\pi^2 b_n) \sin(n\pi t).$$

We now plug into the left hand side of the differential equation.

$$2x_p'' + 18\pi^2 x = -12a_3\pi \sin(3\pi t) - 18\pi^2 a_3 t \cos(3\pi t) + 12b_3\pi \cos(3\pi t) - 18\pi^2 b_3 t \sin(3\pi t)+$$

$$+ 18\pi^2 a_3 t \cos(3\pi t) \qquad\qquad + 18\pi^2 b_3 t \sin(3\pi t)+$$

$$+ \sum_{\substack{n=1 \\ n \text{ odd} \\ n \neq 3}}^{\infty} (-2n^2\pi^2 b_n + 18\pi^2 b_n) \sin(n\pi t).$$

If we simplify we obtain

$$2x_p'' + 18\pi^2 x = -12a_3\pi \sin(3\pi t) + 12b_3\pi \cos(3\pi t) + \sum_{\substack{n=1 \\ n \text{ odd} \\ n \neq 3}}^{\infty} (-2n^2\pi^2 b_n + 18\pi^2 b_n) \sin(n\pi t).$$

This series has to equal to the series for $F(t)$. We equate the coefficients and solve for a_3 and b_n.

$$a_3 = \frac{4/(3\pi)}{-12\pi} = \frac{-1}{9\pi^2},$$

$$b_3 = 0,$$

$$b_n = \frac{4}{n\pi(18\pi^2 - 2n^2\pi^2)} = \frac{2}{\pi^3 n(9 - n^2)} \qquad \text{for } n \text{ odd and } n \neq 3.$$

That is,

$$x_p(t) = \frac{-1}{9\pi^2} t \cos(3\pi t) + \sum_{\substack{n=1 \\ n \text{ odd} \\ n \neq 3}}^{\infty} \frac{2}{\pi^3 n(9 - n^2)} \sin(n\pi t).$$

When $c > 0$, you will not have to worry about pure resonance. That is, there will never be any conflicts and you do not need to multiply any terms by t. There is a corresponding concept of practical resonance and it is very similar to the ideas we already explored in chapter 2. We will not go into details here.

4.5.3   Exercises

Exercise 4.5.2: *Let $F(t) = \frac{1}{2} + \sum_{n=1}^{\infty} \frac{1}{n^2} \cos(n\pi t)$. Find the steady periodic solution to $x'' + 2x = F(t)$. Express your solution as a Fourier series.*

Exercise 4.5.3: *Let $F(t) = \sum_{n=1}^{\infty} \frac{1}{n^3} \sin(n\pi t)$. Find the steady periodic solution to $x'' + x' + x = F(t)$. Express your solution as a Fourier series.*

Exercise 4.5.4: *Let $F(t) = \sum_{n=1}^{\infty} \frac{1}{n^2} \cos(n\pi t)$. Find the steady periodic solution to $x'' + 4x = F(t)$. Express your solution as a Fourier series.*

Exercise 4.5.5: *Let $F(t) = t$ for $-1 < t < 1$ and extended periodically. Find the steady periodic solution to $x'' + x = F(t)$. Express your solution as a series.*

Exercise 4.5.6: *Let $F(t) = t$ for $-1 < t < 1$ and extended periodically. Find the steady periodic solution to $x'' + \pi^2 x = F(t)$. Express your solution as a series.*

Exercise 4.5.101: *Let $F(t) = \sin(2\pi t) + 0.1 \cos(10\pi t)$. Find the steady periodic solution to $x'' + \sqrt{2}\, x = F(t)$. Express your solution as a Fourier series.*

Exercise 4.5.102: *Let $F(t) = \sum_{n=1}^{\infty} e^{-n} \cos(2nt)$. Find the steady periodic solution to $x'' + 3x = F(t)$. Express your solution as a Fourier series.*

Exercise 4.5.103: *Let $F(t) = |t|$ for $-1 \leq t \leq 1$ extended periodically. Find the steady periodic solution to $x'' + \sqrt{3}\, x = F(t)$. Express your solution as a series.*

Exercise 4.5.104: *Let $F(t) = |t|$ for $-1 \leq t \leq 1$ extended periodically. Find the steady periodic solution to $x'' + \pi^2 x = F(t)$. Express your solution as a series.*

4.6 PDEs, separation of variables, and the heat equation

Note: 2 lectures, §9.5 in [EP], §10.5 in [BD]

Let us recall that a *partial differential equation* or *PDE* is an equation containing the partial derivatives with respect to *several* independent variables. Solving PDEs will be our main application of Fourier series.

A PDE is said to be *linear* if the dependent variable and its derivatives appear at most to the first power and in no functions. We will only talk about linear PDEs. Together with a PDE, we usually have specified some *boundary conditions*, where the value of the solution or its derivatives is specified along the boundary of a region, and/or some *initial conditions* where the value of the solution or its derivatives is specified for some initial time. Sometimes such conditions are mixed together and we will refer to them simply as *side conditions*.

We will study three specific partial differential equations, each one representing a more general class of equations. First, we will study the *heat equation*, which is an example of a *parabolic PDE*. Next, we will study the *wave equation*, which is an example of a *hyperbolic PDE*. Finally, we will study the *Laplace equation*, which is an example of an *elliptic PDE*. Each of our examples will illustrate behavior that is typical for the whole class.

4.6.1 Heat on an insulated wire

Let us first study the heat equation. Suppose that we have a wire (or a thin metal rod) of length L that is insulated except at the endpoints. Let x denote the position along the wire and let t denote time. See Figure 4.13.

Figure 4.13: Insulated wire.

Let $u(x, t)$ denote the temperature at point x at time t. The equation governing this setup is the so-called *one-dimensional heat equation*:

$$\frac{\partial u}{\partial t} = k \frac{\partial^2 u}{\partial x^2},$$

where $k > 0$ is a constant (the *thermal conductivity* of the material). That is, the change in heat at a specific point is proportional to the second derivative of the heat along the wire. This makes sense;

if at a fixed t the graph of the heat distribution has a maximum (the graph is concave down), then heat flows away from the maximum. And vice-versa.

We will generally use a more convenient notation for partial derivatives. We will write u_t instead of $\frac{\partial u}{\partial t}$, and we will write u_{xx} instead of $\frac{\partial^2 u}{\partial x^2}$. With this notation the heat equation becomes

$$u_t = k u_{xx}.$$

For the heat equation, we must also have some boundary conditions. We assume that the ends of the wire are either exposed and touching some body of constant heat, or the ends are insulated. For example, if the ends of the wire are kept at temperature 0, then we must have the conditions

$$u(0, t) = 0 \qquad \text{and} \qquad u(L, t) = 0.$$

If, on the other hand, the ends are also insulated we get the conditions

$$u_x(0, t) = 0 \qquad \text{and} \qquad u_x(L, t) = 0.$$

In other words, heat is not flowing in nor out of the wire at the ends. We always have two conditions along the x axis as there are two derivatives in the x direction. These side conditions are called *homogeneous* (that is, u or a derivative of u is set to zero).

Furthermore, suppose that we know the initial temperature distribution at time $t = 0$. That is,

$$u(x, 0) = f(x),$$

for some known function $f(x)$. This initial condition is not a homogeneous side condition.

4.6.2   Separation of variables

The heat equation is linear as u and its derivatives do not appear to any powers or in any functions. Thus the principle of superposition still applies for the heat equation (without side conditions). If u_1 and u_2 are solutions and c_1, c_2 are constants, then $u = c_1 u_1 + c_2 u_2$ is also a solution.

Exercise **4.6.1**: *Verify the principle of superposition for the heat equation.*

Superposition also preserves some of the side conditions. In particular, if u_1 and u_2 are solutions that satisfy $u(0, t) = 0$ and $u(L, t) = 0$, and c_1, c_2 are constants, then $u = c_1 u_1 + c_2 u_2$ is still a solution that satisfies $u(0, t) = 0$ and $u(L, t) = 0$. Similarly for the side conditions $u_x(0, t) = 0$ and $u_x(L, t) = 0$. In general, superposition preserves all homogeneous side conditions.

The method of *separation of variables* is to try to find solutions that are sums or products of functions of one variable. For example, for the heat equation, we try to find solutions of the form

$$u(x, t) = X(x)T(t).$$

That the desired solution we are looking for is of this form is too much to hope for. What is perfectly reasonable to ask, however, is to find enough "building-block" solutions of the form $u(x,t) = X(x)T(t)$ using this procedure so that the desired solution to the PDE is somehow constructed from these building blocks by the use of superposition.

Let us try to solve the heat equation

$$u_t = ku_{xx} \quad \text{with} \quad u(0,t) = 0, \quad u(L,t) = 0, \quad \text{and} \quad u(x,0) = f(x).$$

Let us guess $u(x,t) = X(x)T(t)$. We plug into the heat equation to obtain

$$X(x)T'(t) = kX''(x)T(t).$$

We rewrite as

$$\frac{T'(t)}{kT(t)} = \frac{X''(x)}{X(x)}.$$

This equation must hold for all x and all t. But the left hand side does not depend on x and the right hand side does not depend on t. Hence, each side must be a constant. Let us call this constant $-\lambda$ (the minus sign is for convenience later). We obtain the two equations

$$\frac{T'(t)}{kT(t)} = -\lambda = \frac{X''(x)}{X(x)}.$$

In other words

$$X''(x) + \lambda X(x) = 0,$$
$$T'(t) + \lambda kT(t) = 0.$$

The boundary condition $u(0,t) = 0$ implies $X(0)T(t) = 0$. We are looking for a nontrivial solution and so we can assume that $T(t)$ is not identically zero. Hence $X(0) = 0$. Similarly, $u(L,t) = 0$ implies $X(L) = 0$. We are looking for nontrivial solutions X of the eigenvalue problem $X'' + \lambda X = 0$, $X(0) = 0$, $X(L) = 0$. We have previously found that the only eigenvalues are $\lambda_n = \frac{n^2\pi^2}{L^2}$, for integers $n \geq 1$, where eigenfunctions are $\sin\left(\frac{n\pi}{L}x\right)$. Hence, let us pick the solutions

$$X_n(x) = \sin\left(\frac{n\pi}{L}x\right).$$

The corresponding T_n must satisfy the equation

$$T_n'(t) + \frac{n^2\pi^2}{L^2}kT_n(t) = 0.$$

By the method of integrating factor, the solution of this problem is

$$T_n(t) = e^{\frac{-n^2\pi^2}{L^2}kt}.$$

It will be useful to note that $T_n(0) = 1$. Our building-block solutions are

$$u_n(x,t) = X_n(x)T_n(t) = \sin\left(\frac{n\pi}{L}x\right)e^{\frac{-n^2\pi^2}{L^2}kt}.$$

We note that $u_n(x,0) = \sin\left(\frac{n\pi}{L}x\right)$. Let us write $f(x)$ as the sine series

$$f(x) = \sum_{n=1}^{\infty} b_n \sin\left(\frac{n\pi}{L}x\right).$$

That is, we find the Fourier series of the odd periodic extension of $f(x)$. We used the sine series as it corresponds to the eigenvalue problem for $X(x)$ above. Finally, we use superposition to write the solution as

$$\boxed{u(x,t) = \sum_{n=1}^{\infty} b_n u_n(x,t) = \sum_{n=1}^{\infty} b_n \sin\left(\frac{n\pi}{L}x\right)e^{\frac{-n^2\pi^2}{L^2}kt}.}$$

Why does this solution work? First note that it is a solution to the heat equation by superposition. It satisfies $u(0,t) = 0$ and $u(L,t) = 0$, because $x = 0$ or $x = L$ makes all the sines vanish. Finally, plugging in $t = 0$, we notice that $T_n(0) = 1$ and so

$$u(x,0) = \sum_{n=1}^{\infty} b_n u_n(x,0) = \sum_{n=1}^{\infty} b_n \sin\left(\frac{n\pi}{L}x\right) = f(x).$$

Example 4.6.1: Suppose that we have an insulated wire of length 1, such that the ends of the wire are embedded in ice (temperature 0). Let $k = 0.003$. Then suppose that initial heat distribution is $u(x,0) = 50\,x\,(1-x)$. See Figure 4.14.

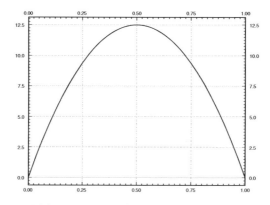

Figure 4.14: Initial distribution of temperature in the wire.

We want to find the temperature function $u(x,t)$. Let us suppose we also want to find when (at what t) does the maximum temperature in the wire drop to one half of the initial maximum of 12.5.

We are solving the following PDE problem:

$$u_t = 0.003\, u_{xx},$$
$$u(0,t) = u(1,t) = 0,$$
$$u(x,0) = 50\,x\,(1-x) \qquad \text{for } 0 < x < 1.$$

We write $f(x) = 50\,x\,(1-x)$ for $0 < x < 1$ as a sine series. That is, $f(x) = \sum_{n=1}^{\infty} b_n \sin(n\pi x)$, where

$$b_n = 2 \int_0^1 50\,x\,(1-x)\sin(n\pi x)\,dx = \frac{200}{\pi^3 n^3} - \frac{200\,(-1)^n}{\pi^3 n^3} = \begin{cases} 0 & \text{if } n \text{ even,} \\ \frac{400}{\pi^3 n^3} & \text{if } n \text{ odd.} \end{cases}$$

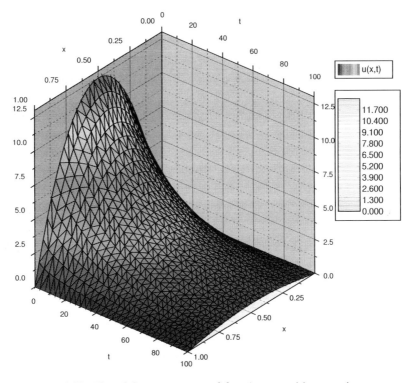

Figure 4.15: Plot of the temperature of the wire at position x at time t.

The solution $u(x,t)$, plotted in Figure 4.15 for $0 \le t \le 100$, is given by the series:

$$u(x,t) = \sum_{\substack{n=1 \\ n\text{ odd}}}^{\infty} \frac{400}{\pi^3 n^3} \sin(n\pi x)\, e^{-n^2 \pi^2 0.003\, t}.$$

Finally, let us answer the question about the maximum temperature. It is relatively easy to see that the maximum temperature will always be at $x = 0.5$, in the middle of the wire. The plot of $u(x, t)$ confirms this intuition.

If we plug in $x = 0.5$ we get

$$u(0.5, t) = \sum_{\substack{n=1 \\ n \text{ odd}}}^{\infty} \frac{400}{\pi^3 n^3} \sin(n\pi\, 0.5)\, e^{-n^2 \pi^2\, 0.003\, t}.$$

For $n = 3$ and higher (remember n is only odd), the terms of the series are insignificant compared to the first term. The first term in the series is already a very good approximation of the function. Hence

$$u(0.5, t) \approx \frac{400}{\pi^3}\, e^{-\pi^2\, 0.003\, t}.$$

The approximation gets better and better as t gets larger as the other terms decay much faster. Let us plot the function $u(0.5, t)$, the temperature at the midpoint of the wire at time t, in Figure 4.16. The figure also plots the approximation by the first term.

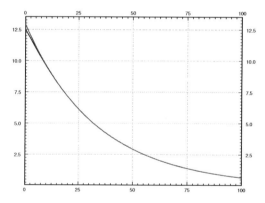

Figure 4.16: Temperature at the midpoint of the wire (the bottom curve), and the approximation of this temperature by using only the first term in the series (top curve).

After $t = 5$ or so it would be hard to tell the difference between the first term of the series for $u(x, t)$ and the real solution $u(x, t)$. This behavior is a general feature of solving the heat equation. If you are interested in behavior for large enough t, only the first one or two terms may be necessary.

Let us get back to the question of when is the maximum temperature one half of the initial maximum temperature. That is, when is the temperature at the midpoint $12.5/2 = 6.25$. We notice on the graph that if we use the approximation by the first term we will be close enough. We solve

$$6.25 = \frac{400}{\pi^3}\, e^{-\pi^2\, 0.003\, t}.$$

That is,

$$t = \frac{\ln \frac{6.25\pi^3}{400}}{-\pi^2 0.003} \approx 24.5.$$

So the maximum temperature drops to half at about $t = 24.5$.

We mention an interesting behavior of the solution to the heat equation. The heat equation "smoothes" out the function $f(x)$ as t grows. For a fixed t, the solution is a Fourier series with coefficients $b_n e^{\frac{-n^2\pi^2}{L^2}kt}$. If $t > 0$, then these coefficients go to zero faster than any $\frac{1}{n^p}$ for any power p. In other words, the Fourier series has infinitely many derivatives everywhere. Thus even if the function $f(x)$ has jumps and corners, then for a fixed $t > 0$, the solution $u(x, t)$ as a function of x is as smooth as we want it to be.

4.6.3 Insulated ends

Now suppose the ends of the wire are insulated. In this case, we are solving the equation

$$u_t = k u_{xx} \quad \text{with} \quad u_x(0, t) = 0, \quad u_x(L, t) = 0, \quad \text{and} \quad u(x, 0) = f(x).$$

Yet again we try a solution of the form $u(x, t) = X(x)T(t)$. By the same procedure as before we plug into the heat equation and arrive at the following two equations

$$X''(x) + \lambda X(x) = 0,$$
$$T'(t) + \lambda k T(t) = 0.$$

At this point the story changes slightly. The boundary condition $u_x(0, t) = 0$ implies $X'(0)T(t) = 0$. Hence $X'(0) = 0$. Similarly, $u_x(L, t) = 0$ implies $X'(L) = 0$. We are looking for nontrivial solutions X of the eigenvalue problem $X'' + \lambda X = 0$, $X'(0) = 0$, $X'(L) = 0$. We have previously found that the only eigenvalues are $\lambda_n = \frac{n^2\pi^2}{L^2}$, for integers $n \geq 0$, where eigenfunctions are $\cos\left(\frac{n\pi}{L} x\right)$ (we include the constant eigenfunction). Hence, let us pick solutions

$$X_n(x) = \cos\left(\frac{n\pi}{L} x\right) \qquad \text{and} \qquad X_0(x) = 1.$$

The corresponding T_n must satisfy the equation

$$T'_n(t) + \frac{n^2\pi^2}{L^2} k T_n(t) = 0.$$

For $n \geq 1$, as before,

$$T_n(t) = e^{\frac{-n^2\pi^2}{L^2}kt}.$$

For $n = 0$, we have $T'_0(t) = 0$ and hence $T_0(t) = 1$. Our building-block solutions will be

$$u_n(x, t) = X_n(x)T_n(t) = \cos\left(\frac{n\pi}{L} x\right) e^{\frac{-n^2\pi^2}{L^2}kt},$$

and
$$u_0(x, t) = 1.$$

We note that $u_n(x, 0) = \cos\left(\frac{n\pi}{L} x\right)$. Let us write f using the cosine series

$$f(x) = \frac{a_0}{2} + \sum_{n=1}^{\infty} a_n \cos\left(\frac{n\pi}{L} x\right).$$

That is, we find the Fourier series of the even periodic extension of $f(x)$.

We use superposition to write the solution as

$$u(x, t) = \frac{a_0}{2} + \sum_{n=1}^{\infty} a_n u_n(x, t) = \frac{a_0}{2} + \sum_{n=1}^{\infty} a_n \cos\left(\frac{n\pi}{L} x\right) e^{\frac{-n^2\pi^2}{L^2} kt}.$$

Example 4.6.2: Let us try the same equation as before, but for insulated ends. We are solving the following PDE problem

$$u_t = 0.003\, u_{xx},$$
$$u_x(0, t) = u_x(1, t) = 0,$$
$$u(x, 0) = 50\, x\,(1 - x) \qquad \text{for } 0 < x < 1.$$

For this problem, we must find the cosine series of $u(x, 0)$. For $0 < x < 1$ we have

$$50\, x\,(1 - x) = \frac{25}{3} + \sum_{\substack{n=2 \\ n \text{ even}}}^{\infty} \left(\frac{-200}{\pi^2 n^2}\right) \cos(n\pi x).$$

The calculation is left to the reader. Hence, the solution to the PDE problem, plotted in Figure 4.17 on the next page, is given by the series

$$u(x, t) = \frac{25}{3} + \sum_{\substack{n=2 \\ n \text{ even}}}^{\infty} \left(\frac{-200}{\pi^2 n^2}\right) \cos(n\pi x)\, e^{-n^2\pi^2\, 0.003\, t}.$$

Note in the graph that the temperature evens out across the wire. Eventually, all the terms except the constant die out, and you will be left with a uniform temperature of $\frac{25}{3} \approx 8.33$ along the entire length of the wire.

4.6.4   Exercises

Exercise 4.6.2: Imagine you have a wire of length 2, with $k = 0.001$ and an initial temperature distribution of $u(x, 0) = 50x$. Suppose that both the ends are embedded in ice (temperature 0). Find the solution as a series.

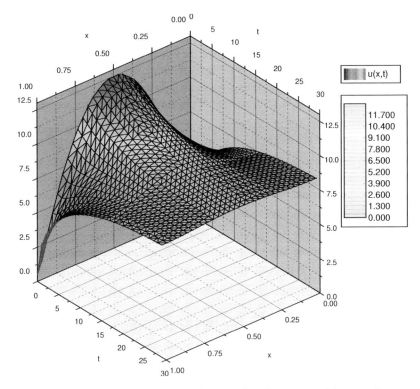

Figure 4.17: Plot of the temperature of the insulated wire at position x at time t.

***Exercise* 4.6.3:** *Find a series solution of*

$$u_t = u_{xx},$$
$$u(0, t) = u(1, t) = 0,$$
$$u(x, 0) = 100 \qquad \textit{for } 0 < x < 1.$$

***Exercise* 4.6.4:** *Find a series solution of*

$$u_t = u_{xx},$$
$$u_x(0, t) = u_x(\pi, t) = 0,$$
$$u(x, 0) = 3 \cos(x) + \cos(3x) \qquad \textit{for } 0 < x < \pi.$$

Exercise **4.6.5***: Find a series solution of*

$$u_t = \frac{1}{3} u_{xx},$$

$$u_x(0, t) = u_x(\pi, t) = 0,$$

$$u(x, 0) = \frac{10x}{\pi} \qquad for \ 0 < x < \pi.$$

Exercise **4.6.6***: Find a series solution of*

$$u_t = u_{xx},$$

$$u(0, t) = 0, \quad u(1, t) = 100,$$

$$u(x, 0) = \sin(\pi x) \qquad for \ 0 < x < 1.$$

Hint: Use the fact that u(x, t) = 100x is a solution satisfying $u_t = u_{xx}$, u(0, t) = 0, u(1, t) = 100. Then use superposition.

Exercise **4.6.7***: Find the* steady state temperature *solution as a function of x alone, by letting t → ∞ in the solution from exercises 4.6.5 and 4.6.6. Verify that it satisfies the equation $u_{xx} = 0$.*

Exercise **4.6.8***: Use separation variables to find a nontrivial solution to $u_{xx} + u_{yy} = 0$, where u(x, 0) = 0 and u(0, y) = 0. Hint: Try u(x, y) = X(x)Y(y).*

Exercise **4.6.9** (challenging)*: Suppose that one end of the wire is insulated (say at x = 0) and the other end is kept at zero temperature. That is, find a series solution of*

$$u_t = k u_{xx},$$

$$u_x(0, t) = u(L, t) = 0,$$

$$u(x, 0) = f(x) \qquad for \ 0 < x < L.$$

Express any coefficients in the series by integrals of f(x).

Exercise **4.6.10** (challenging)*: Suppose that the wire is circular and insulated, so there are no ends. You can think of this as simply connecting the two ends and making sure the solution matches up at the ends. That is, find a series solution of*

$$u_t = k u_{xx},$$

$$u(0, t) = u(L, t), \qquad u_x(0, t) = u_x(L, t),$$

$$u(x, 0) = f(x) \qquad for \ 0 < x < L.$$

Express any coefficients in the series by integrals of f(x).

Exercise 4.6.101: *Find a series solution of*

$$u_t = 3u_{xx},$$
$$u(0,t) = u(\pi,t) = 0,$$
$$u(x,0) = 5\sin(x) + 2\sin(5x) \qquad for\ 0 < x < \pi.$$

Exercise 4.6.102: *Find a series solution of*

$$u_t = 0.1u_{xx},$$
$$u_x(0,t) = u_x(\pi,t) = 0,$$
$$u(x,0) = 1 + 2\cos(x) \qquad for\ 0 < x < \pi.$$

Exercise 4.6.103: *Use separation of variables to find a nontrivial solution to* $u_{xt} = u_{xx}$.

Exercise 4.6.104: *Use separation of variables (Hint: try* $u(x,t) = X(x) + T(t)$*) to find a nontrivial solution to* $u_x + u_t = u$.

4.7   One dimensional wave equation

Note: 1 lecture, §9.6 in [EP], §10.7 in [BD]

Imagine we have a tensioned guitar string of length L. Suppose we only consider vibrations in one direction. That is, let x denote the position along the string, let t denote time, and let y denote the displacement of the string from the rest position. See Figure 4.18.

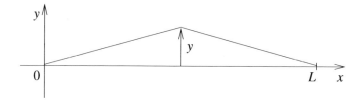

Figure 4.18: Vibrating string.

The equation that governs this setup is the so-called *one-dimensional wave equation*:

$$\boxed{y_{tt} = a^2 y_{xx},}$$

for some constant $a > 0$. Assume that the ends of the string are fixed in place:

$$y(0, t) = 0 \qquad \text{and} \qquad y(L, t) = 0.$$

Note that we have two conditions along the x axis as there are two derivatives in the x direction.

There are also two derivatives along the t direction and hence we need two further conditions here. We need to know the initial position and the initial velocity of the string. That is,

$$y(x, 0) = f(x) \qquad \text{and} \qquad y_t(x, 0) = g(x),$$

for some known functions $f(x)$ and $g(x)$.

As the equation is again linear, superposition works just as it did for the heat equation. And again we will use separation of variables to find enough building-block solutions to get the overall solution. There is one change however. It will be easier to solve two separate problems and add their solutions.

The two problems we will solve are

$$
\begin{aligned}
& w_{tt} = a^2 w_{xx}, \\
& w(0, t) = w(L, t) = 0, \\
& w(x, 0) = 0 && \text{for } 0 < x < L, \\
& w_t(x, 0) = g(x) && \text{for } 0 < x < L.
\end{aligned}
\qquad (4.10)
$$

and

$$z_{tt} = a^2 z_{xx},$$
$$z(0, t) = z(L, t) = 0,$$
$$z(x, 0) = f(x) \qquad \text{for } 0 < x < L,$$
$$z_t(x, 0) = 0 \qquad \text{for } 0 < x < L. \tag{4.11}$$

The principle of superposition implies that $y = w + z$ solves the wave equation and furthermore $y(x, 0) = w(x, 0) + z(x, 0) = f(x)$ and $y_t(x, 0) = w_t(x, 0) + z_t(x, 0) = g(x)$. Hence, y is a solution to

$$y_{tt} = a^2 y_{xx},$$
$$y(0, t) = y(L, t) = 0,$$
$$y(x, 0) = f(x) \qquad \text{for } 0 < x < L,$$
$$y_t(x, 0) = g(x) \qquad \text{for } 0 < x < L. \tag{4.12}$$

The reason for all this complexity is that superposition only works for homogeneous conditions such as $y(0, t) = y(L, t) = 0$, $y(x, 0) = 0$, or $y_t(x, 0) = 0$. Therefore, we will be able to use the idea of separation of variables to find many building-block solutions solving all the homogeneous conditions. We can then use them to construct a solution solving the remaining nonhomogeneous condition.

Let us start with (4.10). We try a solution of the form $w(x, t) = X(x)T(t)$ again. We plug into the wave equation to obtain

$$X(x)T''(t) = a^2 X''(x)T(t).$$

Rewriting we get

$$\frac{T''(t)}{a^2 T(t)} = \frac{X''(x)}{X(x)}.$$

Again, left hand side depends only on t and the right hand side depends only on x. Therefore, both equal a constant, which we will denote by $-\lambda$.

$$\frac{T''(t)}{a^2 T(t)} = -\lambda = \frac{X''(x)}{X(x)}.$$

We solve to get two ordinary differential equations

$$X''(x) + \lambda X(x) = 0,$$
$$T''(t) + \lambda a^2 T(t) = 0.$$

The conditions $0 = w(0, t) = X(0)T(t)$ implies $X(0) = 0$ and $w(L, t) = 0$ implies that $X(L) = 0$. Therefore, the only nontrivial solutions for the first equation are when $\lambda = \lambda_n = \frac{n^2 \pi^2}{L^2}$ and they are

$$X_n(x) = \sin\left(\frac{n\pi}{L} x\right).$$

The general solution for T for this particular λ_n is

$$T_n(t) = A \cos\left(\frac{n\pi a}{L} t\right) + B \sin\left(\frac{n\pi a}{L} t\right).$$

We also have the condition that $w(x, 0) = 0$ or $X(x)T(0) = 0$. This implies that $T(0) = 0$, which in turn forces $A = 0$. It is convenient to pick $B = \frac{L}{n\pi a}$ (you will see why in a moment) and hence

$$T_n(t) = \frac{L}{n\pi a} \sin\left(\frac{n\pi a}{L} t\right).$$

Our building-block solutions are

$$w_n(x, t) = \frac{L}{n\pi a} \sin\left(\frac{n\pi}{L} x\right) \sin\left(\frac{n\pi a}{L} t\right).$$

We differentiate in t, that is

$$\frac{\partial w_n}{\partial t}(x, t) = \sin\left(\frac{n\pi}{L} x\right) \cos\left(\frac{n\pi a}{L} t\right).$$

Hence,

$$\frac{\partial w_n}{\partial t}(x, 0) = \sin\left(\frac{n\pi}{L} x\right).$$

We expand $g(x)$ in terms of these sines as

$$g(x) = \sum_{n=1}^{\infty} b_n \sin\left(\frac{n\pi}{L} x\right).$$

Using superposition we can just write down the solution to (4.10) as a series

$$w(x, t) = \sum_{n=1}^{\infty} b_n w_n(x, t) = \sum_{n=1}^{\infty} b_n \frac{L}{n\pi a} \sin\left(\frac{n\pi}{L} x\right) \sin\left(\frac{n\pi a}{L} t\right).$$

Exercise 4.7.1: *Check that $w(x, 0) = 0$ and $w_t(x, 0) = g(x)$.*

Similarly we proceed to solve (4.11). We again try $z(x, y) = X(x)T(t)$. The procedure works exactly the same at first. We obtain

$$X''(x) + \lambda X(x) = 0,$$
$$T''(t) + \lambda a^2 T(t) = 0.$$

and the conditions $X(0) = 0$, $X(L) = 0$. So again $\lambda = \lambda_n = \frac{n^2 \pi^2}{L^2}$ and

$$X_n(x) = \sin\left(\frac{n\pi}{L} x\right).$$

This time the condition on T is $T'(0) = 0$. Thus we get that $B = 0$ and we take

$$T_n(t) = \cos\left(\frac{n\pi a}{L} t\right).$$

Our building-block solution will be

$$z_n(x, t) = \sin\left(\frac{n\pi}{L} x\right) \cos\left(\frac{n\pi a}{L} t\right).$$

We expand $f(x)$ in terms of these sines as

$$f(x) = \sum_{n=1}^{\infty} c_n \sin\left(\frac{n\pi}{L} x\right).$$

And we write down the solution to (4.11) as a series

$$z(x, t) = \sum_{n=1}^{\infty} c_n z_n(x, t) = \sum_{n=1}^{\infty} c_n \sin\left(\frac{n\pi}{L} x\right) \cos\left(\frac{n\pi a}{L} t\right).$$

Exercise 4.7.2: *Fill in the details in the derivation of the solution of* (4.11). *Check that the solution satisfies all the side conditions.*

Putting these two solutions together, let us state the result as a theorem.

Theorem 4.7.1. *Take the equation*

$$\begin{aligned} y_{tt} &= a^2 y_{xx}, \\ y(0, t) &= y(L, t) = 0, \\ y(x, 0) &= f(x) \qquad \text{for } 0 < x < L, \\ y_t(x, 0) &= g(x) \qquad \text{for } 0 < x < L, \end{aligned} \qquad (4.13)$$

where

$$f(x) = \sum_{n=1}^{\infty} c_n \sin\left(\frac{n\pi}{L} x\right),$$

and

$$g(x) = \sum_{n=1}^{\infty} b_n \sin\left(\frac{n\pi}{L} x\right).$$

Then the solution $y(x,t)$ can be written as a sum of the solutions of (4.10) *and* (4.11). *In other words,*

$$\begin{aligned} y(x, t) &= \sum_{n=1}^{\infty} b_n \frac{L}{n\pi a} \sin\left(\frac{n\pi}{L} x\right) \sin\left(\frac{n\pi a}{L} t\right) + c_n \sin\left(\frac{n\pi}{L} x\right) \cos\left(\frac{n\pi a}{L} t\right) \\ &= \sum_{n=1}^{\infty} \sin\left(\frac{n\pi}{L} x\right) \left[b_n \frac{L}{n\pi a} \sin\left(\frac{n\pi a}{L} t\right) + c_n \cos\left(\frac{n\pi a}{L} t\right) \right]. \end{aligned}$$

Figure 4.19: Plucked string.

Example 4.7.1: Let us try a simple example of a plucked string. Suppose that a string of length 2 is plucked in the middle such that it has the initial shape given in Figure 4.19. That is

$$f(x) = \begin{cases} 0.1\,x & \text{if } 0 \le x \le 1, \\ 0.1\,(2 - x) & \text{if } 1 < x \le 2. \end{cases}$$

The string starts at rest ($g(x) = 0$). Suppose that $a = 1$ in the wave equation for simplicity.

We leave it to the reader to compute the sine series of $f(x)$. The series will be

$$f(x) = \sum_{n=1}^{\infty} \frac{0.8}{n^2\pi^2} \sin\left(\frac{n\pi}{2}\right) \sin\left(\frac{n\pi}{2} x\right).$$

Note that $\sin\left(\frac{n\pi}{2}\right)$ is the sequence $1, 0, -1, 0, 1, 0, -1, \ldots$ for $n = 1, 2, 3, 4, \ldots$. Therefore,

$$f(x) = \frac{0.8}{\pi^2} \sin\left(\frac{\pi}{2} x\right) - \frac{0.8}{9\pi^2} \sin\left(\frac{3\pi}{2} x\right) + \frac{0.8}{25\pi^2} \sin\left(\frac{5\pi}{2} x\right) - \cdots$$

The solution $y(x, t)$ is given by

$$\begin{aligned}
y(x, t) &= \sum_{n=1}^{\infty} \frac{0.8}{n^2\pi^2} \sin\left(\frac{n\pi}{2}\right) \sin\left(\frac{n\pi}{2} x\right) \cos\left(\frac{n\pi}{2} t\right) \\
&= \sum_{m=1}^{\infty} \frac{0.8(-1)^{m+1}}{(2m-1)^2\pi^2} \sin\left(\frac{(2m-1)\pi}{2} x\right) \cos\left(\frac{(2m-1)\pi}{2} t\right) \\
&= \frac{0.8}{\pi^2} \sin\left(\frac{\pi}{2} x\right) \cos\left(\frac{\pi}{2} t\right) - \frac{0.8}{9\pi^2} \sin\left(\frac{3\pi}{2} x\right) \cos\left(\frac{3\pi}{2} t\right) + \frac{0.8}{25\pi^2} \sin\left(\frac{5\pi}{2} x\right) \cos\left(\frac{5\pi}{2} t\right) - \cdots
\end{aligned}$$

A plot for $0 < t < 3$ is given in Figure 4.20 on the facing page. Notice that unlike the heat equation, the solution does not become "smoother," the "sharp edges" remain. We will see the reason for this behavior in the next section where we derive the solution to the wave equation in a different way.

Make sure you understand what the plot, such as the one in the figure, is telling you. For each fixed t, you can think of the function $y(x, t)$ as just a function of x. This function gives you the shape of the string at time t.

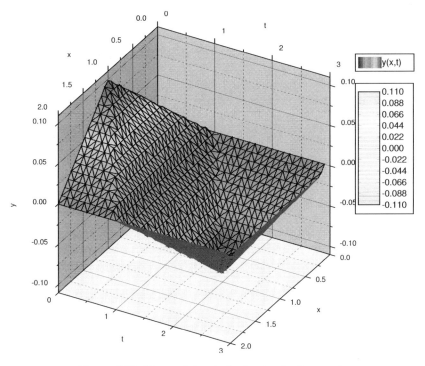

Figure 4.20: Shape of the plucked string for $0 < t < 3$.

4.7.1 Exercises

Exercise **4.7.3***: Solve*

$$
\begin{aligned}
&y_{tt} = 9y_{xx}, \\
&y(0,t) = y(1,t) = 0, \\
&y(x,0) = \sin(3\pi x) + \tfrac{1}{4}\sin(6\pi x) \qquad \text{for } 0 < x < 1, \\
&y_t(x,0) = 0 \qquad\qquad\qquad\qquad\quad \text{for } 0 < x < 1.
\end{aligned}
$$

Exercise **4.7.4***: Solve*

$$
\begin{aligned}
&y_{tt} = 4y_{xx}, \\
&y(0,t) = y(1,t) = 0, \\
&y(x,0) = \sin(3\pi x) + \tfrac{1}{4}\sin(6\pi x) \qquad \text{for } 0 < x < 1, \\
&y_t(x,0) = \sin(9\pi x) \qquad\qquad\qquad\; \text{for } 0 < x < 1.
\end{aligned}
$$

Exercise **4.7.5***: Derive the solution for a general plucked string of length L, where we raise the string some distance b at the midpoint and let go, and for any constant a (in the equation $y_{tt} = a^2 y_{xx}$).*

Exercise **4.7.6**: *Imagine that a stringed musical instrument falls on the floor. Suppose that the length of the string is 1 and $a = 1$. When the musical instrument hits the ground the string was in rest position and hence $y(x, 0) = 0$. However, the string was moving at some velocity at impact $(t = 0)$, say $y_t(x, 0) = -1$. Find the solution $y(x, t)$ for the shape of the string at time t.*

Exercise **4.7.7** (challenging): *Suppose that you have a vibrating string and that there is air resistance proportional to the velocity. That is, you have*

$$y_{tt} = a^2 y_{xx} - k y_t,$$
$$y(0, t) = y(1, t) = 0,$$
$$y(x, 0) = f(x) \qquad \text{for } 0 < x < 1,$$
$$y_t(x, 0) = 0 \qquad \text{for } 0 < x < 1.$$

Suppose that $0 < k < 2\pi a$. Derive a series solution to the problem. Any coefficients in the series should be expressed as integrals of $f(x)$.

Exercise **4.7.101**: *Solve*

$$y_{tt} = y_{xx},$$
$$y(0, t) = y(\pi, t) = 0,$$
$$y(x, 0) = \sin(x) \qquad \text{for } 0 < x < \pi,$$
$$y_t(x, 0) = \sin(x) \qquad \text{for } 0 < x < \pi.$$

Exercise **4.7.102**: *Solve*

$$y_{tt} = 25 y_{xx},$$
$$y(0, t) = y(2, t) = 0,$$
$$y(x, 0) = 0 \qquad \text{for } 0 < x < 2,$$
$$y_t(x, 0) = \sin(\pi t) + 0.1 \sin(2\pi t) \qquad \text{for } 0 < x < 2.$$

Exercise **4.7.103**: *Solve*

$$y_{tt} = 2 y_{xx},$$
$$y(0, t) = y(\pi, t) = 0,$$
$$y(x, 0) = x \qquad \text{for } 0 < x < \pi,$$
$$y_t(x, 0) = 0 \qquad \text{for } 0 < x < \pi.$$

Exercise **4.7.104**: *Let's see what happens when $a = 0$. Find a solution to $y_{tt} = 0$, $y(0, t) = y(\pi, t) = 0$, $y(x, 0) = \sin(2x)$, $y_t(x, 0) = \sin(x)$.*

4.8 D'Alembert solution of the wave equation

Note: 1 lecture, different from §9.6 in [EP], part of §10.7 in [BD]

We have solved the wave equation by using Fourier series. But it is often more convenient to use the so-called *d'Alembert solution to the wave equation*[*]. While this solution can be derived using Fourier series as well, it is really an awkward use of those concepts. It is easier and more instructive to derive this solution by making a correct change of variables to get an equation that can be solved by simple integration.

Suppose we have the wave equation

$$y_{tt} = a^2 y_{xx}. \tag{4.14}$$

We wish to solve the equation (4.14) given the conditions

$$
\begin{aligned}
y(0,t) &= y(L,t) = 0 &&\text{for all } t, \\
y(x,0) &= f(x) && 0 < x < L, \\
y_t(x,0) &= g(x) && 0 < x < L.
\end{aligned}
\tag{4.15}
$$

4.8.1 Change of variables

We will transform the equation into a simpler form where it can be solved by simple integration. We change variables to $\xi = x - at$, $\eta = x + at$. The chain rule says:

$$
\frac{\partial}{\partial x} = \frac{\partial \xi}{\partial x}\frac{\partial}{\partial \xi} + \frac{\partial \eta}{\partial x}\frac{\partial}{\partial \eta} = \frac{\partial}{\partial \xi} + \frac{\partial}{\partial \eta},
$$

$$
\frac{\partial}{\partial t} = \frac{\partial \xi}{\partial t}\frac{\partial}{\partial \xi} + \frac{\partial \eta}{\partial t}\frac{\partial}{\partial \eta} = -a\frac{\partial}{\partial \xi} + a\frac{\partial}{\partial \eta}.
$$

We compute

$$
y_{xx} = \frac{\partial^2 y}{\partial x^2} = \left(\frac{\partial}{\partial \xi} + \frac{\partial}{\partial \eta}\right)\left(\frac{\partial y}{\partial \xi} + \frac{\partial y}{\partial \eta}\right) = \frac{\partial^2 y}{\partial \xi^2} + 2\frac{\partial^2 y}{\partial \xi \partial \eta} + \frac{\partial^2 y}{\partial \eta^2},
$$

$$
y_{tt} = \frac{\partial^2 y}{\partial t^2} = \left(-a\frac{\partial}{\partial \xi} + a\frac{\partial}{\partial \eta}\right)\left(-a\frac{\partial y}{\partial \xi} + a\frac{\partial y}{\partial \eta}\right) = a^2\frac{\partial^2 y}{\partial \xi^2} - 2a^2\frac{\partial^2 y}{\partial \xi \partial \eta} + a^2\frac{\partial^2 y}{\partial \eta^2}.
$$

In the above computations, we used the fact from calculus that $\frac{\partial^2 y}{\partial \xi \partial \eta} = \frac{\partial^2 y}{\partial \eta \partial \xi}$. We plug what we got into the wave equation,

$$
0 = a^2 y_{xx} - y_{tt} = 4a^2\frac{\partial^2 y}{\partial \xi \partial \eta} = 4a^2 y_{\xi\eta}.
$$

---

[*]Named after the French mathematician Jean le Rond d'Alembert (1717–1783).

Therefore, the wave equation (4.14) transforms into $y_{\xi\eta} = 0$. It is easy to find the general solution to this equation by integrating twice. Keeping ξ constant, we integrate with respect to η first[*] and notice that the constant of integration depends on ξ; for each ξ we might get a different constant of integration. We get $y_\xi = C(\xi)$. Next, we integrate with respect to ξ and notice that the constant of integration must depend on η. Thus, $y = \int C(\xi)\,d\xi + B(\eta)$. The solution must, therefore, be of the following form for some functions $A(\xi)$ and $B(\eta)$:

$$y = A(\xi) + B(\eta) = A(x - at) + B(x + at).$$

The solution is a superposition of two functions (waves) travelling at speed a in opposite directions. The coordinates ξ and η are called the *characteristic coordinates*, and a similar technique can be applied to more complicated hyperbolic PDE.

4.8.2   D'Alembert's formula

We know what any solution must look like, but we need to solve for the given side conditions. We will just give the formula and see that it works. First let $F(x)$ denote the *odd extension* of $f(x)$, and let $G(x)$ denote the *odd extension* of $g(x)$. Define

$$A(x) = \frac{1}{2}F(x) - \frac{1}{2a}\int_0^x G(s)\,ds, \qquad B(x) = \frac{1}{2}F(x) + \frac{1}{2a}\int_0^x G(s)\,ds.$$

We claim this $A(x)$ and $B(x)$ give the solution. Explicitly, the solution is $y(x,t) = A(x-at) + B(x+at)$ or in other words:

$$\boxed{\begin{aligned}
y(x,t) &= \frac{1}{2}F(x-at) - \frac{1}{2a}\int_0^{x-at} G(s)\,ds + \frac{1}{2}F(x+at) + \frac{1}{2a}\int_0^{x+at} G(s)\,ds \\
&= \frac{F(x-at) + F(x+at)}{2} + \frac{1}{2a}\int_{x-at}^{x+at} G(s)\,ds.
\end{aligned}}$$

$$(4.16)$$

Let us check that the d'Alembert formula really works.

$$y(x,0) = \frac{1}{2}F(x) - \frac{1}{2a}\int_0^x G(s)\,ds + \frac{1}{2}F(x) + \frac{1}{2a}\int_0^x G(s)\,ds = F(x).$$

So far so good. Assume for simplicity F is differentiable. By the fundamental theorem of calculus we have

$$y_t(x,t) = \frac{-a}{2}F'(x-at) + \frac{1}{2}G(x-at) + \frac{a}{2}F'(x+at) + \frac{1}{2}G(x+at).$$

So

$$y_t(x,0) = \frac{-a}{2}F'(x) + \frac{1}{2}G(x) + \frac{a}{2}F'(x) + \frac{1}{2}G(x) = G(x).$$

---

[*]We can just as well integrate with ξ first, if we wish.

Yay! We're smoking now. OK, now the boundary conditions. Note that $F(x)$ and $G(x)$ are odd. Also $\int_0^x G(s)\,ds$ is an even function of x because $G(x)$ is odd (to see this fact, do the substitution $s = -v$). So

$$y(0, t) = \frac{1}{2}F(-at) - \frac{1}{2a}\int_0^{-at} G(s)\,ds + \frac{1}{2}F(at) + \frac{1}{2a}\int_0^{at} G(s)\,ds$$

$$= \frac{-1}{2}F(at) - \frac{1}{2a}\int_0^{at} G(s)\,ds + \frac{1}{2}F(at) + \frac{1}{2a}\int_0^{at} G(s)\,ds = 0.$$

Note that $F(x)$ and $G(x)$ are $2L$ periodic. We compute

$$y(L, t) = \frac{1}{2}F(L - at) - \frac{1}{2a}\int_0^{L-at} G(s)\,ds + \frac{1}{2}F(L + at) + \frac{1}{2a}\int_0^{L+at} G(s)\,ds$$

$$= \frac{1}{2}F(-L - at) - \frac{1}{2a}\int_0^{L} G(s)\,ds - \frac{1}{2a}\int_0^{-at} G(s)\,ds +$$

$$+ \frac{1}{2}F(L + at) + \frac{1}{2a}\int_0^{L} G(s)\,ds + \frac{1}{2a}\int_0^{at} G(s)\,ds$$

$$= \frac{-1}{2}F(L + at) - \frac{1}{2a}\int_0^{at} G(s)\,ds + \frac{1}{2}F(L + at) + \frac{1}{2a}\int_0^{at} G(s)\,ds = 0.$$

And voilà, it works.

Example 4.8.1: D'Alembert says that the solution is a superposition of two functions (waves) moving in the opposite direction at "speed" a. To get an idea of how it works, let us work out an example. Consider the simpler setup

$$y_{tt} = y_{xx},$$
$$y(0, t) = y(1, t) = 0,$$
$$y(x, 0) = f(x),$$
$$y_t(x, 0) = 0.$$

Here $f(x)$ is an impulse of height 1 centered at $x = 0.5$:

$$f(x) = \begin{cases} 0 & \text{if} \quad 0 \le x < 0.45, \\ 20\,(x - 0.45) & \text{if} \quad 0.45 \le x < 0.5, \\ 20\,(0.55 - x) & \text{if} \quad 0.5 \le x < 0.55, \\ 0 & \text{if} \quad 0.55 \le x \le 1. \end{cases}$$

The graph of this impulse is the top left plot in Figure 4.21 on the next page.

Let $F(x)$ be the odd periodic extension of $f(x)$. Then from (4.16) we know that the solution is given as

$$y(x, t) = \frac{F(x - t) + F(x + t)}{2}.$$

It is not hard to compute specific values of $y(x, t)$. For example, to compute $y(0.1, 0.6)$ we notice $x - t = -0.5$ and $x + t = 0.7$. Now $F(-0.5) = -f(0.5) = -20\,(0.55 - 0.5) = -1$ and $F(0.7) = f(0.7) = 0$. Hence $y(0.1, 0.6) = \frac{-1+0}{2} = -0.5$. As you can see the d'Alembert solution is much easier to actually compute and to plot than the Fourier series solution. See Figure 4.21 for plots of the solution y for several different t.

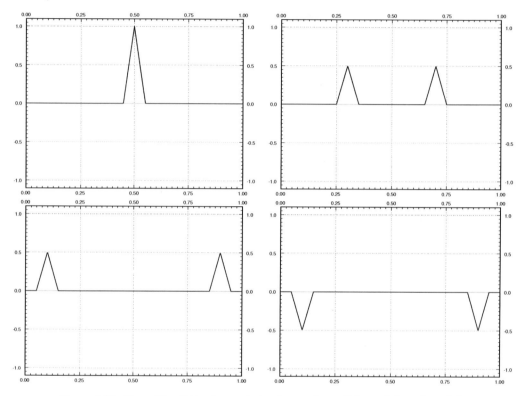

Figure 4.21: Plot of the d'Alembert solution for $t = 0$, $t = 0.2$, $t = 0.4$, and $t = 0.6$.

4.8.3 Another way to solve for the side conditions

It is perhaps easier and more useful to memorize the procedure rather than the formula itself. The important thing to remember is that a solution to the wave equation is a superposition of two waves traveling in opposite directions. That is,

$$y(x, t) = A(x - at) + B(x + at).$$

If you think about it, the exact formulas for A and B are not hard to guess once you realize what kind of side conditions $y(x, t)$ is supposed to satisfy. Let us give the formula again, but slightly differently. Best approach is to do this in stages. When $g(x) = 0$ (and hence $G(x) = 0$) we have the solution

$$\frac{F(x - at) + F(x + at)}{2}.$$

On the other hand, when $f(x) = 0$ (and hence $F(x) = 0$), we let

$$H(x) = \int_0^x G(s)\,ds.$$

The solution in this case is

$$\frac{1}{2a}\int_{x-at}^{x+at} G(s)\,ds = \frac{-H(x - at) + H(x + at)}{2a}.$$

By superposition we get a solution for the general side conditions (4.15) (when neither $f(x)$ nor $g(x)$ are identically zero).

$$y(x, t) = \frac{F(x - at) + F(x + at)}{2} + \frac{-H(x - at) + H(x + at)}{2a}. \tag{4.17}$$

Do note the minus sign before the H, and the a in the second denominator.

Exercise **4.8.1:** *Check that the new formula (4.17) satisfies the side conditions (4.15).*

Warning: Make sure you use the odd extensions $F(x)$ and $G(x)$, when you have formulas for $f(x)$ and $g(x)$. The thing is, those formulas in general hold only for $0 < x < L$, and are not usually equal to $F(x)$ and $G(x)$ for other x.

4.8.4   Exercises

Exercise **4.8.2:** *Using the d'Alembert solution solve* $y_{tt} = 4y_{xx}$, $0 < x < \pi$, $t > 0$, $y(0, t) = y(\pi, t) = 0$, $y(x, 0) = \sin x$, *and* $y_t(x, 0) = \sin x$. *Hint: Note that* $\sin x$ *is the odd extension of* $y(x, 0)$ *and* $y_t(x, 0)$.

Exercise **4.8.3:** *Using the d'Alembert solution solve* $y_{tt} = 2y_{xx}$, $0 < x < 1$, $t > 0$, $y(0, t) = y(1, t) = 0$, $y(x, 0) = \sin^5(\pi x)$, *and* $y_t(x, 0) = \sin^3(\pi x)$.

Exercise **4.8.4:** *Take* $y_{tt} = 4y_{xx}$, $0 < x < \pi$, $t > 0$, $y(0, t) = y(\pi, t) = 0$, $y(x, 0) = x(\pi - x)$, *and* $y_t(x, 0) = 0$. *a) Solve using the d'Alembert formula. Hint: You can use the sine series for* $y(x, 0)$. *b) Find the solution as a function of* x *for a fixed* $t = 0.5$, $t = 1$, *and* $t = 2$. *Do not use the sine series here.*

Exercise **4.8.5:** *Derive the d'Alembert solution for* $y_{tt} = a^2 y_{xx}$, $0 < x < \pi$, $t > 0$, $y(0, t) = y(\pi, t) = 0$, $y(x, 0) = f(x)$, *and* $y_t(x, 0) = 0$, *using the Fourier series solution of the wave equation, by applying an appropriate trigonometric identity.*

Exercise **4.8.6:** *The d'Alembert solution still works if there are no boundary conditions and the initial condition is defined on the whole real line. Suppose that $y_{tt} = y_{xx}$ (for all x on the real line and $t \geq 0$), $y(x, 0) = f(x)$, and $y_t(x, 0) = 0$, where*

$$f(x) = \begin{cases} 0 & \text{if} & x < -1, \\ x + 1 & \text{if} & -1 \leq x < 0, \\ -x + 1 & \text{if} & 0 \leq x < 1, \\ 0 & \text{if} & 1 < x. \end{cases}$$

Solve using the d'Alembert solution. That is, write down a piecewise definition for the solution. Then sketch the solution for $t = 0$, $t = \frac{1}{2}$, $t = 1$, and $t = 2$.

Exercise **4.8.101:** *Using the d'Alembert solution solve $y_{tt} = 9y_{xx}$, $0 < x < 1$, $t > 0$, $y(0, t) = y(1, t) = 0$, $y(x, 0) = \sin(2\pi x)$, and $y_t(x, 0) = \sin(3\pi x)$.*

Exercise **4.8.102:** *Take $y_{tt} = 4y_{xx}$, $0 < x < 1$, $t > 0$, $y(0, t) = y(1, t) = 0$, $y(x, 0) = x - x^2$, and $y_t(x, 0) = 0$. Using the D'Alembert solution find the solution at a) $t = 0.1$, b) $t = \frac{1}{2}$, c) $t = 1$. You may have to split your answer up by cases.*

Exercise **4.8.103:** *Take $y_{tt} = 100y_{xx}$, $0 < x < 4$, $t > 0$, $y(0, t) = y(4, t) = 0$, $y(x, 0) = F(x)$, and $y_t(x, 0) = 0$. Suppose that $F(0) = 0$, $F(1) = 2$, $F(2) = 3$, $F(3) = 1$. Using the D'Alembert solution find a) $y(1, 1)$, b) $y(4, 3)$, c) $y(3, 9)$.*

4.9 Steady state temperature and the Laplacian

Note: 1 lecture, §9.7 in [EP], §10.8 in [BD]

Suppose we have an insulated wire, a plate, or a 3-dimensional object. We apply certain fixed temperatures on the ends of the wire, the edges of the plate, or on all sides of the 3-dimensional object. We wish to find out what is the *steady state temperature* distribution. That is, we wish to know what will be the temperature after long enough period of time.

We are really looking for a solution to the heat equation that is not dependent on time. Let us first solve the problem in one space variable. We are looking for a function u that satisfies

$$u_t = ku_{xx},$$

but such that $u_t = 0$ for all x and t. Hence, we are looking for a function of x alone that satisfies $u_{xx} = 0$. It is easy to solve this equation by integration and we see that $u = Ax + B$ for some constants A and B.

Suppose we have an insulated wire, and we apply constant temperature T_1 at one end (say where $x = 0$) and T_2 on the other end (at $x = L$ where L is the length of the wire). Then our steady state solution is

$$u(x) = \frac{T_2 - T_1}{L}x + T_1.$$

This solution agrees with our common sense intuition with how the heat should be distributed in the wire. So in one dimension, the steady state solutions are basically just straight lines.

Things are more complicated in two or more space dimensions. Let us restrict to two space dimensions for simplicity. The heat equation in two space variables is

$$u_t = k(u_{xx} + u_{yy}), \tag{4.18}$$

or more commonly written as $u_t = k\Delta u$ or $u_t = k\nabla^2 u$. Here the Δ and ∇^2 symbols mean $\frac{\partial^2}{\partial x^2} + \frac{\partial^2}{\partial y^2}$. We will use Δ from now on. The reason for using such a notation is that you can define Δ to be the right thing for any number of space dimensions and then the heat equation is always $u_t = k\Delta u$. The operator Δ is called the *Laplacian*.

OK, now that we have notation out of the way, let us see what does an equation for the steady state solution look like. We are looking for a solution to (4.18) that does not depend on t, or in other words $u_t = 0$. Hence we are looking for a function $u(x, y)$ such that

$$\boxed{\Delta u = u_{xx} + u_{yy} = 0.}$$

This equation is called the *Laplace equation*[*]. Solutions to the Laplace equation are called *harmonic functions* and have many nice properties and applications far beyond the steady state heat problem.

Harmonic functions in two variables are no longer just linear (plane graphs). For example, you can check that the functions $x^2 - y^2$ and xy are harmonic. However, if you remember your

---

[*]Named after the French mathematician Pierre-Simon, marquis de Laplace (1749–1827).

multi-variable calculus we note that if u_{xx} is positive, u is concave up in the x direction, then u_{yy} must be negative and u must be concave down in the y direction. Therefore, a harmonic function can never have any "hilltop" or "valley" on the graph. This observation is consistent with our intuitive idea of steady state heat distribution; the hottest or coldest spot will not be inside.

Commonly the Laplace equation is part of a so-called *Dirichlet problem*[*]. That is, we have a region in the xy-plane and we specify certain values along the boundaries of the region. We then try to find a solution u defined on this region such that u agrees with the values we specified on the boundary.

For simplicity, we consider a rectangular region. Also for simplicity we specify boundary values to be zero at 3 of the four edges and only specify an arbitrary function at one edge. As we still have the principle of superposition, we can use this simpler solution to derive the general solution for arbitrary boundary values by solving 4 different problems, one for each edge, and adding those solutions together. This setup is left as an exercise.

We wish to solve the following problem. Let h and w be the height and width of our rectangle, with one corner at the origin and lying in the first quadrant.

$$\Delta u = 0, \tag{4.19}$$
$$u(0, y) = 0 \quad \text{for } 0 < y < h, \tag{4.20}$$
$$u(x, h) = 0 \quad \text{for } 0 < x < w, \tag{4.21}$$
$$u(w, y) = 0 \quad \text{for } 0 < y < h, \tag{4.22}$$
$$u(x, 0) = f(x) \quad \text{for } 0 < x < w. \tag{4.23}$$

The method we apply is separation of variables. Again, we will come up with enough building-block solutions satisfying all the homogeneous boundary conditions (all conditions except (4.23)). We notice that superposition still works for the equation and all the homogeneous conditions. Therefore, we can use the Fourier series for $f(x)$ to solve the problem as before.

We try $u(x, y) = X(x)Y(y)$. We plug u into the equation to get

$$X''Y + XY'' = 0.$$

We put the Xs on one side and the Ys on the other to get

$$-\frac{X''}{X} = \frac{Y''}{Y}.$$

The left hand side only depends on x and the right hand side only depends on y. Therefore, there is some constant λ such that $\lambda = \frac{-X''}{X} = \frac{Y''}{Y}$. And we get two equations

$$X'' + \lambda X = 0,$$
$$Y'' - \lambda Y = 0.$$

---

[*]Named after the German mathematician Johann Peter Gustav Lejeune Dirichlet (1805–1859).

Furthermore, the homogeneous boundary conditions imply that $X(0) = X(w) = 0$ and $Y(h) = 0$. Taking the equation for X we have already seen that we have a nontrivial solution if and only if $\lambda = \lambda_n = \frac{n^2 \pi^2}{w^2}$ and the solution is a multiple of

$$X_n(x) = \sin\left(\frac{n\pi}{w}x\right).$$

For these given λ_n, the general solution for Y (one for each n) is

$$Y_n(y) = A_n \cosh\left(\frac{n\pi}{w}y\right) + B_n \sinh\left(\frac{n\pi}{w}y\right). \tag{4.24}$$

We only have one condition on Y_n and hence we can pick one of A_n or B_n to be something convenient. It will be useful to have $Y_n(0) = 1$, so we let $A_n = 1$. Setting $Y_n(h) = 0$ and solving for B_n we get that

$$B_n = \frac{-\cosh\left(\frac{n\pi h}{w}\right)}{\sinh\left(\frac{n\pi h}{w}\right)}.$$

After we plug the A_n and B_n we into (4.24) and simplify, we find

$$Y_n(y) = \frac{\sinh\left(\frac{n\pi(h-y)}{w}\right)}{\sinh\left(\frac{n\pi h}{w}\right)}.$$

We define $u_n(x, y) = X_n(x)Y_n(y)$. And note that u_n satisfies (4.19)–(4.22).

Observe that

$$u_n(x, 0) = X_n(x)Y_n(0) = \sin\left(\frac{n\pi}{w}x\right).$$

Suppose

$$f(x) = \sum_{n=1}^{\infty} b_n \sin\left(\frac{n\pi x}{w}\right).$$

Then we get a solution of (4.19)–(4.23) of the following form.

$$u(x, y) = \sum_{n=1}^{\infty} b_n u_n(x, y) = \sum_{n=1}^{\infty} b_n \sin\left(\frac{n\pi}{w}x\right)\left(\frac{\sinh\left(\frac{n\pi(h-y)}{w}\right)}{\sinh\left(\frac{n\pi h}{w}\right)}\right).$$

As u_n satisfies (4.19)–(4.22) and any linear combination (finite or infinite) of u_n must also satisfy (4.19)–(4.22), we see that u must satisfy (4.19)–(4.22). By plugging in $y = 0$ it is easy to see that u satisfies (4.23) as well.

Example 4.9.1: Suppose that we take $w = h = \pi$ and we let $f(x) = \pi$. We compute the sine series for the function π (we will get the square wave). We find that for $0 < x < \pi$ we have

$$f(x) = \sum_{\substack{n=1 \\ n \text{ odd}}}^{\infty} \frac{4}{n} \sin(nx).$$

Therefore the solution $u(x, y)$, see Figure 4.22, to the corresponding Dirichlet problem is given as

$$u(x, y) = \sum_{\substack{n=1 \\ n \text{ odd}}}^{\infty} \frac{4}{n} \sin(nx) \left(\frac{\sinh(n(\pi - y))}{\sinh(n\pi)} \right).$$

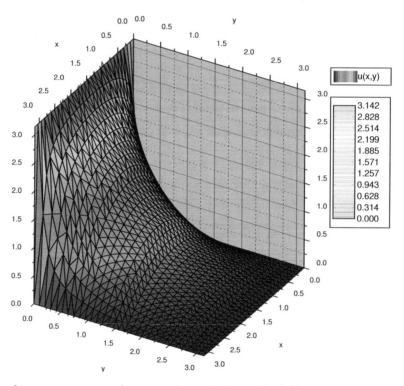

Figure 4.22: Steady state temperature of a square plate with three sides held at zero and one side held at π.

This scenario corresponds to the steady state temperature on a square plate of width π with 3 sides held at 0 degrees and one side held at π degrees. If we have arbitrary initial data on all sides, then we solve four problems, each using one piece of nonhomogeneous data. Then we use the principle of superposition to add up all four solutions to have a solution to the original problem.

A different way to visualize solutions of the Laplace equation is to take a wire and bend it so that it corresponds to the graph of the temperature above the boundary of your region. Cut a rubber sheet in the shape of your region—a square in our case—and stretch it fixing the edges of the sheet to the wire. The rubber sheet is a good approximation of the graph of the solution to the Laplace equation with the given boundary data.

4.9.1 Exercises

Exercise 4.9.1: *Let R be the region described by* $0 < x < \pi$ *and* $0 < y < \pi$. *Solve the problem*

$$\Delta u = 0, \quad u(x, 0) = \sin x, \quad u(x, \pi) = 0, \quad u(0, y) = 0, \quad u(\pi, y) = 0.$$

Exercise 4.9.2: *Let R be the region described by* $0 < x < 1$ *and* $0 < y < 1$. *Solve the problem*

$$u_{xx} + u_{yy} = 0,$$
$$u(x, 0) = \sin(\pi x) - \sin(2\pi x), \quad u(x, 1) = 0,$$
$$u(0, y) = 0, \quad u(1, y) = 0.$$

Exercise 4.9.3: *Let R be the region described by* $0 < x < 1$ *and* $0 < y < 1$. *Solve the problem*

$$u_{xx} + u_{yy} = 0,$$
$$u(x, 0) = u(x, 1) = u(0, y) = u(1, y) = C.$$

for some constant C. Hint: Guess, then check your intuition.

Exercise 4.9.4: *Let R be the region described by* $0 < x < \pi$ *and* $0 < y < \pi$. *Solve*

$$\Delta u = 0, \quad u(x, 0) = 0, \quad u(x, \pi) = \pi, \quad u(0, y) = y, \quad u(\pi, y) = y.$$

Hint: Try a solution of the form $u(x, y) = X(x) + Y(y)$ *(different separation of variables).*

Exercise 4.9.5: *Use the solution of Exercise 4.9.4 to solve*

$$\Delta u = 0, \quad u(x, 0) = \sin x, \quad u(x, \pi) = \pi, \quad u(0, y) = y, \quad u(\pi, y) = y.$$

Hint: Use superposition.

Exercise 4.9.6: *Let R be the region described by* $0 < x < w$ *and* $0 < y < h$. *Solve the problem*

$$u_{xx} + u_{yy} = 0,$$
$$u(x, 0) = 0, \quad u(x, h) = f(x),$$
$$u(0, y) = 0, \quad u(w, y) = 0.$$

The solution should be in series form using the Fourier series coefficients of $f(x)$.

Exercise 4.9.7: *Let R be the region described by* $0 < x < w$ *and* $0 < y < h$. *Solve the problem*

$$u_{xx} + u_{yy} = 0,$$
$$u(x, 0) = 0, \quad u(x, h) = 0,$$
$$u(0, y) = f(y), \quad u(w, y) = 0.$$

The solution should be in series form using the Fourier series coefficients of $f(y)$.

Exercise **4.9.8:** *Let R be the region described by* $0 < x < w$ *and* $0 < y < h$. *Solve the problem*

$$u_{xx} + u_{yy} = 0,$$
$$u(x, 0) = 0, \quad u(x, h) = 0,$$
$$u(0, y) = 0, \quad u(w, y) = f(y).$$

The solution should be in series form using the Fourier series coefficients of $f(y)$.

Exercise **4.9.9:** *Let R be the region described by* $0 < x < 1$ *and* $0 < y < 1$. *Solve the problem*

$$u_{xx} + u_{yy} = 0,$$
$$u(x, 0) = \sin(9\pi x), \quad u(x, 1) = \sin(2\pi x),$$
$$u(0, y) = 0, \quad u(1, y) = 0.$$

Hint: Use superposition.

Exercise **4.9.10:** *Let R be the region described by* $0 < x < 1$ *and* $0 < y < 1$. *Solve the problem*

$$u_{xx} + u_{yy} = 0,$$
$$u(x, 0) = \sin(\pi x), \quad u(x, 1) = \sin(\pi x),$$
$$u(0, y) = \sin(\pi y), \quad u(1, y) = \sin(\pi y).$$

Hint: Use superposition.

Exercise **4.9.11** (challenging)*: Using only your intuition find* $u(1/2, 1/2)$, *for the problem* $\Delta u = 0$, *where* $u(0, y) = u(1, y) = 100$ *for* $0 < y < 1$, *and* $u(x, 0) = u(x, 1) = 0$ *for* $0 < x < 1$. *Explain.*

Exercise **4.9.101:** *Let R be the region described by* $0 < x < 1$ *and* $0 < y < 1$. *Solve the problem*

$$\Delta u = 0, \quad u(x, 0) = \sum_{n=1}^{\infty} \frac{1}{n^2} \sin(n\pi x), \quad u(x, 1) = 0, \quad u(0, y) = 0, \quad u(1, y) = 0.$$

Exercise **4.9.102:** *Let R be the region described by* $0 < x < 1$ *and* $0 < y < 2$. *Solve the problem*

$$\Delta u = 0, \quad u(x, 0) = 0.1 \sin(\pi x), \quad u(x, 2) = 0, \quad u(0, y) = 0, \quad u(1, y) = 0.$$

4.10 Dirichlet problem in the circle and the Poisson kernel

Note: 2 lectures, §9.7 in [EP], §10.8 in [BD]

4.10.1 Laplace in polar coordinates

A more natural setting for the Laplace equation $\Delta u = 0$ is the circle rather than the square. On the other hand, what makes the problem somewhat more difficult is that we need polar coordinates.

Recall that the polar coordinates for the (x, y)-plane are (r, θ):

$$x = r \cos \theta, \quad y = r \sin \theta,$$

where $r \geq 0$ and $-\pi < \theta \leq \pi$. So (x, y) is distance r from the origin at angle θ.

Now that we know our coordinates, let us give the problem we wish to solve. We have a circular region of radius 1, and we are interested in the Dirichlet problem for the Laplace equation for this region. Let $u(r, \theta)$ denote the temperature at the point (r, θ) in polar coordinates. We have the problem:

$$\begin{aligned} \Delta u &= 0, & \text{for } r < 1, \\ u(1, \theta) &= g(\theta), & \text{for } -\pi < \theta \leq \pi. \end{aligned} \tag{4.25}$$

The first issue we face is that we do not know what the Laplacian is in polar coordinates. Normally we would find u_{xx} and u_{yy} in terms of the derivatives in r and θ. We would need to solve for r and θ in terms of x and y. While this is certainly possible, it happens to be more convenient to work in reverse. Let us instead compute derivatives in r and θ in terms of derivatives in x and y and then solve. The computations are easier this way. First

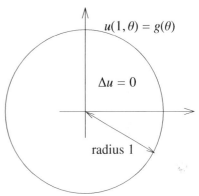

$$x_r = \cos \theta, \qquad x_\theta = -r \sin \theta, \qquad y_r = \sin \theta, \qquad y_\theta = r \cos \theta.$$

Next by chain rule we obtain

$$u_r = u_x x_r + u_y y_r = \cos(\theta) u_x + \sin(\theta) u_y,$$
$$u_{rr} = \cos(\theta)(u_{xx} x_r + u_{xy} y_r) + \sin(\theta)(u_{yx} x_r + u_{yy} y_r) = \cos^2(\theta) u_{xx} + 2 \cos(\theta) \sin(\theta) u_{xy} + \sin^2(\theta) u_{yy}.$$

Similarly for the θ derivative. Note that we have to use product rule for the second derivative.

$$u_\theta = u_x x_\theta + u_y y_\theta = -r \sin(\theta) u_x + r \cos(\theta) u_y,$$
$$u_{\theta\theta} = -r \cos(\theta) u_x - r \sin(\theta)(u_{xx} x_\theta + u_{xy} y_\theta) - r \sin(\theta) u_y + r \cos(\theta)(u_{yx} x_\theta + u_{yy} y_\theta)$$
$$= -r \cos(\theta) u_x - r \sin(\theta) u_y + r^2 \sin^2(\theta) u_{xx} - r^2 2 \sin(\theta) \cos(\theta) u_{xy} + r^2 \cos^2(\theta) u_{yy}.$$

Let us now try to solve for $u_{xx} + u_{yy}$. We start with $\frac{1}{r^2}u_{\theta\theta}$ to get rid of those pesky r^2. If we add u_{rr} and use the fact that $\cos^2(\theta) + \sin^2(\theta) = 1$, we get

$$\frac{1}{r^2}u_{\theta\theta} + u_{rr} = u_{xx} + u_{yy} - \frac{1}{r}\cos(\theta)u_x - \frac{1}{r}\sin(\theta)u_y.$$

We're not quite there yet, but all we are lacking is $\frac{1}{r}u_r$. Adding it we obtain the *Laplacian in polar coordinates*:

$$\boxed{\frac{1}{r^2}u_{\theta\theta} + \frac{1}{r}u_r + u_{rr} = u_{xx} + u_{yy} = \Delta u.}$$

Notice that the Laplacian in polar coordinates no longer has constant coefficients.

4.10.2   Series solution

Let us separate variables as usual. That is let us try $u(r, \theta) = R(r)\Theta(\theta)$. Then

$$0 = \Delta u = \frac{1}{r^2}R\Theta'' + \frac{1}{r}R'\Theta + R''\Theta.$$

Let us put R on one side and Θ on the other and conclude that both sides must be constant.

$$\frac{1}{r^2}R\Theta'' = -\left(\frac{1}{r}R' + R''\right)\Theta.$$

$$\frac{\Theta''}{\Theta} = -\frac{rR' + r^2R''}{R} = -\lambda.$$

We get two equations:

$$\Theta'' + \lambda\Theta = 0,$$
$$r^2R'' + rR' - \lambda R = 0.$$

Let us first focus on Θ. We know that $u(r, \theta)$ ought to be 2π-periodic in θ, that is, $u(r, \theta) = u(r, \theta+2\pi)$. Therefore, the solution to $\Theta'' + \lambda\Theta = 0$ must be 2π-periodic. We conclude that $\lambda = n^2$ for a nonnegative integer $n = 0, 1, 2, 3, \ldots$. The equation becomes $\Theta'' + n^2\Theta = 0$. When $n = 0$ the equation is just $\Theta'' = 0$, so we have the general solution $A\theta + B$. As Θ is periodic, $A = 0$. For convenience let us write this solution as

$$\Theta_0 = \frac{a_0}{2}$$

for some constant a_0. For positive n, the solution to $\Theta'' + n^2\Theta = 0$ is

$$\Theta_n = a_n\cos(n\theta) + b_n\sin(n\theta),$$

for some constants a_n and b_n.

Next, we consider the equation for R,

$$r^2 R'' + r R' - n^2 R = 0.$$

This equation appeared in exercises before—we solved it in Exercise 2.1.6 and Exercise 2.1.7 on page 66. The idea is to try a solution r^s and if that does not work out try a solution of the form $r^s \ln r$. When $n = 0$ we obtain

$$R_0 = A r^0 + B r^0 \ln r = A + B \ln r,$$

and if $n > 0$, we get

$$R_n = A r^n + B r^{-n}.$$

The function $u(r, \theta)$ must be finite at the origin, that is, when $r = 0$. Therefore, $B = 0$ in both cases. Let us set $A = 1$ in both cases as well, the constants in Θ_n will pick up the slack so we do not lose anything. Therefore let

$$R_0 = 1, \qquad \text{and} \qquad R_n = r^n.$$

Hence our building block solutions are

$$u_0(r, \theta) = \frac{a_0}{2}, \qquad\qquad u_n(r, \theta) = a_n r^n \cos(n\theta) + b_n r^n \sin(n\theta).$$

Putting everything together our solution is:

$$\boxed{u(r, \theta) = \frac{a_0}{2} + \sum_{n=1}^{\infty} a_n r^n \cos(n\theta) + b_n r^n \sin(n\theta).}$$

We look at the boundary condition in (4.25),

$$g(\theta) = u(1, \theta) = \frac{a_0}{2} + \sum_{n=1}^{\infty} a_n \cos(n\theta) + b_n \sin(n\theta).$$

Therefore, the solution to (4.25) is to expand $g(\theta)$, which is a 2π-periodic function, as a Fourier series, and then the nth coordinate is multiplied by r^n. In other words, to compute a_n and b_n from the formula we can, as usual, compute

$$a_n = \frac{1}{\pi} \int_{-\pi}^{\pi} g(\theta) \cos(n\theta) \, d\theta, \qquad \text{and} \qquad b_n = \frac{1}{\pi} \int_{-\pi}^{\pi} g(\theta) \sin(n\theta) \, d\theta.$$

Example 4.10.1: Suppose we wish to solve

$$\Delta u = 0, \qquad 0 \le r < 1, \qquad -\pi < \theta \le \pi,$$
$$u(1, \theta) = \cos(10\,\theta), \qquad -\pi < \theta \le \pi.$$

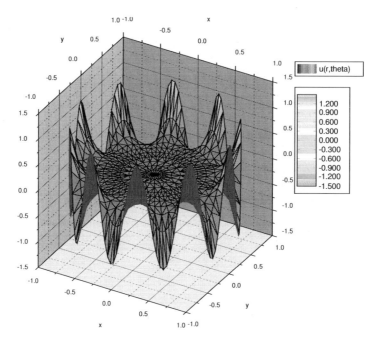

Figure 4.23: The solution of the Dirichlet problem in the disc with $\cos(10\,\theta)$ *as boundary data.*

The solution is

$$u(r,\theta) = r^{10}\cos(10\,\theta).$$

See the plot in Figure 4.23. The thing to notice in this example is that the effect of a high frequency is mostly felt at the boundary. In the middle of the disc, the solution is very close to zero. That is because r^{10} rather small when r is close to 0.

Example 4.10.2: Let us solve a more difficult problem. Suppose we have a long rod with circular cross section of radius 1 and we wish to solve the steady state heat problem. If the rod is long enough we simply need to solve the Laplace equation in two dimensions. Let us put the center of the rod at the origin and we have exactly the region we are currently studying—a circle of radius 1. For the boundary conditions, suppose in Cartesian coordinates x and y, the temperature is fixed at 0 when $y < 0$ and at $2y$ when $y > 0$.

We set the problem up. As $y = r\sin(\theta)$, then on the circle of radius 1 we have $2y = 2\sin(\theta)$. So

$$\Delta u = 0, \qquad 0 \le r < 1, \quad -\pi < \theta \le \pi,$$

$$u(1,\theta) = \begin{cases} 2\sin(\theta) & \text{if} \quad 0 \le \theta \le \pi, \\ 0 & \text{if} \quad -\pi < \theta < 0. \end{cases}$$

We must now compute the Fourier series for the boundary condition. By now the reader has plentiful experience in computing Fourier series and so we simply state that

$$u(1, \theta) = \frac{2}{\pi} + \sin(\theta) + \sum_{n=1}^{\infty} \frac{-4}{\pi(4n^2 - 1)} \cos(2n\theta).$$

Exercise 4.10.1: *Compute the series for $u(1, \theta)$ and verify that it really is what we have just claimed. Hint: Be careful, make sure not to divide by zero.*

We now simply write the solution (see Figure 4.24) by multiplying by r^n in the right places.

$$u(r, \theta) = \frac{2}{\pi} + r\sin(\theta) + \sum_{n=1}^{\infty} \frac{-4r^{2n}}{\pi(4n^2 - 1)} \cos(2n\theta).$$

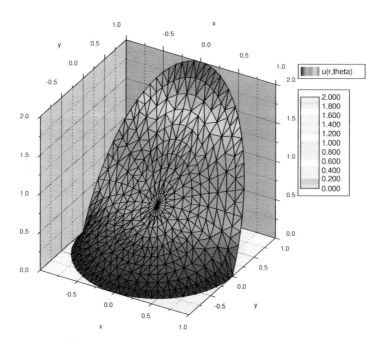

Figure 4.24: The solution of the Dirichlet problem with boundary data 0 for $y < 0$ and $2y$ for $y > 0$.

4.10.3   Poisson kernel

There is another way to solve the Dirichlet problem with the help of an integral kernel. That is, we will find a function $P(r, \theta, \alpha)$ called the *Poisson kernel*[*] such that

$$u(r, \theta) = \frac{1}{2\pi} \int_{-\pi}^{\pi} P(r, \theta, \alpha)\, g(\alpha)\, d\alpha.$$

While the integral will generally not be solvable analytically, it can be evaluated numerically. In fact, unless the boundary data is given as a Fourier series already, it may be much easier to numerically evaluate this formula as there is only one integral to evaluate.

The formula also has theoretical applications. For instance, as $P(r, \theta, \alpha)$ will have infinitely many derivatives, then via differentiating under the integral we find that the solution $u(r, \theta)$ has infinitely many derivatives, at least when inside the circle, $r < 1$. By infinitely many derivatives what you should think of is that $u(r, \theta)$ has "no corners" and all of its partial derivatives of all orders exist and also have "no corners."

We will compute the formula for $P(r, \theta, \alpha)$ from the series solution, and this idea can be applied anytime you have a convenient series solution where the coefficients are obtained via integration. Hence you can apply this reasoning to obtain such integral kernels for other equations, such as the heat equation. The computation is long and tedious, but not overly difficult. Since the ideas are often applied in similar contexts, it is good to understand how this computation works.

What we do is start with the series solution and replace the coefficients with the integrals that compute them. Then we try to write everything as a single integral. We must use a different dummy variable for the integration and hence we use α instead of θ.

$$
\begin{aligned}
u(r, \theta) &= \frac{a_0}{2} + \sum_{n=1}^{\infty} a_n r^n \cos(n\theta) + b_n r^n \sin(n\theta) \\
&= \frac{1}{2\pi} \int_{-\pi}^{\pi} g(\alpha)\, d\alpha \\
&\quad + \sum_{n=1}^{\infty} \left(\frac{1}{\pi} \int_{-\pi}^{\pi} g(\alpha) \cos(n\alpha)\, d\alpha \right) r^n \cos(n\theta) + \left(\frac{1}{\pi} \int_{-\pi}^{\pi} g(\alpha) \sin(n\alpha)\, d\alpha \right) r^n \sin(n\theta) \\
&= \frac{1}{2\pi} \int_{-\pi}^{\pi} \left[g(\alpha) + 2 \sum_{n=1}^{\infty} g(\alpha) \cos(n\alpha) r^n \cos(n\theta) + g(\alpha) \sin(n\alpha) r^n \sin(n\theta) \right] d\alpha \\
&= \frac{1}{2\pi} \int_{-\pi}^{\pi} \left[1 + 2 \sum_{n=1}^{\infty} r^n (\cos(n\alpha) \cos(n\theta) + \sin(n\alpha) \sin(n\theta)) \right] g(\alpha)\, d\alpha
\end{aligned}
$$

OK, so we have what we wanted, the expression in the parentheses is the Poisson kernel, $P(r, \theta, \alpha)$. However, we can do a lot better. It is still given as a series, and we would really like to have a nice

---

[*]Named for the French mathematician Siméon Denis Poisson (1781–1840).

simple expression for it. We must work a little harder. The trick is to rewrite everything in terms of complex exponentials. Let us work just on the kernel.

$$P(r, \theta, \alpha) = 1 + 2 \sum_{n=1}^{\infty} r^n (\cos(n\alpha) \cos(n\theta) + \sin(n\alpha) \sin(n\theta))$$

$$= 1 + 2 \sum_{n=1}^{\infty} r^n \cos(n(\theta - \alpha))$$

$$= 1 + \sum_{n=1}^{\infty} r^n (e^{in(\theta - \alpha)} + e^{-in(\theta - \alpha)})$$

$$= 1 + \sum_{n=1}^{\infty} (re^{i(\theta - \alpha)})^n + \sum_{n=1}^{\infty} (re^{-i(\theta - \alpha)})^n.$$

In the above expression we recognize the *geometric series*. That is, recall from calculus that as long as $|z| < 1$, then

$$\sum_{n=1}^{\infty} z^n = \frac{z}{1 - z}.$$

Note that n starts at 1 and that is why we have the z in the numerator. It is the standard geometric series multiplied by z. Let us continue with the computation.

$$P(r, \theta, \alpha) = 1 + \sum_{n=1}^{\infty} (re^{i(\theta - \alpha)})^n + \sum_{n=1}^{\infty} (re^{-i(\theta - \alpha)})^n$$

$$= 1 + \frac{re^{i(\theta - \alpha)}}{1 - re^{i(\theta - \alpha)}} + \frac{re^{-i(\theta - \alpha)}}{1 - re^{-i(\theta - \alpha)}}$$

$$= \frac{(1 - re^{i(\theta - \alpha)})(1 - re^{-i(\theta - \alpha)}) + (1 - re^{-i(\theta - \alpha)})re^{i(\theta - \alpha)} + (1 - re^{i(\theta - \alpha)})re^{-i(\theta - \alpha)}}{(1 - re^{i(\theta - \alpha)})(1 - re^{-i(\theta - \alpha)})}$$

$$= \frac{1 - r^2}{1 - re^{i(\theta - \alpha)} - re^{-i(\theta - \alpha)} + r^2}$$

$$= \frac{1 - r^2}{1 - 2r \cos(\theta - \alpha) + r^2}.$$

Now that's a formula we can live with. The solution to the Dirichlet problem using the Poisson kernel is

$$\boxed{u(r, \theta) = \frac{1}{2\pi} \int_{-\pi}^{\pi} \frac{1 - r^2}{1 - 2r \cos(\theta - \alpha) + r^2} g(\alpha) \, d\alpha.}$$

Sometimes the formula for the Poisson kernel is given together with the constant $\frac{1}{2\pi}$, in which case we should of course not leave it in front of the integral. Also, often the limits of the integral are given as 0 to 2π; everything inside is 2π-periodic in α, so this does not change the integral.

Let us not leave the Poisson kernel without explaining its geometric meaning. Let s be the distance from (r, θ) to $(1, \alpha)$. You may recall from calculus that this distance s in polar coordinates is given precisely by the square root of $1 - 2r\cos(\theta - \alpha) + r^2$. That is, the Poisson kernel is really the formula

$$\frac{1 - r^2}{s^2}.$$

One final note we make about the formula is that it is really a weighted average of the boundary values. First let us look at what happens at the origin, that is when $r = 0$.

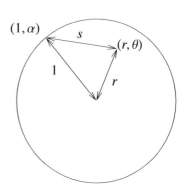

$$u(0, 0) = \frac{1}{2\pi} \int_{-\pi}^{\pi} \frac{1 - 0^2}{1 - 2(0)\cos(\theta - \alpha) + 0^2} g(\alpha)\, d\alpha$$

$$= \frac{1}{2\pi} \int_{-\pi}^{\pi} g(\alpha)\, d\alpha.$$

So $u(0, 0)$ is precisely the average value of $g(\theta)$ and therefore the average value of u on the boundary. This is a general feature of harmonic functions, the value at some point p is equal to the average of the values on a circle centered at p.

What the formula says is that the value of the solution at any point in the circle is a weighted average of the boundary data $g(\theta)$. The kernel is bigger when $(1, \alpha)$ is closer to (r, θ). Therefore when computing $u(r, \theta)$ we give more weight to the values $g(\alpha)$ when $(1, \alpha)$ is closer to (r, θ) and less weight to the values $g(\alpha)$ when $(1, \alpha)$ far from (r, θ).

4.10.4   Exercises

Exercise 4.10.2: *Using series solve $\Delta u = 0$, $u(1, \theta) = |\theta|$, for $-\pi < \theta \leq \pi$.*

Exercise 4.10.3: *Using series solve $\Delta u = 0$, $u(1, \theta) = g(\theta)$ for the following data. Hint: trig identities.*

  a) $g(\theta) = 1/2 + 3\sin(\theta) + \cos(3\theta)$ *b) $g(\theta) = 3\cos(3\theta) + 3\sin(3\theta) + \sin(9\theta)$*
  c) $g(\theta) = 2\cos(\theta + 1)$ *d) $g(\theta) = \sin^2(\theta)$*

Exercise 4.10.4: *Using the Poisson kernel, give the solution to $\Delta u = 0$, where $u(1, \theta)$ is zero for θ outside the interval $[-\pi/4, \pi/4]$ and $u(1, \theta)$ is 1 for θ on the interval $[-\pi/4, \pi/4]$.*

Exercise 4.10.5: *a) Draw a graph for the Poisson kernel as a function of α when $r = 1/2$ and $\theta = 0$. b) Describe what happens to the graph when you make r bigger (as it approaches 1). c) Knowing that the solution $u(r, \theta)$ is the weighted average of $g(\theta)$ with Poisson kernel as the weight, explain what your answer to part b means.*

Exercise 4.10.6: *Take the function $g(\theta)$ to be the function $xy = \cos\theta \sin\theta$ on the boundary. Use the series solution to find a solution to the Dirichlet problem $\Delta u = 0$, $u(1, \theta) = g(\theta)$. Now convert the solution to Cartesian coordinates x and y. Is this solution surprising? Hint: use your trig identities.*

***Exercise* 4.10.7:** *Carry out the computation we needed in the separation of variables and solve* $r^2 R'' + rR' - n^2 R = 0$, *for* $n = 0, 1, 2, 3, \ldots$.

***Exercise* 4.10.8** (challenging)**:** *Derive the series solution to the Dirichlet problem if the region is a circle of radius ρ rather than 1. That is, solve $\Delta u = 0$, $u(\rho, \theta) = g(\theta)$.*

***Exercise* 4.10.9** (challenging)**:** *a) Find the solution for $\Delta u = 0$, $u(1, \theta) = x^2 y^3 + 5x^2$. Write the answer in Cartesian coordinates.*
b) Now solve $\Delta u = 0$, $u(1, \theta) = x^k y^\ell$. Write the solution in Cartesian coordinates.
c) Suppose you have a polynomial $P(x, y) = \sum_{j=0}^{m} \sum_{k=0}^{n} c_{j,k} x^j y^k$, solve $\Delta u = 0$, $u(1, \theta) = P(x, y)$ (that is, write down the formula for the answer). Write the answer in Cartesian coordinates.
Notice the answer is again a polynomial in x and y. See also Exercise 4.10.6.

***Exercise* 4.10.101:** *Using series solve $\Delta u = 0$, $u(1, \theta) = 1 + \sum_{n=1}^{\infty} \frac{1}{n^2} \sin(n\theta)$.*

***Exercise* 4.10.102:** *Using the series solution find the solution to $\Delta u = 0$, $u(1, \theta) = 1 - \cos(\theta)$. Express the solution in Cartesian coordinates (that is, using x and y).*

***Exercise* 4.10.103:** *a) Try and guess a solution to $\Delta u = -1$, $u(1, \theta) = 0$. Hint: try a solution that only depends on r. Also first, don't worry about the boundary condition. b) Now solve $\Delta u = -1$, $u(1, \theta) = \sin(2\theta)$ using superposition.*

***Exercise* 4.10.104** (challenging)**:** *Derive the Poisson kernel solution if the region is a circle of radius ρ rather than 1. That is, solve $\Delta u = 0$, $u(\rho, \theta) = g(\theta)$.*

Chapter 5

Eigenvalue problems

5.1 Sturm-Liouville problems

Note: 2 lectures, §10.1 in [EP], §11.2 in [BD]

5.1.1 Boundary value problems

We have encountered several different eigenvalue problems such as:

$$X''(x) + \lambda X(x) = 0$$

with different boundary conditions

$$
\begin{array}{llll}
X(0) = 0 & X(L) = 0 & \text{(Dirichlet) or,} \\
X'(0) = 0 & X'(L) = 0 & \text{(Neumann) or,} \\
X'(0) = 0 & X(L) = 0 & \text{(Mixed) or,} \\
X(0) = 0 & X'(L) = 0 & \text{(Mixed), \ldots}
\end{array}
$$

For example for the insulated wire, Dirichlet conditions correspond to applying a zero temperature at the ends, Neumann means insulating the ends, etc.... Other types of endpoint conditions also arise naturally, such as the *Robin boundary conditions*

$$hX(0) - X'(0) = 0 \qquad hX(L) + X'(L) = 0,$$

for some constant h. These conditions come up when the ends are immersed in some medium.

Boundary problems came up in the study of the heat equation $u_t = ku_{xx}$ when we were trying to solve the equation by the method of separation of variables. In the computation we encountered a certain eigenvalue problem and found the eigenfunctions $X_n(x)$. We then found the *eigenfunction decomposition* of the initial temperature $f(x) = u(x, 0)$ in terms of the eigenfunctions

$$f(x) = \sum_{n=1}^{\infty} c_n X_n(x).$$

Once we had this decomposition and found suitable $T_n(t)$ such that $T_n(0) = 1$ and $T_n(t)X(x)$ were solutions, the solution to the original problem including the initial condition could be written as

$$u(x,t) = \sum_{n=1}^{\infty} c_n T_n(t) X_n(x).$$

We will try to solve more general problems using this method. First, we will study second order linear equations of the form

$$\frac{d}{dx}\left(p(x)\frac{dy}{dx}\right) - q(x)y + \lambda r(x)y = 0. \tag{5.1}$$

Essentially any second order linear equation of the form $a(x)y'' + b(x)y' + c(x)y + \lambda d(x)y = 0$ can be written as (5.1) after multiplying by a proper factor.

Example 5.1.1: [Bessel] Put the following equation into the form (5.1):

$$x^2 y'' + xy' + \left(\lambda x^2 - n^2\right)y = 0.$$

Multiply both sides by $\frac{1}{x}$ to obtain

$$\frac{1}{x}\left(x^2 y'' + xy' + \left(\lambda x^2 - n^2\right)y\right) = xy'' + y' + \left(\lambda x - \frac{n^2}{x}\right)y = \frac{d}{dx}\left(x\frac{dy}{dx}\right) - \frac{n^2}{x}y + \lambda xy = 0.$$

The so-called *Sturm-Liouville problem*[*] is to seek nontrivial solutions to

$$\boxed{\begin{aligned} &\frac{d}{dx}\left(p(x)\frac{dy}{dx}\right) - q(x)y + \lambda r(x)y = 0, \qquad a < x < b, \\ &\alpha_1 y(a) - \alpha_2 y'(a) = 0, \\ &\beta_1 y(b) + \beta_2 y'(b) = 0. \end{aligned}} \tag{5.2}$$

In particular, we seek λs that allow for nontrivial solutions. The λs that admit nontrivial solutions are called the *eigenvalues* and the corresponding nontrivial solutions are called *eigenfunctions*. The constants α_1 and α_2 should not be both zero, same for β_1 and β_2.

Theorem 5.1.1. *Suppose $p(x)$, $p'(x)$, $q(x)$ and $r(x)$ are continuous on $[a,b]$ and suppose $p(x) > 0$ and $r(x) > 0$ for all x in $[a,b]$. Then the Sturm-Liouville problem (5.2) has an increasing sequence of eigenvalues*

$$\lambda_1 < \lambda_2 < \lambda_3 < \cdots$$

such that

$$\lim_{n \to \infty} \lambda_n = +\infty$$

and such that to each λ_n there is (up to a constant multiple) a single eigenfunction $y_n(x)$.
    Moreover, if $q(x) \geq 0$ and $\alpha_1, \alpha_2, \beta_1, \beta_2 \geq 0$, then $\lambda_n \geq 0$ for all n.

---

[*]Named after the French mathematicians Jacques Charles François Sturm (1803–1855) and Joseph Liouville (1809–1882).

Problems satisfying the hypothesis of the theorem are called *regular Sturm-Liouville problems* and we will only consider such problems here. That is, a regular problem is one where $p(x)$, $p'(x)$, $q(x)$ and $r(x)$ are continuous, $p(x) > 0$, $r(x) > 0$, $q(x) \geq 0$, and $\alpha_1, \alpha_2, \beta_1, \beta_2 \geq 0$. Note: Be careful about the signs. Also be careful about the inequalities for r and p, they must be strict for all x!

When zero is an eigenvalue, we usually start labeling the eigenvalues at 0 rather than at 1 for convenience.

Example 5.1.2: The problem $y'' + \lambda y$, $0 < x < L$, $y(0) = 0$, and $y(L) = 0$ is a regular Sturm-Liouville problem. $p(x) = 1$, $q(x) = 0$, $r(x) = 1$, and we have $p(x) = 1 > 0$ and $r(x) = 1 > 0$. The eigenvalues are $\lambda_n = \frac{n^2 \pi^2}{L^2}$ and eigenfunctions are $y_n(x) = \sin(\frac{n\pi}{L} x)$. All eigenvalues are nonnegative as predicted by the theorem.

Exercise 5.1.1: Find eigenvalues and eigenfunctions for

$$y'' + \lambda y = 0, \quad y'(0) = 0, \quad y'(1) = 0.$$

Identify the $p, q, r, \alpha_j, \beta_j$. Can you use the theorem to make the search for eigenvalues easier? (Hint: Consider the condition $-y'(0) = 0$)

Example 5.1.3: Find eigenvalues and eigenfunctions of the problem

$$y'' + \lambda y = 0, \quad 0 < x < 1,$$
$$hy(0) - y'(0) = 0, \quad y'(1) = 0, \quad h > 0.$$

These equations give a regular Sturm-Liouville problem.

Exercise 5.1.2: Identify $p, q, r, \alpha_j, \beta_j$ in the example above.

First note that $\lambda \geq 0$ by Theorem 5.1.1. Therefore, the general solution (without boundary conditions) is

$$y(x) = A \cos(\sqrt{\lambda}\, x) + B \sin(\sqrt{\lambda}\, x) \qquad \text{if } \lambda > 0,$$
$$y(x) = Ax + B \qquad\qquad\qquad\qquad\quad \text{if } \lambda = 0.$$

Let us see if $\lambda = 0$ is an eigenvalue: We must satisfy $0 = hB - A$ and $A = 0$, hence $B = 0$ (as $h > 0$), therefore, 0 is not an eigenvalue (no nonzero solution, so no eigenfunction).

Now let us try $\lambda > 0$. We plug in the boundary conditions.

$$0 = hA - \sqrt{\lambda}\, B,$$
$$0 = -A \sqrt{\lambda} \sin(\sqrt{\lambda}) + B \sqrt{\lambda} \cos(\sqrt{\lambda}).$$

If $A = 0$, then $B = 0$ and vice-versa, hence both are nonzero. So $B = \frac{hA}{\sqrt{\lambda}}$, and $0 = -A \sqrt{\lambda} \sin(\sqrt{\lambda}) + \frac{hA}{\sqrt{\lambda}} \sqrt{\lambda} \cos(\sqrt{\lambda})$. As $A \neq 0$ we get

$$0 = - \sqrt{\lambda} \sin(\sqrt{\lambda}) + h \cos(\sqrt{\lambda}),$$

or

$$\frac{h}{\sqrt{\lambda}} = \tan\ \sqrt{\lambda}.$$

Now use a computer to find λ_n. There are tables available, though using a computer or a graphing calculator is far more convenient nowadays. Easiest method is to plot the functions h/x and $\tan x$ and see for which x they intersect. There is an infinite number of intersections. Denote by $\sqrt{\lambda_1}$ the first intersection, by $\sqrt{\lambda_2}$ the second intersection, etc.... For example, when $h = 1$, we get that $\sqrt{\lambda_1} \approx 0.86$, $\sqrt{\lambda_2} \approx 3.43$, ... That is $\lambda_1 \approx 0.74$, $\lambda_2 \approx 11.73$, ... A plot for $h = 1$ is given in Figure 5.1. The appropriate eigenfunction (let $A = 1$ for convenience, then $B = h/\sqrt{\lambda}$) is

$$y_n(x) = \cos(\sqrt{\lambda_n}\ x) + \frac{h}{\sqrt{\lambda_n}}\ \sin(\sqrt{\lambda_n}\ x).$$

When $h = 1$ we get (approximately)

$$y_1(x) \approx \cos(0.86\ x) + \frac{1}{0.86}\ \sin(0.86\ x), \qquad y_2(x) \approx \cos(3.43\ x) + \frac{1}{3.43}\ \sin(3.43\ x), \qquad \ldots.$$

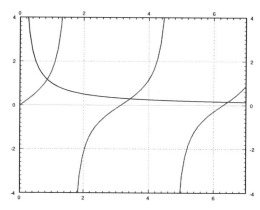

Figure 5.1: Plot of $\frac{1}{x}$ and $\tan x$.

5.1.2   Orthogonality

We have seen the notion of orthogonality before. For example, we have shown that $\sin(nx)$ are orthogonal for distinct n on $[0, \pi]$. For general Sturm-Liouville problems we will need a more general setup. Let $r(x)$ be a *weight function* (any function, though generally we will assume it is positive) on $[a, b]$. Two functions $f(x)$, $g(x)$ are said to be *orthogonal* with respect to the weight function $r(x)$ when

$$\int_a^b f(x) g(x) r(x) \, dx = 0.$$

In this setting, we define the *inner product* as

$$\langle f, g \rangle \stackrel{\text{def}}{=} \int_a^b f(x) g(x) r(x) \, dx,$$

and then say f and g are orthogonal whenever $\langle f, g \rangle = 0$. The results and concepts are again analogous to finite dimensional linear algebra.

The idea of the given inner product is that those x where $r(x)$ is greater have more weight. Nontrivial (nonconstant) $r(x)$ arise naturally, for example from a change of variables. Hence, you could think of a change of variables such that $d\xi = r(x) \, dx$.

We have the following orthogonality property of eigenfunctions of a regular Sturm-Liouville problem.

Theorem 5.1.2. *Suppose we have a regular Sturm-Liouville problem*

$$\frac{d}{dx} \left(p(x) \frac{dy}{dx} \right) - q(x) y + \lambda r(x) y = 0,$$

$$\alpha_1 y(a) - \alpha_2 y'(a) = 0,$$

$$\beta_1 y(b) + \beta_2 y'(b) = 0.$$

Let y_j and y_k be two distinct eigenfunctions for two distinct eigenvalues λ_j and λ_k. Then

$$\int_a^b y_j(x) \, y_k(x) \, r(x) \, dx = 0,$$

that is, y_j and y_k are orthogonal with respect to the weight function r.

Proof is very similar to the analogous theorem from § 4.1. It can also be found in many books including, for example, Edwards and Penney [EP].

5.1.3   Fredholm alternative

We also have the *Fredholm alternative* theorem we talked about before for all regular Sturm-Liouville problems. We state it here for completeness.

Theorem 5.1.3 (Fredholm alternative). *Suppose that we have a regular Sturm-Liouville problem. Then either*

$$\frac{d}{dx} \left(p(x) \frac{dy}{dx} \right) - q(x) y + \lambda r(x) y = 0,$$

$$\alpha_1 y(a) - \alpha_2 y'(a) = 0,$$

$$\beta_1 y(b) + \beta_2 y'(b) = 0,$$

has a nonzero solution, or

$$\frac{d}{dx}\left(p(x)\frac{dy}{dx}\right) - q(x)y + \lambda r(x)y = f(x),$$

$$\alpha_1 y(a) - \alpha_2 y'(a) = 0,$$

$$\beta_1 y(b) + \beta_2 y'(b) = 0,$$

has a unique solution for any $f(x)$ continuous on $[a, b]$.

This theorem is used in much the same way as we did before in § 4.4. It is used when solving more general nonhomogeneous boundary value problems. The theorem does not help us solve the problem, but it tells us when a unique solution exists, so that we know when to spend time looking for it. To solve the problem we decompose $f(x)$ and $y(x)$ in terms of the eigenfunctions of the homogeneous problem, and then solve for the coefficients of the series for $y(x)$.

5.1.4   Eigenfunction series

What we want to do with the eigenfunctions once we have them is to compute the *eigenfunction decomposition* of an arbitrary function $f(x)$. That is, we wish to write

$$f(x) = \sum_{n=1}^{\infty} c_n y_n(x), \tag{5.3}$$

where $y_n(x)$ the eigenfunctions. We wish to find out if we can represent any function $f(x)$ in this way, and if so, we wish to calculate c_n (and of course we would want to know if the sum converges). OK, so imagine we could write $f(x)$ as (5.3). We will assume convergence and the ability to integrate the series term by term. Because of orthogonality we have

$$\langle f, y_m \rangle = \int_a^b f(x)\, y_m(x)\, r(x)\, dx$$

$$= \sum_{n=1}^{\infty} c_n \int_a^b y_n(x)\, y_m(x)\, r(x)\, dx$$

$$= c_m \int_a^b y_m(x)\, y_m(x)\, r(x)\, dx = c_m \langle y_m, y_m \rangle.$$

Hence,

$$\boxed{c_m = \frac{\langle f, y_m \rangle}{\langle y_m, y_m \rangle} = \frac{\int_a^b f(x)\, y_m(x)\, r(x)\, dx}{\int_a^b \left(y_m(x)\right)^2 r(x)\, dx}.} \tag{5.4}$$

Note that y_m are known up to a constant multiple, so we could have picked a scalar multiple of an eigenfunction such that $\langle y_m, y_m \rangle = 1$ (if we had an arbitrary eigenfunction $\tilde{y}_m$, divide it by

$\sqrt{\langle \tilde{y}_m, \tilde{y}_m \rangle}$). When $\langle y_m, y_m \rangle = 1$ we have the simpler form $c_m = \langle f, y_m \rangle$ as we did for the Fourier series. The following theorem holds more generally, but the statement given is enough for our purposes.

Theorem 5.1.4. *Suppose f is a piecewise smooth continuous function on $[a, b]$. If $y_1, y_2, \ldots$ are the eigenfunctions of a regular Sturm-Liouville problem, then there exist real constants $c_1, c_2, \ldots$ given by (5.4) such that (5.3) converges and holds for $a < x < b$.*

Example 5.1.4: Take the simple Sturm-Liouville problem

$$y'' + \lambda y = 0, \quad 0 < x < \frac{\pi}{2},$$

$$y(0) = 0, \quad y'\left(\frac{\pi}{2}\right) = 0.$$

The above is a regular problem and furthermore we know by Theorem 5.1.1 on page 248 that $\lambda \geq 0$.

Suppose $\lambda = 0$, then the general solution is $y(x) = Ax + B$, we plug in the initial conditions to get $0 = y(0) = B$, and $0 = y'(\frac{\pi}{2}) = A$, hence $\lambda = 0$ is not an eigenvalue. The general solution, therefore, is

$$y(x) = A\cos(\sqrt{\lambda}\,x) + B\sin(\sqrt{\lambda}\,x).$$

Plugging in the boundary conditions we get $0 = y(0) = A$ and $0 = y'(\frac{\pi}{2}) = \sqrt{\lambda}\,B\cos(\sqrt{\lambda}\,\frac{\pi}{2})$. B cannot be zero and hence $\cos(\sqrt{\lambda}\,\frac{\pi}{2}) = 0$. This means that $\sqrt{\lambda}\,\frac{\pi}{2}$ must be an odd integral multiple of $\frac{\pi}{2}$, i.e. $(2n - 1)\frac{\pi}{2} = \sqrt{\lambda_n}\,\frac{\pi}{2}$. Hence

$$\lambda_n = (2n - 1)^2.$$

We can take $B = 1$. Hence our eigenfunctions are

$$y_n(x) = \sin((2n - 1)x).$$

Finally we compute

$$\int_0^{\frac{\pi}{2}} \left(\sin((2n - 1)x)\right)^2 dx = \frac{\pi}{4}.$$

So any piecewise smooth function on $[0, \frac{\pi}{2}]$ can be written as

$$f(x) = \sum_{n=1}^{\infty} c_n \sin((2n - 1)x),$$

where

$$c_n = \frac{\langle f, y_n \rangle}{\langle y_n, y_n \rangle} = \frac{\int_0^{\frac{\pi}{2}} f(x)\,\sin((2n - 1)x)\,dx}{\int_0^{\frac{\pi}{2}} \left(\sin((2n - 1)x)\right)^2 dx} = \frac{4}{\pi}\int_0^{\frac{\pi}{2}} f(x)\,\sin((2n - 1)x)\,dx.$$

Note that the series converges to an odd 2π-periodic (not π-periodic!) extension of $f(x)$.

***Exercise* 5.1.3** (challenging)*: In the above example, the function is defined on $0 < x < \frac{\pi}{2}$, yet the series converges to an odd 2π-periodic extension of $f(x)$. Find out how is the extension defined for $\frac{\pi}{2} < x < \pi$.*

5.1.5   Exercises

Exercise 5.1.4: *Find eigenvalues and eigenfunctions of*

$$y'' + \lambda y = 0, \quad y(0) - y'(0) = 0, \quad y(1) = 0.$$

Exercise 5.1.5: *Expand the function $f(x) = x$ on $0 \le x \le 1$ using the eigenfunctions of the system*

$$y'' + \lambda y = 0, \quad y'(0) = 0, \quad y(1) = 0.$$

Exercise 5.1.6: *Suppose that you had a Sturm-Liouville problem on the interval $[0, 1]$ and came up with $y_n(x) = \sin(\gamma n x)$, where $\gamma > 0$ is some constant. Decompose $f(x) = x$, $0 < x < 1$ in terms of these eigenfunctions.*

Exercise 5.1.7: *Find eigenvalues and eigenfunctions of*

$$y^{(4)} + \lambda y = 0, \quad y(0) = 0, \quad y'(0) = 0, \quad y(1) = 0, \quad y'(1) = 0.$$

This problem is not a Sturm-Liouville problem, but the idea is the same.

Exercise 5.1.8 (more challenging)*: Find eigenvalues and eigenfunctions for*

$$\frac{d}{dx}(e^x y') + \lambda e^x y = 0, \quad y(0) = 0, \quad y(1) = 0.$$

Hint: First write the system as a constant coefficient system to find general solutions. Do note that Theorem 5.1.1 on page 248 guarantees $\lambda \ge 0$.

Exercise 5.1.101: *Find eigenvalues and eigenfunctions of*

$$y'' + \lambda y = 0, \quad y(-1) = 0, \quad y(1) = 0.$$

Exercise 5.1.102: *Put the following problems into the standard form for Sturm-Liouville problems, that is, find $p(x)$, $q(x)$, $r(x)$, α_1, α_2, β_1, and β_2, and decide if the problems are regular or not.*
a) $xy'' + \lambda y = 0$ for $0 < x < 1$, $y(0) = 0$, $y(1) = 0$,
b)* $(1 + x^2)y'' + 2xy' + (\lambda - x^2)y = 0$ for $-1 < x < 1$, $y(-1) = 0$, $y(1) + y'(1) = 0$.

---

*In a previous version of the book, a typo rendered the equation as $(1 + x^2)y'' - 2xy' + (\lambda - x^2)y = 0$ ending up with something harder than intended. Try this equation for a further challenge.

5.2 Application of eigenfunction series

Note: 1 lecture, §10.2 in [EP], exercises in §11.2 in [BD]

The eigenfunction series can arise even from higher order equations. Suppose we have an elastic beam (say made of steel). We will study the transversal vibrations of the beam. That is, assume the beam lies along the x axis and let $y(x, t)$ measure the displacement of the point x on the beam at time t. See Figure 5.2.

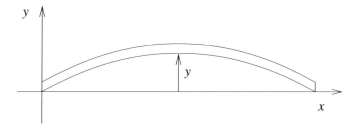

Figure 5.2: Transversal vibrations of a beam.

The equation that governs this setup is

$$a^4 \frac{\partial^4 y}{\partial x^4} + \frac{\partial^2 y}{\partial t^2} = 0,$$

for some constant $a > 0$.

Suppose the beam is of length 1 simply supported (hinged) at the ends. The beam is displaced by some function $f(x)$ at time $t = 0$ and then let go (initial velocity is 0). Then y satisfies:

$$
\begin{aligned}
&a^4 y_{xxxx} + y_{tt} = 0 \quad (0 < x < 1, t > 0), \\
&y(0, t) = y_{xx}(0, t) = 0, \\
&y(1, t) = y_{xx}(1, t) = 0, \\
&y(x, 0) = f(x), \quad y_t(x, 0) = 0.
\end{aligned}
\tag{5.5}
$$

Again we try $y(x, t) = X(x)T(t)$ and plug in to get $a^4 X^{(4)} T + X T'' = 0$ or

$$\frac{X^{(4)}}{X} = \frac{-T''}{a^4 T} = \lambda.$$

We note that we want $T'' + \lambda a^4 T = 0$. Let us assume that $\lambda > 0$. We can argue that we expect vibration and not exponential growth nor decay in the t direction (there is no friction in our model for instance). Similarly $\lambda = 0$ will not occur.

Exercise 5.2.1: *Try to justify $\lambda > 0$ just from the equations.*

Write $\omega^4 = \lambda$, so that we do not need to write the fourth root all the time. For X we get the equation $X^{(4)} - \omega^4 X = 0$. The general solution is

$$X(x) = Ae^{\omega x} + Be^{-\omega x} + C\sin(\omega x) + D\cos(\omega x).$$

Now $0 = X(0) = A + B + D, 0 = X''(0) = \omega^2(A + B - D)$. Hence, $D = 0$ and $A + B = 0$, or $B = -A$. So we have

$$X(x) = Ae^{\omega x} - Ae^{-\omega x} + C\sin(\omega x).$$

Also $0 = X(1) = A(e^\omega - e^{-\omega}) + C\sin\omega$, and $0 = X''(1) = A\omega^2(e^\omega - e^{-\omega}) - C\omega^2\sin\omega$. This means that $C\sin\omega = 0$ and $A(e^\omega - e^{-\omega}) = 2A\sinh\omega = 0$. If $\omega > 0$, then $\sinh\omega \neq 0$ and so $A = 0$. This means that $C \neq 0$ otherwise λ is not an eigenvalue. Also ω must be an integer multiple of π. Hence $\omega = n\pi$ and $n \geq 1$ (as $\omega > 0$). We can take $C = 1$. So the eigenvalues are $\lambda_n = n^4\pi^4$ and the eigenfunctions are $\sin(n\pi x)$.

Now $T'' + n^4\pi^4 a^4 T = 0$. The general solution is $T(t) = A\sin(n^2\pi^2 a^2 t) + B\cos(n^2\pi^2 a^2 t)$. But $T'(0) = 0$ and hence we must have $A = 0$ and we can take $B = 1$ to make $T(0) = 1$ for convenience. So our solutions are $T_n(t) = \cos(n^2\pi^2 a^2 t)$.

As the eigenfunctions are just sines again, we can decompose the function $f(x)$ on $0 < x < 1$ using the sine series. We find numbers b_n such that for $0 < x < 1$ we have

$$f(x) = \sum_{n=1}^{\infty} b_n \sin(n\pi x).$$

Then the solution to (5.5) is

$$y(x, t) = \sum_{n=1}^{\infty} b_n X_n(x) T_n(t) = \sum_{n=1}^{\infty} b_n \sin(n\pi x)\cos(n^2\pi^2 a^2 t).$$

The point is that $X_n T_n$ is a solution that satisfies all the homogeneous conditions (that is, all conditions except the initial position). And since and $T_n(0) = 1$, we have

$$y(x, 0) = \sum_{n=1}^{\infty} b_n X_n(x) T_n(0) = \sum_{n=1}^{\infty} b_n X_n(x) = \sum_{n=1}^{\infty} b_n \sin(n\pi x) = f(x).$$

So $y(x, t)$ solves (5.5).

The natural (circular) frequencies of the system are $n^2\pi^2 a^2$. These frequencies are all integer multiples of the fundamental frequency $\pi^2 a^2$, so we get a nice musical note. The exact frequencies and their amplitude are what we call the timbre of the note.

The timbre of a beam is different than for a vibrating string where we get "more" of the lower frequencies since we get all integer multiples, $1, 2, 3, 4, 5, \ldots$. For a steel beam we get only the square multiples $1, 4, 9, 16, 25, \ldots$. That is why when you hit a steel beam you hear a very pure sound. The sound of a xylophone or vibraphone is, therefore, very different from a guitar or piano.

Example 5.2.1: Let us assume that $f(x) = \frac{x(x-1)}{10}$. On $0 < x < 1$ we have (you know how to do this by now)

$$f(x) = \sum_{\substack{n=1 \\ n \text{ odd}}}^{\infty} \frac{4}{5\pi^3 n^3} \sin(n\pi x).$$

Hence, the solution to (5.5) with the given initial position $f(x)$ is

$$y(x, t) = \sum_{\substack{n=1 \\ n \text{ odd}}}^{\infty} \frac{4}{5\pi^3 n^3} \sin(n\pi x) \cos(n^2 \pi^2 a^2 t).$$

5.2.1 Exercises

Exercise 5.2.2: Suppose you have a beam of length 5 with free ends. Let y be the transverse deviation of the beam at position x on the beam ($0 < x < 5$). You know that the constants are such that this satisfies the equation $y_{tt} + 4y_{xxxx} = 0$. Suppose you know that the initial shape of the beam is the graph of $x(5 - x)$, and the initial velocity is uniformly equal to 2 (same for each x) in the positive y direction. Set up the equation together with the boundary and initial conditions. Just set up, do not solve.

Exercise 5.2.3: Suppose you have a beam of length 5 with one end free and one end fixed (the fixed end is at $x = 5$). Let u be the longitudinal deviation of the beam at position x on the beam ($0 < x < 5$). You know that the constants are such that this satisfies the equation $u_{tt} = 4u_{xx}$. Suppose you know that the initial displacement of the beam is $\frac{x-5}{50}$, and the initial velocity is $\frac{-(x-5)}{100}$ in the positive u direction. Set up the equation together with the boundary and initial conditions. Just set up, do not solve.

Exercise 5.2.4: Suppose the beam is L units long, everything else kept the same as in (5.5). What is the equation and the series solution?

Exercise 5.2.5: Suppose you have

$$a^4 y_{xxxx} + y_{tt} = 0 \quad (0 < x < 1, t > 0),$$
$$y(0, t) = y_{xx}(0, t) = 0,$$
$$y(1, t) = y_{xx}(1, t) = 0,$$
$$y(x, 0) = f(x), \quad y_t(x, 0) = g(x).$$

That is, you have also an initial velocity. Find a series solution. Hint: Use the same idea as we did for the wave equation.

Exercise 5.2.101: Suppose you have a beam of length 1 with hinged ends. Let y be the transverse deviation of the beam at position x on the beam ($0 < x < 1$). You know that the constants are such that this satisfies the equation $y_{tt} + 4y_{xxxx} = 0$. Suppose you know that the initial shape of the beam is the graph of $\sin(\pi x)$, and the initial velocity is 0. Solve for y.

Exercise **5.2.102:** *Suppose you have a beam of length 10 with two fixed ends. Let y be the transverse deviation of the beam at position x on the beam (0 < x < 10). You know that the constants are such that this satisfies the equation $y_{tt} + 9y_{xxxx} = 0$. Suppose you know that the initial shape of the beam is the graph of $\sin(\pi x)$, and the initial velocity is uniformly equal to $x(10 - x)$. Set up the equation together with the boundary and initial conditions. Just set up, do not solve.*

5.3 Steady periodic solutions

Note: 1–2 lectures, §10.3 in [EP], not in [BD]

5.3.1 Forced vibrating string.

Suppose that we have a guitar string of length L. We have studied the wave equation problem in this case, where x was the position on the string, t was time and y was the displacement of the string. See Figure 5.3.

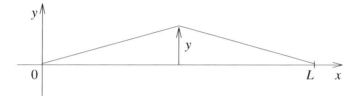

Figure 5.3: Vibrating string.

The problem is governed by the equations

$$
\begin{aligned}
&y_{tt} = a^2 y_{xx}, \\
&y(0, t) = 0, \qquad y(L, t) = 0, \\
&y(x, 0) = f(x), \quad y_t(x, 0) = g(x).
\end{aligned}
\tag{5.6}
$$

We saw previously that the solution is of the form

$$
y = \sum_{n=1}^{\infty} \left(A_n \cos\left(\frac{n\pi a}{L} t\right) + B_n \sin\left(\frac{n\pi a}{L} t\right) \right) \sin\left(\frac{n\pi}{L} x\right),
$$

where A_n and B_n were determined by the initial conditions. The natural frequencies of the system are the (circular) frequencies $\frac{n\pi a}{L}$ for integers $n \geq 1$.

But these are free vibrations. What if there is an external force acting on the string. Let us assume say air vibrations (noise), for example a second string. Or perhaps a jet engine. For simplicity, assume nice pure sound and assume the force is uniform at every position on the string. Let us say $F(t) = F_0 \cos(\omega t)$ as force per unit mass. Then our wave equation becomes (remember force is mass times acceleration)

$$
y_{tt} = a^2 y_{xx} + F_0 \cos(\omega t),
\tag{5.7}
$$

with the same boundary conditions of course.

We want to find the solution here that satisfies the above equation and

$$y(0, t) = 0, \qquad y(L, t) = 0, \qquad y(x, 0) = 0, \qquad y_t(x, 0) = 0. \qquad (5.8)$$

That is, the string is initially at rest. First we find a particular solution y_p of (5.7) that satisfies $y(0, t) = y(L, t) = 0$. We define the functions f and g as

$$f(x) = -y_p(x, 0), \qquad g(x) = -\frac{\partial y_p}{\partial t}(x, 0).$$

We then find solution y_c of (5.6). If we add the two solutions, we find that $y = y_c + y_p$ solves (5.7) with the initial conditions.

***Exercise* 5.3.1:** *Check that $y = y_c + y_p$ solves (5.7) and the side conditions (5.8).*

So the big issue here is to find the particular solution y_p. We look at the equation and we make an educated guess

$$y_p(x, t) = X(x) \cos(\omega t).$$

We plug in to get

$$-\omega^2 X \cos(\omega t) = a^2 X'' \cos(\omega t) + F_0 \cos(\omega t),$$

or $-\omega^2 X = a^2 X'' + F_0$ after canceling the cosine. We know how to find a general solution to this equation (it is a nonhomogeneous constant coefficient equation). The general solution is

$$X(x) = A \cos\left(\frac{\omega}{a} x\right) + B \sin\left(\frac{\omega}{a} x\right) - \frac{F_0}{\omega^2}.$$

The endpoint conditions imply $X(0) = X(L) = 0$. So

$$0 = X(0) = A - \frac{F_0}{\omega^2},$$

or $A = \frac{F_0}{\omega^2}$, and also

$$0 = X(L) = \frac{F_0}{\omega^2} \cos\left(\frac{\omega L}{a}\right) + B \sin\left(\frac{\omega L}{a}\right) - \frac{F_0}{\omega^2}.$$

Assuming that $\sin(\frac{\omega L}{a})$ is not zero we can solve for B to get

$$B = \frac{-F_0\left(\cos\left(\frac{\omega L}{a}\right) - 1\right)}{\omega^2 \sin\left(\frac{\omega L}{a}\right)}. \qquad (5.9)$$

Therefore,

$$X(x) = \frac{F_0}{\omega^2}\left(\cos\left(\frac{\omega}{a} x\right) - \frac{\cos\left(\frac{\omega L}{a}\right) - 1}{\sin\left(\frac{\omega L}{a}\right)} \sin\left(\frac{\omega}{a} x\right) - 1\right).$$

The particular solution y_p we are looking for is

$$y_p(x,t) = \frac{F_0}{\omega^2}\left(\cos\left(\frac{\omega}{a}x\right) - \frac{\cos\left(\frac{\omega L}{a}\right) - 1}{\sin\left(\frac{\omega L}{a}\right)}\sin\left(\frac{\omega}{a}x\right) - 1\right)\cos(\omega t).$$

***Exercise* 5.3.2:** *Check that y_p works.*

Now we get to the point that we skipped. Suppose $\sin(\frac{\omega L}{a}) = 0$. What this means is that ω is equal to one of the natural frequencies of the system, i.e. a multiple of $\frac{\pi a}{L}$. We notice that if ω is not equal to a multiple of the base frequency, but is very close, then the coefficient B in (5.9) seems to become very large. But let us not jump to conclusions just yet. When $\omega = \frac{n\pi a}{L}$ for n even, then $\cos(\frac{\omega L}{a}) = 1$ and hence we really get that $B = 0$. So resonance occurs only when both $\cos(\frac{\omega L}{a}) = -1$ and $\sin(\frac{\omega L}{a}) = 0$. That is when $\omega = \frac{n\pi a}{L}$ for *odd n*.

We could again solve for the resonance solution if we wanted to, but it is, in the right sense, the limit of the solutions as ω gets close to a resonance frequency. In real life, pure resonance never occurs anyway.

The above calculation explains why a string will begin to vibrate if the identical string is plucked close by. In the absence of friction this vibration would get louder and louder as time goes on. On the other hand, you are unlikely to get large vibration if the forcing frequency is not close to a resonance frequency even if you have a jet engine running close to the string. That is, the amplitude will not keep increasing unless you tune to just the right frequency.

Similar resonance phenomena occur when you break a wine glass using human voice (yes this is possible, but not easy[*]) if you happen to hit just the right frequency. Remember a glass has much purer sound, i.e. it is more like a vibraphone, so there are far fewer resonance frequencies to hit.

When the forcing function is more complicated, you decompose it in terms of the Fourier series and apply the above result. You may also need to solve the above problem if the forcing function is a sine rather than a cosine, but if you think about it, the solution is almost the same.

Example 5.3.1: Let us do the computation for specific values. Suppose $F_0 = 1$ and $\omega = 1$ and $L = 1$ and $a = 1$. Then

$$y_p(x,t) = \left(\cos(x) - \frac{\cos(1) - 1}{\sin(1)}\sin(x) - 1\right)\cos(t).$$

Write $B = \frac{\cos(1)-1}{\sin(1)}$ for simplicity.
   Then plug in $t = 0$ to get

$$f(x) = -y_p(x,0) = -\cos x + B\sin x + 1,$$

and after differentiating in t we see that $g(x) = -\frac{\partial y_p}{\partial t}(x,0) = 0$.

---

[*]*Mythbusters*, episode 31, Discovery Channel, originally aired may 18th 2005.

Hence to find y_c we need to solve the problem

$$y_{tt} = y_{xx},$$
$$y(0, t) = 0, \quad y(1, t) = 0,$$
$$y(x, 0) = -\cos x + B \sin x + 1,$$
$$y_t(x, 0) = 0.$$

Note that the formula that we use to define $y(x, 0)$ is not odd, hence it is not a simple matter of plugging in to apply the D'Alembert formula directly! You must define F to be the odd, 2-periodic extension of $y(x, 0)$. Then our solution would look like

$$y(x, t) = \frac{F(x + t) + F(x - t)}{2} + \left(\cos(x) - \frac{\cos(1) - 1}{\sin(1)} \sin(x) - 1\right)\cos(t). \tag{5.10}$$

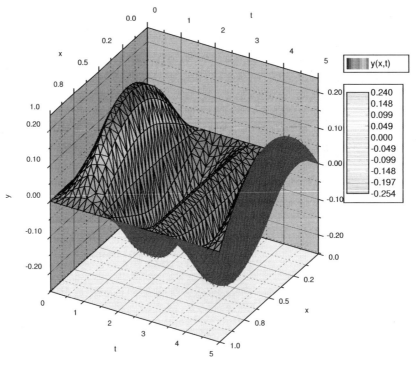

Figure 5.4: Plot of $y(x, t) = \frac{F(x+t)+F(x-t)}{2} + \left(\cos(x) - \frac{\cos(1)-1}{\sin(1)} \sin(x) - 1\right)\cos(t).$

It is not hard to compute specific values for an odd extension of a function and hence (5.10) is a wonderful solution to the problem. For example it is very easy to have a computer do it, unlike a series solution. A plot is given in Figure 5.4.

5.3.2 Underground temperature oscillations

Let $u(x, t)$ be the temperature at a certain location at depth x underground at time t. See Figure 5.5.

The temperature u satisfies the heat equation $u_t = ku_{xx}$, where k is the diffusivity of the soil. We know the temperature at the surface $u(0, t)$ from weather records. Let us assume for simplicity that

$$u(0, t) = T_0 + A_0 \cos(\omega t),$$

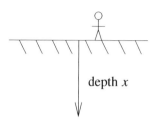

where T_0 is the yearly mean temperature, and $t = 0$ is midsummer (you can put negative sign above to make it midwinter if you wish). A_0 gives the typical variation for the year. That is, the hottest temperature is $T_0 + A_0$ and the coldest is $T_0 - A_0$. For simplicity, we will assume that $T_0 = 0$. The frequency ω is picked depending on the units of t, such that when $t = 1$ year, then $\omega t = 2\pi$. For example if t is in years, then $\omega = 2\pi$.

Figure 5.5: Underground temperature.

It seems reasonable that the temperature at depth x will also oscillate with the same frequency. This, in fact, will be the steady periodic solution, independent of the initial conditions. So we are looking for a solution of the form

$$u(x, t) = V(x) \cos(\omega t) + W(x) \sin(\omega t).$$

for the problem

$$u_t = ku_{xx}, \qquad u(0, t) = A_0 \cos(\omega t). \tag{5.11}$$

We will employ the complex exponential here to make calculations simpler. Suppose we have a complex valued function

$$h(x, t) = X(x) e^{i\omega t}.$$

We will look for an h such that $\operatorname{Re} h = u$. To find an h, whose real part satisfies (5.11), we look for an h such that

$$h_t = kh_{xx}, \qquad h(0, t) = A_0 e^{i\omega t}. \tag{5.12}$$

Exercise 5.3.3: *Suppose h satisfies* (5.12). *Use Euler's formula for the complex exponential to check that* $u = \operatorname{Re} h$ *satisfies* (5.11).

Substitute h into (5.12).

$$i\omega X e^{i\omega t} = kX'' e^{i\omega t}.$$

Hence,

$$kX'' - i\omega X = 0,$$

or

$$X'' - \alpha^2 X = 0,$$

where $\alpha = \pm \sqrt{\frac{i\omega}{k}}$. Note that $\pm \sqrt{i} = \pm\frac{1+i}{\sqrt{2}}$ so you could simplify to $\alpha = \pm(1+i)\sqrt{\frac{\omega}{2k}}$. Hence the general solution is

$$X(x) = Ae^{-(1+i)\sqrt{\frac{\omega}{2k}}\,x} + Be^{(1+i)\sqrt{\frac{\omega}{2k}}\,x}.$$

We assume that an $X(x)$ that solves the problem must be bounded as $x \to \infty$ since $u(x,t)$ should be bounded (we are not worrying about the earth core!). If you use Euler's formula to expand the complex exponentials, you will note that the second term will be unbounded (if $B \neq 0$), while the first term is always bounded. Hence $B = 0$.

Exercise 5.3.4: *Use Euler's formula to show that $e^{(1+i)\sqrt{\frac{\omega}{2k}}\,x}$ is unbounded as $x \to \infty$, while $e^{-(1+i)\sqrt{\frac{\omega}{2k}}\,x}$ is bounded as $x \to \infty$.*

Furthermore, $X(0) = A_0$ since $h(0,t) = A_0 e^{i\omega t}$. Thus $A = A_0$. This means that

$$h(x,t) = A_0 e^{-(1+i)\sqrt{\frac{\omega}{2k}}\,x} e^{i\omega t} = A_0 e^{-(1+i)\sqrt{\frac{\omega}{2k}}\,x + i\omega t} = A_0 e^{-\sqrt{\frac{\omega}{2k}}\,x} e^{i(\omega t - \sqrt{\frac{\omega}{2k}}\,x)}.$$

We will need to get the real part of h, so we apply Euler's formula to get

$$h(x,t) = A_0 e^{-\sqrt{\frac{\omega}{2k}}\,x}\left(\cos\left(\omega t - \sqrt{\frac{\omega}{2k}}\,x\right) + i\sin\left(\omega t - \sqrt{\frac{\omega}{2k}}\,x\right)\right).$$

Then finally

$$u(x,t) = \operatorname{Re} h(x,t) = A_0 e^{-\sqrt{\frac{\omega}{2k}}\,x}\cos\left(\omega t - \sqrt{\frac{\omega}{2k}}\,x\right).$$

Yay!

Notice the phase is different at different depths. At depth x the phase is delayed by $x\sqrt{\frac{\omega}{2k}}$. For example in cgs units (centimeters-grams-seconds) we have $k = 0.005$ (typical value for soil), $\omega = \frac{2\pi}{\text{seconds in a year}} = \frac{2\pi}{31{,}557{,}341} \approx 1.99 \times 10^{-7}$. Then if we compute where the phase shift $x\sqrt{\frac{\omega}{2k}} = \pi$ we find the depth in centimeters where the seasons are reversed. That is, we get the depth at which summer is the coldest and winter is the warmest. We get approximately 700 centimeters, which is approximately 23 feet below ground.

Be careful not to jump to conclusions. The temperature swings decay rapidly as you dig deeper. The amplitude of the temperature swings is $A_0 e^{-\sqrt{\frac{\omega}{2k}}x}$. This function decays *very* quickly as x (the depth) grows. Let us again take typical parameters as above. We will also assume that our surface temperature swing is $\pm 15°$ Celsius, that is, $A_0 = 15$. Then the maximum temperature variation at 700 centimeters is only $\pm 0.66°$ Celsius.

You need not dig very deep to get an effective "refrigerator," with nearly constant temperature. That is why wines are kept in a cellar; you need consistent temperature. The temperature differential could also be used for energy. A home could be heated or cooled by taking advantage of the above fact. Even without the earth core you could heat a home in the winter and cool it in the summer. The earth core makes the temperature higher the deeper you dig, although you need to dig somewhat deep to feel a difference. We did not take that into account above.

5.3.3 Exercises

***Exercise* 5.3.5:** *Suppose that the forcing function for the vibrating string is $F_0 \sin(\omega t)$. Derive the particular solution y_p.*

***Exercise* 5.3.6:** *Take the forced vibrating string. Suppose that $L = 1$, $a = 1$. Suppose that the forcing function is the square wave that is 1 on the interval $0 < x < 1$ and -1 on the interval $-1 < x < 0$. Find the particular solution. Hint: You may want to use result of Exercise 5.3.5.*

***Exercise* 5.3.7:** *The units are cgs (centimeters-grams-seconds). For $k = 0.005$, $\omega = 1.991 \times 10^{-7}$, $A_0 = 20$. Find the depth at which the temperature variation is half (± 10 degrees) of what it is on the surface.*

***Exercise* 5.3.8:** *Derive the solution for underground temperature oscillation without assuming that $T_0 = 0$.*

***Exercise* 5.3.101:** *Take the forced vibrating string. Suppose that $L = 1$, $a = 1$. Suppose that the forcing function is a sawtooth, that is $|x| - \frac{1}{2}$ on $-1 < x < 1$ extended periodically. Find the particular solution.*

***Exercise* 5.3.102:** *The units are cgs (centimeters-grams-seconds). For $k = 0.01$, $\omega = 1.991 \times 10^{-7}$, $A_0 = 25$. Find the depth at which the summer is again the hottest point.*

Chapter 6

The Laplace transform

6.1 The Laplace transform

Note: 1.5–2 lectures, §10.1 in [EP], §6.1 and parts of §6.2 in [BD]

6.1.1 The transform

In this chapter we will discuss the Laplace transform[*]. The Laplace transform turns out to be a very efficient method to solve certain ODE problems. In particular, the transform can take a differential equation and turn it into an algebraic equation. If the algebraic equation can be solved, applying the inverse transform gives us our desired solution. The Laplace transform also has applications in the analysis of electrical circuits, NMR spectroscopy, signal processing, and elsewhere. Finally, understanding the Laplace transform will also help with understanding the related Fourier transform, which, however, requires more understanding of complex numbers. We will not cover the Fourier transform.

The Laplace transform also gives a lot of insight into the nature of the equations we are dealing with. It can be seen as converting between the time and the frequency domain. For example, take the standard equation

$$mx''(t) + cx'(t) + kx(t) = f(t).$$

We can think of t as time and $f(t)$ as incoming signal. The Laplace transform will convert the equation from a differential equation in time to an algebraic (no derivatives) equation, where the new independent variable s is the frequency.

We can think of the *Laplace transform* as a black box. It eats functions and spits out functions in a new variable. We write $\mathcal{L}\{f(t)\} = F(s)$ for the Laplace transform of $f(t)$. It is common to write lower case letters for functions in the time domain and upper case letters for functions in the

---

[*]Just like the Laplace equation and the Laplacian, the Laplace transform is also named after Pierre-Simon, marquis de Laplace (1749–1827).

frequency domain. We use the same letter to denote that one function is the Laplace transform of the other. For example $F(s)$ is the Laplace transform of $f(t)$. Let us define the transform.

$$\mathcal{L}\{f(t)\} = F(s) \overset{\text{def}}{=} \int_0^\infty e^{-st} f(t)\, dt.$$

We note that we are only considering $t \geq 0$ in the transform. Of course, if we think of t as time there is no problem, we are generally interested in finding out what will happen in the future (Laplace transform is one place where it is safe to ignore the past). Let us compute some simple transforms.

Example 6.1.1: Suppose $f(t) = 1$, then

$$\mathcal{L}\{1\} = \int_0^\infty e^{-st}\, dt = \left[\frac{e^{-st}}{-s}\right]_{t=0}^\infty = \lim_{h\to\infty}\left[\frac{e^{-st}}{-s}\right]_{t=0}^h = \lim_{h\to\infty}\left(\frac{e^{-sh}}{-s} - \frac{1}{-s}\right) = \frac{1}{s}.$$

The limit (the improper integral) only exists if $s > 0$. So $\mathcal{L}\{1\}$ is only defined for $s > 0$.

Example 6.1.2: Suppose $f(t) = e^{-at}$, then

$$\mathcal{L}\{e^{-at}\} = \int_0^\infty e^{-st} e^{-at}\, dt = \int_0^\infty e^{-(s+a)t}\, dt = \left[\frac{e^{-(s+a)t}}{-(s+a)}\right]_{t=0}^\infty = \frac{1}{s+a}.$$

The limit only exists if $s + a > 0$. So $\mathcal{L}\{e^{-at}\}$ is only defined for $s + a > 0$.

Example 6.1.3: Suppose $f(t) = t$, then using integration by parts

$$\begin{aligned}
\mathcal{L}\{t\} &= \int_0^\infty e^{-st} t\, dt \\
&= \left[\frac{-te^{-st}}{s}\right]_{t=0}^\infty + \frac{1}{s}\int_0^\infty e^{-st}\, dt \\
&= 0 + \frac{1}{s}\left[\frac{e^{-st}}{-s}\right]_{t=0}^\infty \\
&= \frac{1}{s^2}.
\end{aligned}$$

Again, the limit only exists if $s > 0$.

Example 6.1.4: A common function is the *unit step function*, which is sometimes called the *Heaviside function*[*]. This function is generally given as

$$u(t) = \begin{cases} 0 & \text{if } t < 0, \\ 1 & \text{if } t \geq 0. \end{cases}$$

---

[*]The function is named after the English mathematician, engineer, and physicist Oliver Heaviside (1850–1925). Only by coincidence is the function "heavy" on "one side."

Let us find the Laplace transform of $u(t - a)$, where $a \geq 0$ is some constant. That is, the function that is 0 for $t < a$ and 1 for $t \geq a$.

$$\mathcal{L}\{u(t - a)\} = \int_0^\infty e^{-st} u(t - a)\, dt = \int_a^\infty e^{-st}\, dt = \left[\frac{e^{-st}}{-s}\right]_{t=a}^\infty = \frac{e^{-as}}{s},$$

where of course $s > 0$ (and $a \geq 0$ as we said before).

By applying similar procedures we can compute the transforms of many elementary functions. Many basic transforms are listed in Table 6.1.

$f(t)$	$\mathcal{L}\{f(t)\}$
C	$\frac{C}{s}$
t	$\frac{1}{s^2}$
t^2	$\frac{2}{s^3}$
t^3	$\frac{6}{s^4}$
t^n	$\frac{n!}{s^{n+1}}$
e^{-at}	$\frac{1}{s+a}$
$\sin(\omega t)$	$\frac{\omega}{s^2+\omega^2}$
$\cos(\omega t)$	$\frac{s}{s^2+\omega^2}$
$\sinh(\omega t)$	$\frac{\omega}{s^2-\omega^2}$
$\cosh(\omega t)$	$\frac{s}{s^2-\omega^2}$
$u(t - a)$	$\frac{e^{-as}}{s}$

Table 6.1: Some Laplace transforms (C, ω, and a are constants).

Exercise 6.1.1: *Verify Table 6.1.*

Since the transform is defined by an integral. We can use the linearity properties of the integral. For example, suppose C is a constant, then

$$\mathcal{L}\{Cf(t)\} = \int_0^\infty e^{-st} Cf(t)\, dt = C \int_0^\infty e^{-st} f(t)\, dt = C\mathcal{L}\{f(t)\}.$$

So we can "pull out" a constant out of the transform. Similarly we have linearity. Since linearity is very important we state it as a theorem.

Theorem 6.1.1 (Linearity of the Laplace transform). *Suppose that A, B, and C are constants, then*

$$\mathcal{L}\{Af(t) + Bg(t)\} = A\mathcal{L}\{f(t)\} + B\mathcal{L}\{g(t)\},$$

and in particular

$$\mathcal{L}\{Cf(t)\} = C\mathcal{L}\{f(t)\}.$$

***Exercise* 6.1.2:** *Verify the theorem. That is, show that* $\mathcal{L}\{Af(t) + Bg(t)\} = A\mathcal{L}\{f(t)\} + B\mathcal{L}\{g(t)\}$.

These rules together with Table 6.1 on the preceding page make it easy to find the Laplace transform of a whole lot of functions already. But be careful. It is a common mistake to think that the Laplace transform of a product is the product of the transforms. In general

$$\mathcal{L}\{f(t)g(t)\} \neq \mathcal{L}\{f(t)\}\mathcal{L}\{g(t)\}.$$

It must also be noted that not all functions have a Laplace transform. For example, the function $\frac{1}{t}$ does not have a Laplace transform as the integral diverges for all s. Similarly, $\tan t$ or e^{t^2} do not have Laplace transforms.

6.1.2   Existence and uniqueness

Let us consider when does the Laplace transform exist in more detail. First let us consider functions of exponential order. The function $f(t)$ is of *exponential order* as t goes to infinity if

$$|f(t)| \leq Me^{ct},$$

for some constants M and c, for sufficiently large t (say for all $t > t_0$ for some t_0). The simplest way to check this condition is to try and compute

$$\lim_{t \to \infty} \frac{f(t)}{e^{ct}}.$$

If the limit exists and is finite (usually zero), then $f(t)$ is of exponential order.

***Exercise* 6.1.3:** *Use L'Hopital's rule from calculus to show that a polynomial is of exponential order. Hint: Note that a sum of two exponential order functions is also of exponential order. Then show that t^n is of exponential order for any n.*

For an exponential order function we have existence and uniqueness of the Laplace transform.

Theorem 6.1.2 (Existence). *Let $f(t)$ be continuous and of exponential order for a certain constant c. Then $F(s) = \mathcal{L}\{f(t)\}$ is defined for all $s > c$.*

The existence is not difficult to see. Let $f(t)$ be of exponential order, that is $|f(t)| \leq Me^{ct}$ for all $t > 0$ (for simplicity $t_0 = 0$). Let $s > c$, or in other words $(c - s) < 0$. By the comparison theorem from calculus, the improper integral defining $\mathcal{L}\{f(t)\}$ exists if the following integral exists

$$\int_0^\infty e^{-st}(Me^{ct})\, dt = M \int_0^\infty e^{(c-s)t}\, dt = M \left[\frac{e^{(c-s)t}}{c-s} \right]_{t=0}^\infty = \frac{M}{c-s}.$$

The transform also exists for some other functions that are not of exponential order, but that will not be relevant to us. Before dealing with uniqueness, let us note that for exponential order functions we obtain that their Laplace transform decays at infinity:

$$\lim_{s \to \infty} F(s) = 0.$$

Theorem 6.1.3 (Uniqueness). *Let $f(t)$ and $g(t)$ be continuous and of exponential order. Suppose that there exists a constant C, such that $F(s) = G(s)$ for all $s > C$. Then $f(t) = g(t)$ for all $t \geq 0$.*

Both theorems hold for piecewise continuous functions as well. Recall that piecewise continuous means that the function is continuous except perhaps at a discrete set of points where it has jump discontinuities like the Heaviside function. Uniqueness, however, does not "see" values at the discontinuities. So we can only conclude that $f(t) = g(t)$ outside of discontinuities. For example, the unit step function is sometimes defined using $u(0) = 1/2$. This new step function, however, has the exact same Laplace transform as the one we defined earlier where $u(0) = 1$.

6.1.3 The inverse transform

As we said, the Laplace transform will allow us to convert a differential equation into an algebraic equation. Once we solve the algebraic equation in the frequency domain we will want to get back to the time domain, as that is what we are interested in. If we have a function $F(s)$, to be able to find $f(t)$ such that $\mathcal{L}\{f(t)\} = F(s)$, we need to first know if such a function is unique. It turns out we are in luck by Theorem 6.1.3. So we can without fear make the following definition.

If $F(s) = \mathcal{L}\{f(t)\}$ for some function $f(t)$. We define the *inverse Laplace transform* as

$$\mathcal{L}^{-1}\{F(s)\} \overset{\text{def}}{=} f(t).$$

There is an integral formula for the inverse, but it is not as simple as the transform itself—it requires complex numbers and path integrals. For us it will suffice to compute the inverse using Table 6.1 on page 269.

Example 6.1.5: Take $F(s) = \frac{1}{s+1}$. Find the inverse Laplace transform.

We look at the table to find

$$\mathcal{L}^{-1}\left\{\frac{1}{s+1}\right\} = e^{-t}.$$

As the Laplace transform is linear, the inverse Laplace transform is also linear. That is,

$$\mathcal{L}^{-1}\{AF(s) + BG(s)\} = A\mathcal{L}^{-1}\{F(s)\} + B\mathcal{L}^{-1}\{G(s)\}.$$

Of course, we also have $\mathcal{L}^{-1}\{AF(s)\} = A\mathcal{L}^{-1}\{F(s)\}$. Let us demonstrate how linearity can be used.

Example 6.1.6: Take $F(s) = \frac{s^2+s+1}{s^3+s}$. Find the inverse Laplace transform.

First we use the *method of partial fractions* to write F in a form where we can use Table 6.1 on page 269. We factor the denominator as $s(s^2 + 1)$ and write

$$\frac{s^2 + s + 1}{s^3 + s} = \frac{A}{s} + \frac{Bs + C}{s^2 + 1}.$$

Putting the right hand side over a common denominator and equating the numerators we get $A(s^2 + 1) + s(Bs + C) = s^2 + s + 1$. Expanding and equating coefficients we obtain $A + B = 1$, $C = 1$, $A = 1$, and thus $B = 0$. In other words,

$$F(s) = \frac{s^2 + s + 1}{s^3 + s} = \frac{1}{s} + \frac{1}{s^2 + 1}.$$

By linearity of the inverse Laplace transform we get

$$\mathcal{L}^{-1} \left\{ \frac{s^2 + s + 1}{s^3 + s} \right\} = \mathcal{L}^{-1} \left\{ \frac{1}{s} \right\} + \mathcal{L}^{-1} \left\{ \frac{1}{s^2 + 1} \right\} = 1 + \sin t.$$

Another useful property is the so-called *shifting property* or the *first shifting property*

$$\boxed{\mathcal{L}\{e^{-at} f(t)\} = F(s + a),}$$

where $F(s)$ is the Laplace transform of $f(t)$.

Exercise 6.1.4: *Derive the first shifting property from the definition of the Laplace transform.*

The shifting property can be used, for example, when the denominator is a more complicated quadratic that may come up in the method of partial fractions. We complete the square and write such quadratics as $(s + a)^2 + b$ and then use the shifting property.

Example 6.1.7: Find $\mathcal{L}^{-1} \left\{ \frac{1}{s^2 + 4s + 8} \right\}$.

First we complete the square to make the denominator $(s + 2)^2 + 4$. Next we find

$$\mathcal{L}^{-1} \left\{ \frac{1}{s^2 + 4} \right\} = \frac{1}{2} \sin(2t).$$

Putting it all together with the shifting property, we find

$$\mathcal{L}^{-1} \left\{ \frac{1}{s^2 + 4s + 8} \right\} = \mathcal{L}^{-1} \left\{ \frac{1}{(s + 2)^2 + 4} \right\} = \frac{1}{2} e^{-2t} \sin(2t).$$

In general, we want to be able to apply the Laplace transform to rational functions, that is functions of the form

$$\frac{F(s)}{G(s)}$$

where $F(s)$ and $G(s)$ are polynomials. Since normally, for the functions that we are considering, the Laplace transform goes to zero as $s \to \infty$, it is not hard to see that the degree of $F(s)$ must be smaller than that of $G(s)$. Such rational functions are called *proper rational functions* and we can always apply the method of partial fractions. Of course this means we need to be able to factor the denominator into linear and quadratic terms, which involves finding the roots of the denominator.

6.1.4 Exercises

Exercise 6.1.5: *Find the Laplace transform of* $3 + t^5 + \sin(\pi t)$.

Exercise 6.1.6: *Find the Laplace transform of* $a + bt + ct^2$ *for some constants* a, b, *and* c.

Exercise 6.1.7: *Find the Laplace transform of* $A\cos(\omega t) + B\sin(\omega t)$.

Exercise 6.1.8: *Find the Laplace transform of* $\cos^2(\omega t)$.

Exercise 6.1.9: *Find the inverse Laplace transform of* $\frac{4}{s^2-9}$.

Exercise 6.1.10: *Find the inverse Laplace transform of* $\frac{2s}{s^2-1}$.

Exercise 6.1.11: *Find the inverse Laplace transform of* $\frac{1}{(s-1)^2(s+1)}$.

Exercise 6.1.12: *Find the Laplace transform of* $f(t) = \begin{cases} t & \text{if } t \geq 1, \\ 0 & \text{if } t < 1. \end{cases}$

Exercise 6.1.13: *Find the inverse Laplace transform of* $\frac{s}{(s^2+s+2)(s+4)}$.

Exercise 6.1.14: *Find the Laplace transform of* $\sin(\omega(t-a))$.

Exercise 6.1.15: *Find the Laplace transform of* $t\sin(\omega t)$. *Hint: Several integrations by parts.*

Exercise 6.1.101: *Find the Laplace transform of* $4(t+1)^2$.

Exercise 6.1.102: *Find the inverse Laplace transform of* $\frac{8}{s^3(s+2)}$.

Exercise 6.1.103: *Find the Laplace transform of* te^{-t} *(Hint: integrate by parts).*

Exercise 6.1.104: *Find the Laplace transform of* $\sin(t)e^{-t}$ *(Hint: integrate by parts).*

6.2   Transforms of derivatives and ODEs

Note: 2 lectures, §7.2–7.3 in [EP], §6.2 and §6.3 in [BD]

6.2.1   Transforms of derivatives

Let us see how the Laplace transform is used for differential equations. First let us try to find the Laplace transform of a function that is a derivative. Suppose $g(t)$ is a differentiable function of exponential order, that is, $|g(t)| \le Me^{ct}$ for some M and c. So $\mathcal{L}\{g(t)\}$ exists, and what is more, $\lim_{t\to\infty} e^{-st}g(t) = 0$ when $s > c$. Then

$$\mathcal{L}\{g'(t)\} = \int_0^\infty e^{-st}g'(t)\,dt = \left[e^{-st}g(t)\right]_{t=0}^\infty - \int_0^\infty (-s)\,e^{-st}g(t)\,dt = -g(0) + s\mathcal{L}\{g(t)\}.$$

We repeat this procedure for higher derivatives. The results are listed in Table 6.2. The procedure also works for piecewise smooth functions, that is functions that are piecewise continuous with a piecewise continuous derivative.

$f(t)$	$\mathcal{L}\{f(t)\} = F(s)$
$g'(t)$	$sG(s) - g(0)$
$g''(t)$	$s^2G(s) - sg(0) - g'(0)$
$g'''(t)$	$s^3G(s) - s^2g(0) - sg'(0) - g''(0)$

Table 6.2: Laplace transforms of derivatives ($G(s) = \mathcal{L}\{g(t)\}$ as usual).

Exercise 6.2.1: *Verify Table 6.2.*

6.2.2   Solving ODEs with the Laplace transform

Notice that the Laplace transform turns differentiation into multiplication by s. Let us see how to apply this fact to differential equations.

Example 6.2.1: Take the equation

$$x''(t) + x(t) = \cos(2t), \quad x(0) = 0, \quad x'(0) = 1.$$

We will take the Laplace transform of both sides. By $X(s)$ we will, as usual, denote the Laplace transform of $x(t)$.

$$\mathcal{L}\{x''(t) + x(t)\} = \mathcal{L}\{\cos(2t)\},$$
$$s^2X(s) - sx(0) - x'(0) + X(s) = \frac{s}{s^2 + 4}.$$

We plug in the initial conditions now—this makes the computations more streamlined—to obtain

$$s^2 X(s) - 1 + X(s) = \frac{s}{s^2 + 4}.$$

We solve for $X(s)$,

$$X(s) = \frac{s}{(s^2 + 1)(s^2 + 4)} + \frac{1}{s^2 + 1}.$$

We use partial fractions (exercise) to write

$$X(s) = \frac{1}{3} \frac{s}{s^2 + 1} - \frac{1}{3} \frac{s}{s^2 + 4} + \frac{1}{s^2 + 1}.$$

Now take the inverse Laplace transform to obtain

$$x(t) = \frac{1}{3} \cos(t) - \frac{1}{3} \cos(2t) + \sin(t).$$

The procedure for linear constant coefficient equations is as follows. We take an ordinary differential equation in the time variable t. We apply the Laplace transform to transform the equation into an algebraic (non differential) equation in the frequency domain. All the $x(t)$, $x'(t)$, $x''(t)$, and so on, will be converted to $X(s)$, $sX(s) - x(0)$, $s^2 X(s) - sx(0) - x'(0)$, and so on. We solve the equation for $X(s)$. Then taking the inverse transform, if possible, we find $x(t)$.

It should be noted that since not every function has a Laplace transform, not every equation can be solved in this manner. Also if the equation is not a linear constant coefficient ODE, then by applying the Laplace transform we may not obtain an algebraic equation.

6.2.3   Using the Heaviside function

Before we move on to more general equations than those we could solve before, we want to consider the Heaviside function. See Figure 6.1 on the next page for the graph.

$$u(t) = \begin{cases} 0 & \text{if } t < 0, \\ 1 & \text{if } t \geq 0. \end{cases}$$

This function is useful for putting together functions, or cutting functions off. Most commonly it is used as $u(t - a)$ for some constant a. This just shifts the graph to the right by a. That is, it is a function that is 0 when $t < a$ and 1 when $t \geq a$. Suppose for example that $f(t)$ is a "signal" and you started receiving the signal $\sin t$ at time $t = \pi$. The function $f(t)$ should then be defined as

$$f(t) = \begin{cases} 0 & \text{if } t < \pi, \\ \sin t & \text{if } t \geq \pi. \end{cases}$$

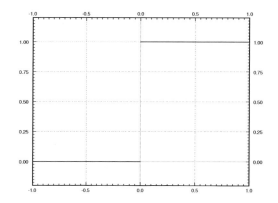

Figure 6.1: Plot of the Heaviside (unit step) function u(t).

Using the Heaviside function, $f(t)$ can be written as

$$f(t) = u(t - \pi) \sin t.$$

Similarly the step function that is 1 on the interval $[1, 2)$ and zero everywhere else can be written as

$$u(t - 1) - u(t - 2).$$

The Heaviside function is useful to define functions defined piecewise. If you want to define $f(t)$ such that $f(t) = t$ when t is in $[0, 1]$, $f(t) = -t + 2$ when t is in $[1, 2]$, and $f(t) = 0$ otherwise, then you can use the expression

$$f(t) = t\left(u(t) - u(t - 1)\right) + (-t + 2)\left(u(t - 1) - u(t - 2)\right).$$

Hence it is useful to know how the Heaviside function interacts with the Laplace transform. We have already seen that

$$\mathcal{L}\{u(t - a)\} = \frac{e^{-as}}{s}.$$

This can be generalized into a *shifting property* or *second shifting property*.

$$\boxed{\mathcal{L}\{f(t - a)\, u(t - a)\} = e^{-as}\mathcal{L}\{f(t)\}.} \tag{6.1}$$

Example 6.2.2: Suppose that the forcing function is not periodic. For example, suppose that we had a mass-spring system

$$x''(t) + x(t) = f(t), \quad x(0) = 0, \quad x'(0) = 0,$$

where $f(t) = 1$ if $1 \leq t < 5$ and zero otherwise. We could imagine a mass-spring system, where a rocket is fired for 4 seconds starting at $t = 1$. Or perhaps an RLC circuit, where the voltage is raised at a constant rate for 4 seconds starting at $t = 1$, and then held steady again starting at $t = 5$.

We can write $f(t) = u(t - 1) - u(t - 5)$. We transform the equation and we plug in the initial conditions as before to obtain

$$s^2 X(s) + X(s) = \frac{e^{-s}}{s} - \frac{e^{-5s}}{s}.$$

We solve for $X(s)$ to obtain

$$X(s) = \frac{e^{-s}}{s(s^2 + 1)} - \frac{e^{-5s}}{s(s^2 + 1)}.$$

We leave it as an exercise to the reader to show that

$$\mathcal{L}^{-1}\left\{\frac{1}{s(s^2 + 1)}\right\} = 1 - \cos t.$$

In other words $\mathcal{L}\{1 - \cos t\} = \frac{1}{s(s^2+1)}$. So using (6.1) we find

$$\mathcal{L}^{-1}\left\{\frac{e^{-s}}{s(s^2 + 1)}\right\} = \mathcal{L}^{-1}\{e^{-s}\mathcal{L}\{1 - \cos t\}\} = (1 - \cos(t - 1))\, u(t - 1).$$

Similarly

$$\mathcal{L}^{-1}\left\{\frac{e^{-5s}}{s(s^2 + 1)}\right\} = \mathcal{L}^{-1}\{e^{-5s}\mathcal{L}\{1 - \cos t\}\} = (1 - \cos(t - 5))\, u(t - 5).$$

Hence, the solution is

$$x(t) = (1 - \cos(t - 1))\, u(t - 1) - (1 - \cos(t - 5))\, u(t - 5).$$

The plot of this solution is given in Figure 6.2.

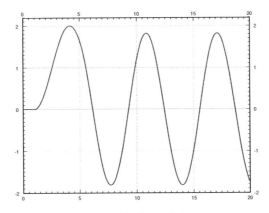

Figure 6.2: Plot of $x(t)$.

6.2.4   Transfer functions

Laplace transform leads to the following useful concept for studying the steady state behavior of a linear system. Suppose we have an equation of the form

$$Lx = f(t),$$

where L is a linear constant coefficient differential operator. Then $f(t)$ is usually thought of as input of the system and $x(t)$ is thought of as the output of the system. For example, for a mass-spring system the input is the forcing function and output is the behavior of the mass. We would like to have an convenient way to study the behavior of the system for different inputs.

Let us suppose that all the initial conditions are zero and take the Laplace transform of the equation, we obtain the equation

$$A(s)X(s) = F(s).$$

Solving for the ratio $X(s)/F(s)$ we obtain the so-called *transfer function* $H(s) = 1/A(s)$.

$$H(s) = \frac{X(s)}{F(s)}.$$

In other words, $X(s) = H(s)F(s)$. We obtain an algebraic dependence of the output of the system based on the input. We can now easily study the steady state behavior of the system given different inputs by simply multiplying by the transfer function.

Example 6.2.3: Given $x'' + \omega_0^2 x = f(t)$, let us find the transfer function (assuming the initial conditions are zero).

First, we take the Laplace transform of the equation.

$$s^2 X(s) + \omega_0^2 X(s) = F(s).$$

Now we solve for the transfer function $X(s)/F(s)$.

$$H(s) = \frac{X(s)}{F(s)} = \frac{1}{s^2 + \omega_0^2}.$$

Let us see how to use the transfer function. Suppose we have the constant input $f(t) = 1$. Hence $F(s) = 1/s$, and

$$X(s) = H(s)F(s) = \frac{1}{s^2 + \omega_0^2} \frac{1}{s}.$$

Taking the inverse Laplace transform of $X(s)$ we obtain

$$x(t) = \frac{1 - \cos(\omega_0 t)}{\omega_0^2}.$$

6.2.5 Transforms of integrals

A feature of Laplace transforms is that it is also able to easily deal with integral equations. That is, equations in which integrals rather than derivatives of functions appear. The basic property, which can be proved by applying the definition and doing integration by parts, is

$$\mathcal{L}\left\{\int_0^t f(\tau)\,d\tau\right\} = \frac{1}{s}F(s).$$

It is sometimes useful (e.g. for computing the inverse transform) to write this as

$$\int_0^t f(\tau)\,d\tau = \mathcal{L}^{-1}\left\{\frac{1}{s}F(s)\right\}.$$

Example 6.2.4: To compute $\mathcal{L}^{-1}\left\{\frac{1}{s(s^2+1)}\right\}$ we could proceed by applying this integration rule.

$$\mathcal{L}^{-1}\left\{\frac{1}{s}\frac{1}{s^2+1}\right\} = \int_0^t \mathcal{L}^{-1}\left\{\frac{1}{s^2+1}\right\}\,d\tau = \int_0^t \sin\tau\,d\tau = 1 - \cos t.$$

Example 6.2.5: An equation containing an integral of the unknown function is called an *integral equation*. For example, take

$$t^2 = \int_0^t e^\tau x(\tau)\,d\tau,$$

where we wish to solve for $x(t)$. We apply the Laplace transform and the shifting property to get

$$\frac{2}{s^3} = \frac{1}{s}\mathcal{L}\{e^t x(t)\} = \frac{1}{s}X(s-1),$$

where $X(s) = \mathcal{L}\{x(t)\}$. Thus

$$X(s-1) = \frac{2}{s^2} \qquad \text{or} \qquad X(s) = \frac{2}{(s+1)^2}.$$

We use the shifting property again

$$x(t) = 2e^{-t}t.$$

6.2.6 Exercises

Exercise 6.2.2: Using the Heaviside function write down the piecewise function that is 0 for $t < 0$, t^2 for t in $[0, 1]$ and t for $t > 1$.

Exercise 6.2.3: Using the Laplace transform solve

$$mx'' + cx' + kx = 0, \quad x(0) = a, \quad x'(0) = b,$$

where $m > 0$, $c > 0$, $k > 0$, and $c^2 - 4km > 0$ (system is overdamped).

Exercise 6.2.4: Using the Laplace transform solve

$$mx'' + cx' + kx = 0, \quad x(0) = a, \quad x'(0) = b,$$

where m > 0, c > 0, k > 0, and $c^2 - 4km < 0$ (system is underdamped).

Exercise 6.2.5: Using the Laplace transform solve

$$mx'' + cx' + kx = 0, \quad x(0) = a, \quad x'(0) = b,$$

where m > 0, c > 0, k > 0, and $c^2 = 4km$ (system is critically damped).

Exercise 6.2.6: Solve $x'' + x = u(t - 1)$ for initial conditions $x(0) = 0$ and $x'(0) = 0$.

Exercise 6.2.7: Show the differentiation of the transform property. Suppose $\mathcal{L}\{f(t)\} = F(s)$, then show

$$\mathcal{L}\{-t f(t)\} = F'(s).$$

Hint: Differentiate under the integral sign.

Exercise 6.2.8: Solve $x''' + x = t^3 u(t - 1)$ for initial conditions $x(0) = 1$ and $x'(0) = 0$, $x''(0) = 0$.

Exercise 6.2.9: Show the second shifting property: $\mathcal{L}\{f(t - a) u(t - a)\} = e^{-as} \mathcal{L}\{f(t)\}$.

Exercise 6.2.10: Let us think of the mass-spring system with a rocket from Example 6.2.2. We noticed that the solution kept oscillating after the rocket stopped running. The amplitude of the oscillation depends on the time that the rocket was fired (for 4 seconds in the example). a) Find a formula for the amplitude of the resulting oscillation in terms of the amount of time the rocket is fired. b) Is there a nonzero time (if so what is it?) for which the rocket fires and the resulting oscillation has amplitude 0 (the mass is not moving)?

Exercise 6.2.11: Define

$$f(t) = \begin{cases} (t - 1)^2 & \text{if } 1 \le t < 2, \\ 3 - t & \text{if } 2 \le t < 3, \\ 0 & \text{otherwise.} \end{cases}$$

a) Sketch the graph of $f(t)$. b) Write down $f(t)$ using the Heaviside function. c) Solve $x'' + x = f(t)$, $x(0) = 0$, $x'(0) = 0$ using Laplace transform.

Exercise 6.2.12: Find the transfer function for $mx'' + cx' + kx = f(t)$ (assuming the initial conditions are zero).

Exercise 6.2.101: Using the Heaviside function $u(t)$, write down the function

$$f(t) = \begin{cases} 0 & \text{if } \quad t < 1, \\ t - 1 & \text{if } 1 \le t < 2, \\ 1 & \text{if } 2 \le t. \end{cases}$$

Exercise 6.2.102: *Solve $x'' - x = (t^2 - 1)u(t - 1)$ for initial conditions $x(0) = 1$, $x'(0) = 2$ using the Laplace transform.*

Exercise 6.2.103: *Find the transfer function for $x' + x = f(t)$ (assuming the initial conditions are zero).*

6.3   Convolution

Note: 1 or 1.5 lectures, §7.2 in [EP], §6.6 in [BD]

6.3.1   The convolution

We said that the Laplace transformation of a product is not the product of the transforms. All hope is not lost however. We simply have to use a different type of a "product." Take two functions $f(t)$ and $g(t)$ defined for $t \geq 0$, and define the *convolution** of $f(t)$ and $g(t)$ as

$$(f * g)(t) \overset{\text{def}}{=} \int_0^t f(\tau) g(t - \tau) \, d\tau. \tag{6.2}$$

As you can see, the convolution of two functions of t is another function of t.

Example 6.3.1: Take $f(t) = e^t$ and $g(t) = t$ for $t \geq 0$. Then

$$(f * g)(t) = \int_0^t e^\tau (t - \tau) \, d\tau = e^t - t - 1.$$

To solve the integral we did one integration by parts.

Example 6.3.2: Take $f(t) = \sin(\omega t)$ and $g(t) = \cos(\omega t)$ for $t \geq 0$. Then

$$(f * g)(t) = \int_0^t \sin(\omega \tau) \, \cos(\omega(t - \tau)) \, d\tau.$$

We apply the identity

$$\cos(\theta) \sin(\psi) = \frac{1}{2} \left(\sin(\theta + \psi) - \sin(\theta - \psi) \right).$$

Hence,

$$\begin{aligned}
(f * g)(t) &= \int_0^t \frac{1}{2} \left(\sin(\omega t) - \sin(\omega t - 2\omega \tau) \right) d\tau \\
&= \left[\frac{1}{2} \tau \sin(\omega t) + \frac{1}{4\omega} \cos(2\omega \tau - \omega t) \right]_{\tau=0}^t \\
&= \frac{1}{2} t \sin(\omega t).
\end{aligned}$$

---

*For those that have seen convolution defined before, you may have seen it defined as $(f * g)(t) = \int_{-\infty}^\infty f(\tau) g(t - \tau) \, d\tau$. This definition agrees with (6.2) if you define $f(t)$ and $g(t)$ to be zero for $t < 0$. When discussing the Laplace transform the definition we gave is sufficient. Convolution does occur in many other applications, however, where you may have to use the more general definition with infinities.

The formula holds only for $t \geq 0$. We assumed that f and g are zero (or simply not defined) for negative t.

The convolution has many properties that make it behave like a product. Let c be a constant and f, g, and h be functions then

$$f * g = g * f,$$
$$(cf) * g = f * (cg) = c(f * g),$$
$$(f * g) * h = f * (g * h).$$

The most interesting property for us, and the main result of this section is the following theorem.

Theorem 6.3.1. *Let $f(t)$ and $g(t)$ be of exponential type, then*

$$\mathcal{L}\{(f * g)(t)\} = \mathcal{L}\left\{\int_0^t f(\tau)g(t - \tau)\, d\tau\right\} = \mathcal{L}\{f(t)\}\mathcal{L}\{g(t)\}.$$

In other words, the Laplace transform of a convolution is the product of the Laplace transforms. The simplest way to use this result is in reverse.

Example 6.3.3: Suppose we have the function of s defined by

$$\frac{1}{(s + 1)s^2} = \frac{1}{s + 1}\frac{1}{s^2}.$$

We recognize the two entries of Table 6.2. That is

$$\mathcal{L}^{-1}\left\{\frac{1}{s + 1}\right\} = e^{-t} \quad \text{and} \quad \mathcal{L}^{-1}\left\{\frac{1}{s^2}\right\} = t.$$

Therefore,

$$\mathcal{L}^{-1}\left\{\frac{1}{s + 1}\frac{1}{s^2}\right\} = \int_0^t \tau e^{-(t-\tau)}\, d\tau = e^{-t} + t - 1.$$

The calculation of the integral involved an integration by parts.

6.3.2 Solving ODEs

The next example demonstrates the full power of the convolution and the Laplace transform. We can give the solution to the forced oscillation problem for any forcing function as a definite integral.

Example 6.3.4: Find the solution to

$$x'' + \omega_0^2 x = f(t), \quad x(0) = 0, \quad x'(0) = 0,$$

for an arbitrary function $f(t)$.

We first apply the Laplace transform to the equation. Denote the transform of $x(t)$ by $X(s)$ and the transform of $f(t)$ by $F(s)$ as usual.

$$s^2 X(s) + \omega_0^2 X(s) = F(s),$$

or in other words

$$X(s) = F(s)\frac{1}{s^2 + \omega_0^2}.$$

We know

$$\mathcal{L}^{-1}\left\{\frac{1}{s^2 + \omega_0^2}\right\} = \frac{\sin(\omega_0 t)}{\omega_0}.$$

Therefore,

$$x(t) = \int_0^t f(\tau)\frac{\sin(\omega_0(t - \tau))}{\omega_0}\, d\tau,$$

or if we reverse the order

$$x(t) = \int_0^t \frac{\sin(\omega_0 \tau)}{\omega_0} f(t - \tau)\, d\tau.$$

Let us notice one more feature of this example. We can now see how Laplace transform handles resonance. Suppose that $f(t) = \cos(\omega_0 t)$. Then

$$x(t) = \int_0^t \frac{\sin(\omega_0 \tau)}{\omega_0}\cos(\omega_0(t - \tau))\, d\tau = \frac{1}{\omega_0}\int_0^t \sin(\omega_0 \tau)\cos(\omega_0(t - \tau))\, d\tau.$$

We have computed the convolution of sine and cosine in Example 6.3.2. Hence

$$x(t) = \left(\frac{1}{\omega_0}\right)\left(\frac{1}{2}t\,\sin(\omega_0 t)\right) = \frac{1}{2\omega_0}t\,\sin(\omega_0 t).$$

Note the t in front of the sine. The solution, therefore, grows without bound as t gets large, meaning we get resonance.

Similarly, we can solve any constant coefficient equation with an arbitrary forcing function $f(t)$ as a definite integral using convolution. A definite integral, rather than a closed form solution, is usually enough for most practical purposes. It is not hard to numerically evaluate a definite integral.

6.3.3 Volterra integral equation

A common integral equation is the *Volterra integral equation*[*]

$$x(t) = f(t) + \int_0^t g(t - \tau)x(\tau)\, d\tau,$$

---

[*]Named for the Italian mathematician Vito Volterra (1860–1940).

where $f(t)$ and $g(t)$ are known functions and $x(t)$ is an unknown we wish to solve for. To find $x(t)$, we apply the Laplace transform to the equation to obtain

$$X(s) = F(s) + G(s)X(s),$$

where $X(s)$, $F(s)$, and $G(s)$ are the Laplace transforms of $x(t)$, $f(t)$, and $g(t)$ respectively. We find

$$X(s) = \frac{F(s)}{1 - G(s)}.$$

To find $x(t)$ we now need to find the inverse Laplace transform of $X(s)$.

Example 6.3.5: Solve

$$x(t) = e^{-t} + \int_0^t \sinh(t - \tau)x(\tau)\,d\tau.$$

We apply Laplace transform to obtain

$$X(s) = \frac{1}{s + 1} + \frac{1}{s^2 - 1}X(s),$$

or

$$X(s) = \frac{\frac{1}{s+1}}{1 - \frac{1}{s^2-1}} = \frac{s - 1}{s^2 - 2} = \frac{s}{s^2 - 2} - \frac{1}{s^2 - 2}.$$

It is not hard to apply Table 6.1 on page 269 to find

$$x(t) = \cosh(\sqrt{2}\,t) - \frac{1}{\sqrt{2}}\sinh(\sqrt{2}\,t).$$

6.3.4 Exercises

Exercise 6.3.1: *Let $f(t) = t^2$ for $t \geq 0$, and $g(t) = u(t - 1)$. Compute $f * g$.*

Exercise 6.3.2: *Let $f(t) = t$ for $t \geq 0$, and $g(t) = \sin t$ for $t \geq 0$. Compute $f * g$.*

Exercise 6.3.3: *Find the solution to*

$$mx'' + cx' + kx = f(t), \quad x(0) = 0, \quad x'(0) = 0,$$

for an arbitrary function $f(t)$, where $m > 0$, $c > 0$, $k > 0$, and $c^2 - 4km > 0$ (system is overdamped). Write the solution as a definite integral.

Exercise 6.3.4: *Find the solution to*

$$mx'' + cx' + kx = f(t), \quad x(0) = 0, \quad x'(0) = 0,$$

for an arbitrary function $f(t)$, where $m > 0$, $c > 0$, $k > 0$, and $c^2 - 4km < 0$ (system is underdamped). Write the solution as a definite integral.

Exercise 6.3.5: *Find the solution to*

$$mx'' + cx' + kx = f(t), \quad x(0) = 0, \quad x'(0) = 0,$$

for an arbitrary function $f(t)$, where $m > 0$, $c > 0$, $k > 0$, and $c^2 = 4km$ (system is critically damped). Write the solution as a definite integral.

Exercise 6.3.6: *Solve*

$$x(t) = e^{-t} + \int_0^t \cos(t - \tau)x(\tau)\, d\tau.$$

Exercise 6.3.7: *Solve*

$$x(t) = \cos t + \int_0^t \cos(t - \tau)x(\tau)\, d\tau.$$

Exercise 6.3.8: *Compute $\mathcal{L}^{-1}\left\{\frac{s}{(s^2+4)^2}\right\}$ using convolution.*

Exercise 6.3.9: *Write down the solution to $x'' - 2x = e^{-t^2}$, $x(0) = 0$, $x'(0) = 0$ as a definite integral. Hint: Do not try to compute the Laplace transform of e^{-t^2}.*

Exercise 6.3.101: *Let $f(t) = \cos t$ for $t \geq 0$, and $g(t) = e^{-t}$. Compute $f * g$.*

Exercise 6.3.102: *Compute $\mathcal{L}^{-1}\left\{\frac{5}{s^4+s^2}\right\}$ using convolution.*

Exercise 6.3.103: *Solve $x'' + x = \sin t$, $x(0) = 0$, $x'(0) = 0$ using convolution.*

Exercise 6.3.104: *Solve $x''' + x' = f(t)$, $x(0) = 0$, $x'(0) = 0$, $x''(0) = 0$ using convolution. Write the result as a definite integral.*

6.4 Dirac delta and impulse response

Note: 1 or 1.5 lecture, §7.6 in [EP], §6.5 in [BD]

6.4.1 Rectangular pulse

Often in applications we study a physical system by putting in a short pulse and then seeing what the system does. The resulting behavior is often called *impulse response*. Let us see what we mean by a pulse. The simplest kind of a pulse is a simple rectangular pulse defined by

$$\varphi(t) = \begin{cases} 0 & \text{if} \quad t < a, \\ M & \text{if} \quad a \leq t < b, \\ 0 & \text{if} \quad b \leq t. \end{cases}$$

See Figure 6.3 for a graph. Notice that

$$\varphi(t) = M(u(t - a) - u(t - b)),$$

where $u(t)$ is the unit step function.

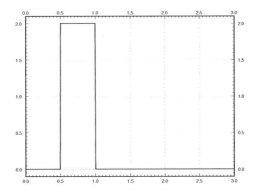

Figure 6.3: Sample square pulse with $a = 0.5$, $b = 1$ and $M = 2$.

Let us take the Laplace transform of a square pulse,

$$\mathcal{L}\{\varphi(t)\} = \mathcal{L}\{M(u(t - a) - u(t - b))\} - M\frac{e^{-as} - e^{-bs}}{s}.$$

For simplicity we let $a = 0$, and it is convenient to set $M = 1/b$ to have

$$\int_0^\infty \varphi(t)\, dt = 1.$$

That is, to have the pulse have "unit mass." For such a pulse we compute

$$\mathcal{L}\{\varphi(t)\} = \mathcal{L}\left\{\frac{u(t) - u(t-b)}{b}\right\} = \frac{1 - e^{-bs}}{bs}.$$

We generally want b to be very small. That is, we wish to have the pulse be very short and very tall. By letting b go to zero we arrive at the concept of the Dirac delta function.

6.4.2   The delta function

The *Dirac delta function*[*] is not exactly a function; it is sometimes called a *generalized function*. We avoid unnecessary details and simply say that it is an object that does not really make sense unless we integrate it. The motivation is that we would like a "function" $\delta(t)$ such that for any continuous function $f(t)$ we have

$$\boxed{\int_{-\infty}^{\infty} \delta(t)f(t)\, dt = f(0).}$$

The formula should hold if we integrate over any interval that contains 0, not just $(-\infty, \infty)$. So $\delta(t)$ is a "function" with all its "mass" at the single point $t = 0$. In other words, for any interval $[c, d]$

$$\int_c^d \delta(t)\, dt = \begin{cases} 1 & \text{if the interval } [c, d] \text{ contains 0, i.e. } c \le 0 \le d, \\ 0 & \text{otherwise.} \end{cases}$$

Unfortunately there is no such function in the classical sense. You could informally think that $\delta(t)$ is zero for $t \ne 0$ and somehow infinite at $t = 0$.

A good way to think about $\delta(t)$ is as a limit of short pulses whose integral is 1. For example, suppose that we have a square pulse $\varphi(t)$ as above with $a = 0$, $M = 1/b$, that is $\varphi(t) = \frac{u(t) - u(t-b)}{b}$. Compute

$$\int_{-\infty}^{\infty} \varphi(t)f(t)\, dt = \int_{-\infty}^{\infty} \frac{u(t) - u(t-b)}{b} f(t)\, dt = \frac{1}{b}\int_0^b f(t)\, dt.$$

If $f(t)$ is continuous at $t = 0$, then for very small b, the function $f(t)$ is approximately equal to $f(0)$ on the interval $[0, b]$. We approximate the integral

$$\frac{1}{b}\int_0^b f(t)\, dt \approx \frac{1}{b}\int_0^b f(0)\, dt = f(0).$$

Therefore,

$$\lim_{b \to 0} \int_{-\infty}^{\infty} \varphi(t)f(t)\, dt = \lim_{b \to 0} \frac{1}{b}\int_0^b f(t)\, dt = f(0).$$

---

[*]Named after the English physicist and mathematician Paul Adrien Maurice Dirac (1902–1984).

Let us therefore accept $\delta(t)$ as an object that is possible to integrate. We often want to shift δ to another point, for example $\delta(t - a)$. In that case we have

$$\int_{-\infty}^{\infty} \delta(t - a) f(t) \, dt = f(a).$$

Note that $\delta(a - t)$ is the same object as $\delta(t - a)$. In other words, the convolution of $\delta(t)$ with $f(t)$ is again $f(t)$,

$$(f * \delta)(t) = \int_{0}^{t} \delta(t - s) f(s) \, ds = f(t).$$

As we can integrate $\delta(t)$, let us compute its Laplace transform.

$$\boxed{\mathcal{L}\{\delta(t - a)\} = \int_{0}^{\infty} e^{-st} \delta(t - a) \, dt = e^{-as}.}$$

In particular,

$$\mathcal{L}\{\delta(t)\} = 1.$$

Remark 6.4.1: Notice that the Laplace transform of $\delta(t - a)$ looks like the Laplace transform of the derivative of the Heaviside function $u(t - a)$, if we could differentiate the Heaviside function. First notice

$$\mathcal{L}\{u(t - a)\} = \frac{e^{-as}}{s}.$$

To obtain what the Laplace transform of the derivative would be we multiply by s, to obtain e^{-as}, which is the Laplace transform of $\delta(t - a)$. We see the same thing using integration,

$$\int_{0}^{t} \delta(s - a) \, ds = u(t - a).$$

So in a certain sense

$$\text{``} \frac{d}{dt}[u(t - a)] = \delta(t - a) \quad \text{''}$$

This line of reasoning allows us to talk about derivatives of functions with jump discontinuities. We can think of the derivative of the Heaviside function $u(t - a)$ as being somehow infinite at a, which is precisely our intuitive understanding of the delta function.

Example 6.4.1: Let us compute $\mathcal{L}^{-1}\left\{\frac{s+1}{s}\right\}$. So far we have always looked at proper rational functions in the s variable. That is, the numerator was always of lower degree than the denominator. Not so with $\frac{s+1}{s}$. We write,

$$\mathcal{L}^{-1}\left\{\frac{s+1}{s}\right\} = \mathcal{L}^{-1}\left\{1 + \frac{1}{s}\right\} = \mathcal{L}^{-1}\{1\} + \mathcal{L}^{-1}\left\{\frac{1}{s}\right\} = \delta(t) + 1.$$

The resulting object is a generalized function and only makes sense when put underneath an integral.

6.4.3   Impulse response

As we said before, in the differential equation $Lx = f(t)$, we think of $f(t)$ as input, and $x(t)$ as the output. Often it is important to find the response to an impulse, and then we use the delta function in place of $f(t)$. The solution to

$$Lx = \delta(t)$$

is called the *impulse response*.

Example 6.4.2: Solve (find the impulse response)

$$x'' + \omega_0^2 x = \delta(t), \quad x(0) = 0, \quad x'(0) = 0. \tag{6.3}$$

We first apply the Laplace transform to the equation. Denote the transform of $x(t)$ by $X(s)$.

$$s^2 X(s) + \omega_0^2 X(s) = 1, \quad \text{and so} \quad X(s) = \frac{1}{s^2 + \omega_0^2}.$$

Taking the inverse Laplace transform we obtain

$$x(t) = \frac{\sin(\omega_0 t)}{\omega_0}.$$

Let us notice something about the above example. We have proved before that when the input was $f(t)$, then the solution to $Lx = f(t)$ was given by

$$x(t) = \int_0^t f(\tau) \frac{\sin(\omega_0(t - \tau))}{\omega_0} \, d\tau.$$

Notice that the solution for an arbitrary input is given as convolution with the impulse response. Let us see why. The key is to notice that for functions $x(t)$ and $f(t)$,

$$(x * f)''(t) = \frac{d^2}{dt^2} \left[\int_0^t f(\tau) x(t - \tau) \, d\tau \right] = \int_0^t f(\tau) x''(t - \tau) \, d\tau = (x'' * f)(t).$$

We simply differentiate twice under the integral[*], the details are left as an exercise. And so if we convolve the entire equation (6.3), the left hand side becomes

$$(x'' + \omega_0^2 x) * f = (x'' * f) + \omega_0^2(x * f) = (x * f)'' + \omega_0^2(x * f).$$

The right hand side becomes

$$(\delta * f)(t) = f(t).$$

Therefore $y(t) = (x * f)(t)$ is the solution to

$$y'' + \omega_0^2 y = f(t).$$

This procedure works in general for other linear equations $Lx = f(t)$. If you determine the impulse response, you also know how to obtain the output $x(t)$ for any input $f(t)$ by simply convolving the impulse response and the input $f(t)$.

---

[*]You should really think of the integral going over $(-\infty, \infty)$ rather than over $[0, t]$ and simply assume that $f(t)$ and $x(t)$ are continuous and zero for negative t.

6.4.4 Three-point beam bending

Let us give another quite different example where delta functions turn up. In this case representing point loads on a steel beam. Suppose we have a beam of length L, resting on two simple supports at the ends. Let x denote the position on the beam, and let $y(x)$ denote the deflection of the beam in the vertical direction. The deflection $y(x)$ satisfies the *Euler-Bernoulli equation**,

$$EI\frac{d^4y}{dx^4} = F(x),$$

where E and I are constants[†] and $F(x)$ is the force applied per unit length at position x. The situation we are interested in is when the force is applied at a single point as in Figure 6.4.

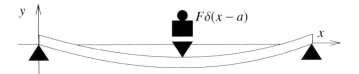

Figure 6.4: Three-point bending.

In this case the equation becomes

$$EI\frac{d^4y}{dx^4} = -F\delta(x-a),$$

where $x = a$ is the point where the mass is applied. F is the force applied and the minus sign indicates that the force is downward, that is, in the negative y direction. The end points of the beam satisfy the conditions,

$$y(0) = 0, \qquad y''(0) = 0,$$
$$y(L) = 0, \qquad y''(L) = 0.$$

See § 5.2, for further information about endpoint conditions applied to beams.

Example 6.4.3: Suppose that length of the beam is 2, and suppose that $EI = 1$ for simplicity. Further suppose that the force $F = 1$ is applied at $x = 1$. That is, we have the equation

$$\frac{d^4y}{dx^4} = -\delta(x-1),$$

---

[*]Named for the Swiss mathematicians Jacob Bernoulli (1654–1705), Daniel Bernoulli —nephew of Jacob— (1700–1782), and Leonhard Paul Euler (1707–1783).

[†]E is the elastic modulus and I is the second moment of area. Let us not worry about the details and simply think of these as some given constants.

and the endpoint conditions are

$$y(0) = 0, \qquad y''(0) = 0, \qquad y(2) = 0, \qquad y''(2) = 0.$$

We could integrate, but using the Laplace transform is even easier. We apply the transform in the x variable rather than the t variable. Let us again denote the transform of $y(x)$ as $Y(s)$.

$$s^4 Y(s) - s^3 y(0) - s^2 y'(0) - s y''(0) - y'''(0) = -e^{-s}.$$

We notice that $y(0) = 0$ and $y''(0) = 0$. Let us call $C_1 = y'(0)$ and $C_2 = y'''(0)$. We solve for $Y(s)$,

$$Y(s) = \frac{-e^{-s}}{s^4} + \frac{C_1}{s^2} + \frac{C_2}{s^4}.$$

We take the inverse Laplace transform utilizing the second shifting property (6.1) to take the inverse of the first term.

$$y(x) = \frac{-(x-1)^3}{6} u(x-1) + C_1 x + \frac{C_2}{6} x^3.$$

We still need to apply two of the endpoint conditions. As the conditions are at $x = 2$ we can simply replace $u(x-1) = 1$ when taking the derivatives. Therefore,

$$0 = y(2) = \frac{-(2-1)^3}{6} + C_1(2) + \frac{C_2}{6} 2^3 = \frac{-1}{6} + 2C_1 + \frac{4}{3} C_2,$$

and

$$0 = y''(2) = \frac{-3 \cdot 2 \cdot (2-1)}{6} + \frac{C_2}{6} 3 \cdot 2 \cdot 2 = -1 + 2C_2.$$

Hence $C_2 = \frac{1}{2}$ and solving for C_1 using the first equation we obtain $C_1 = \frac{-1}{4}$. Our solution for the beam deflection is

$$y(x) = \frac{-(x-1)^3}{6} u(x-1) - \frac{x}{4} + \frac{x^3}{12}.$$

6.4.5 Exercises

Exercise 6.4.1: *Solve (find the impulse response) $x'' + x' + x = \delta(t)$, $x(0) = 0$, $x'(0) = 0$.*

Exercise 6.4.2: *Solve (find the impulse response) $x'' + 2x' + x = \delta(t)$, $x(0) = 0$, $x'(0) = 0$.*

Exercise 6.4.3: *A pulse can come later and can be bigger. Solve $x'' + 4x = 4\delta(t-1)$, $x(0) = 0$, $x'(0) = 0$.*

Exercise 6.4.4: *Suppose that $f(t)$ and $g(t)$ are differentiable functions and suppose that $f(t) = g(t) = 0$ for all $t \le 0$. Show that*

$$(f * g)'(t) = (f' * g)(t) = (f * g')(t).$$

Exercise 6.4.5: *Suppose that $Lx = \delta(t)$, $x(0) = 0$, $x'(0) = 0$, has the solution $x = e^{-t}$ for $t > 0$. Find the solution to $Lx = t^2$, $x(0) = 0$, $x'(0) = 0$ for $t > 0$.*

Exercise 6.4.6: *Compute $\mathcal{L}^{-1}\left\{\frac{s^2+s+1}{s^2}\right\}$.*

Exercise 6.4.7 (challenging)*: Solve Example 6.4.3 via integrating 4 times in the x variable.*

Exercise 6.4.8: *Suppose we have a beam of length 1 simply supported at the ends and suppose that force $F = 1$ is applied at $x = \frac{3}{4}$ in the downward direction. Suppose that $EI = 1$ for simplicity. Find the beam deflection $y(x)$.*

Exercise 6.4.101: *Solve (find the impulse response) $x'' = \delta(t)$, $x(0) = 0$, $x'(0) = 0$.*

Exercise 6.4.102: *Solve (find the impulse response) $x' + ax = \delta(t)$, $x(0) = 0$, $x'(0) = 0$.*

Exercise 6.4.103: *Suppose that $Lx = \delta(t)$, $x(0) = 0$, $x'(0) = 0$, has the solution $x(t) = \cos(t)$ for $t > 0$. Find (in closed form) the solution to $Lx = \sin(t)$, $x(0) = 0$, $x'(0) = 0$ for $t > 0$.*

Exercise 6.4.104: *Compute $\mathcal{L}^{-1}\left\{\frac{s^2}{s^2+1}\right\}$.*

Exercise 6.4.105: *Compute $\mathcal{L}^{-1}\left\{\frac{3s^2e^{-s}+2}{s^2}\right\}$.*

Chapter 7

Power series methods

7.1 Power series

Note: 1 or 1.5 lecture, §8.1 in [EP], §5.1 in [BD]

Many functions can be written in terms of a power series

$$\sum_{k=0}^{\infty} a_k(x - x_0)^k.$$

If we assume that a solution of a differential equation is written as a power series, then perhaps we can use a method reminiscent of undetermined coefficients. That is, we will try to solve for the numbers a_k. Before we can carry out this process, let us review some results and concepts about power series.

7.1.1 Definition

As we said, a *power series* is an expression such as

$$\sum_{k=0}^{\infty} a_k(x - x_0)^k = a_0 + a_1(x - x_0) + a_2(x - x_0)^2 + a_3(x - x_0)^3 + \cdots, \tag{7.1}$$

where $a_0, a_1, a_2, \ldots, a_k, \ldots$ and x_0 are constants. Let

$$S_n(x) = \sum_{k=0}^{n} a_k(x - x_0)^k = a_0 + a_1(x - x_0) + a_2(x - x_0)^2 + a_3(x - x_0)^3 + \cdots + a_n(x - x_0)^n,$$

denote the so-called *partial sum*. If for some x, the limit

$$\lim_{n \to \infty} S_n(x) = \lim_{n \to \infty} \sum_{k=0}^{n} a_k(x - x_0)^k$$

exists, then we say that the series (7.1) *converges* at x. Note that for $x = x_0$, the series always converges to a_0. When (7.1) converges at any other point $x \neq x_0$, we say that (7.1) is a *convergent power series*. In this case we write

$$\sum_{k=0}^{\infty} a_k(x - x_0)^k = \lim_{n \to \infty} \sum_{k=0}^{n} a_k(x - x_0)^k.$$

If the series does not converge for any point $x \neq x_0$, we say that the series is *divergent*.

Example 7.1.1: The series

$$\sum_{k=0}^{\infty} \frac{1}{k!} x^k = 1 + x + \frac{x^2}{2} + \frac{x^3}{6} + \cdots$$

is convergent for any x. Recall that $k! = 1 \cdot 2 \cdot 3 \cdots k$ is the factorial. By convention we define $0! = 1$. In fact, you may recall that this series converges to e^x.

We say that (7.1) *converges absolutely* at x whenever the limit

$$\lim_{n \to \infty} \sum_{k=0}^{n} |a_k| |x - x_0|^k$$

exists. That is, the series $\sum_{k=0}^{\infty} |a_k| |x - x_0|^k$ is convergent. If (7.1) converges absolutely at x, then it converges at x. However, the opposite implication is not true.

Example 7.1.2: The series

$$\sum_{k=1}^{\infty} \frac{1}{k} x^k$$

converges absolutely for all x in the interval $(-1, 1)$. It converges at $x = -1$, as $\sum_{k=1}^{\infty} \frac{(-1)^k}{k}$ converges (conditionally) by the alternating series test. But the power series does not converge absolutely at $x = -1$, because $\sum_{k=1}^{\infty} \frac{1}{k}$ does not converge. The series diverges at $x = 1$.

7.1.2   Radius of convergence

If a power series converges absolutely at some x_1, then for all x such that $|x - x_0| \leq |x_1 - x_0|$ (that is, x is closer than x_1 to x_0) we have $\left| a_k(x - x_0)^k \right| \leq \left| a_k(x_1 - x_0)^k \right|$ for all k. As the numbers $\left| a_k(x_1 - x_0)^k \right|$ sum to some finite limit, summing smaller positive numbers $\left| a_k(x - x_0)^k \right|$ must also have a finite limit. Therefore, the series must converge absolutely at x. We have the following result.

Theorem 7.1.1. *For a power series (7.1), there exists a number ρ (we allow $\rho = \infty$) called the* radius of convergence *such that the series converges absolutely on the interval $(x_0 - \rho, x_0 + \rho)$ and diverges for $x < x_0 - \rho$ and $x > x_0 + \rho$. We write $\rho = \infty$ if the series converges for all x.*

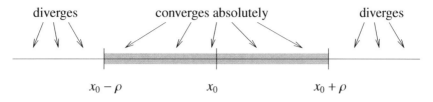

Figure 7.1: Convergence of a power series.

See Figure 7.1. In Example 7.1.1 the radius of convergence is $\rho = \infty$ as the series converges everywhere. In Example 7.1.2 the radius of convergence is $\rho = 1$. We note that $\rho = 0$ is another way of saying that the series is divergent.

A useful test for convergence of a series is the *ratio test*. Suppose that

$$\sum_{k=0}^{\infty} c_k$$

is a series such that the limit

$$L = \lim_{n \to \infty} \left| \frac{c_{k+1}}{c_k} \right|$$

exists. Then the series converges absolutely if $L < 1$ and diverges if $L > 1$.

Let us apply this test to the series (7.1). That is we let $c_k = a_k(x - x_0)^k$ in the test. Compute

$$L = \lim_{n \to \infty} \left| \frac{c_{k+1}}{c_k} \right| = \lim_{n \to \infty} \left| \frac{a_{k+1}(x - x_0)^{k+1}}{a_k(x - x_0)^k} \right| = \lim_{n \to \infty} \left| \frac{a_{k+1}}{a_k} \right| |x - x_0|.$$

Define A by

$$A = \lim_{n \to \infty} \left| \frac{a_{k+1}}{a_k} \right|.$$

Then if $1 > L = A|x - x_0|$ the series (7.1) converges absolutely. If $A = 0$, then the series always converges. If $A > 0$, then the series converges absolutely if $|x - x_0| < 1/A$, and diverges if $|x - x_0| > 1/A$. That is, the radius of convergence is $1/A$. Let us summarize.

Theorem 7.1.2. *Let*

$$\sum_{k=0}^{\infty} a_k(x - x_0)^k$$

be a power series such that

$$A = \lim_{n \to \infty} \left| \frac{a_{k+1}}{a_k} \right|$$

exists. If $A = 0$, then the radius of convergence of the series is ∞. Otherwise the radius of convergence is $1/A$.

Example 7.1.3: Suppose we have the series

$$\sum_{k=0}^{\infty} 2^{-k}(x-1)^k.$$

First we compute,

$$A = \lim_{k\to\infty}\left|\frac{a_{k+1}}{a_k}\right| = \lim_{k\to\infty}\left|\frac{2^{-k-1}}{2^{-k}}\right| = 2^{-1} = 1/2.$$

Therefore the radius of convergence is 2, and the series converges absolutely on the interval $(-1, 3)$.

The ratio test does not always apply. That is the limit of $\left|\frac{a_{k+1}}{a_k}\right|$ might not exist. There exist more sophisticated ways of finding the radius of convergence, but those would be beyond the scope of this chapter.

7.1.3   Analytic functions

Functions represented by power series are called *analytic functions*. Not every function is analytic, although the majority of the functions you have seen in calculus are.

An analytic function $f(x)$ is equal to its *Taylor series** near a point x_0. That is, for x near x_0 we have

$$f(x) = \sum_{k=0}^{\infty} \frac{f^{(k)}(x_0)}{k!}(x-x_0)^k, \tag{7.2}$$

where $f^{(k)}(x_0)$ denotes the kth derivative of $f(x)$ at the point x_0.

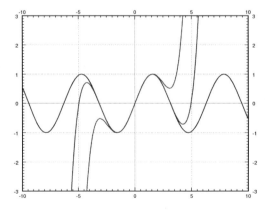

Figure 7.2: The sine function and its Taylor approximations around $x_0 = 0$ of 5th and 9th degree.

---

*Named after the English mathematician Sir Brook Taylor (1685–1731).

For example, sine is an analytic function and its Taylor series around $x_0 = 0$ is given by

$$\sin(x) = \sum_{n=0}^{\infty} \frac{(-1)^n}{(2n+1)!} x^{2n+1}.$$

In Figure 7.2 on the facing page we plot $\sin(x)$ and the truncations of the series up to degree 5 and 9. You can see that the approximation is very good for x near 0, but gets worse for x further away from 0. This is what happens in general. To get a good approximation far away from x_0 you need to take more and more terms of the Taylor series.

7.1.4 Manipulating power series

One of the main properties of power series that we will use is that we can differentiate them term by term. That is, suppose that $\sum a_k(x - x_0)^k$ is a convergent power series. Then for x in the radius of convergence we have

$$\frac{d}{dx}\left[\sum_{k=0}^{\infty} a_k(x - x_0)^k\right] = \sum_{k=1}^{\infty} ka_k(x - x_0)^{k-1}.$$

Notice that the term corresponding to $k = 0$ disappeared as it was constant. The radius of convergence of the differentiated series is the same as that of the original.

Example 7.1.4: Let us show that the exponential $y = e^x$ solves $y' = y$. First write

$$y = e^x = \sum_{k=0}^{\infty} \frac{1}{k!} x^k.$$

Now differentiate

$$y' = \sum_{k=1}^{\infty} k \frac{1}{k!} x^{k-1} = \sum_{k=1}^{\infty} \frac{1}{(k-1)!} x^{k-1}.$$

We *reindex* the series by simply replacing k with $k + 1$. The series does not change, what changes is simply how we write it. After reindexing the series starts at $k = 0$ again.

$$\sum_{k=1}^{\infty} \frac{1}{(k-1)!} x^{k-1} = \sum_{k+1=1}^{\infty} \frac{1}{((k+1)-1)!} x^{(k+1)-1} = \sum_{k=0}^{\infty} \frac{1}{k!} x^k.$$

That was precisely the power series for e^x that we started with, so we showed that $\frac{d}{dx}[e^x] = e^x$.

Convergent power series can be added and multiplied together, and multiplied by constants using the following rules. First, we can add series by adding term by term,

$$\left(\sum_{k=0}^{\infty} a_k(x - x_0)^k\right) + \left(\sum_{k=0}^{\infty} b_k(x - x_0)^k\right) = \sum_{k=0}^{\infty} (a_k + b_k)(x - x_0)^k.$$

We can multiply by constants,

$$\alpha \left(\sum_{k=0}^{\infty} a_k (x - x_0)^k \right) = \sum_{k=0}^{\infty} \alpha a_k (x - x_0)^k.$$

We can also multiply series together,

$$\left(\sum_{k=0}^{\infty} a_k (x - x_0)^k \right) \left(\sum_{k=0}^{\infty} b_k (x - x_0)^k \right) = \sum_{k=0}^{\infty} c_k (x - x_0)^k,$$

where $c_k = a_0 b_k + a_1 b_{k-1} + \cdots + a_k b_0$. The radius of convergence of the sum or the product is at least the minimum of the radii of convergence of the two series involved.

7.1.5   Power series for rational functions

Polynomials are simply finite power series. That is, a polynomial is a power series where the a_k are zero for all k large enough. We can always expand a polynomial as a power series about any point x_0 by writing the polynomial as a polynomial in $(x - x_0)$. For example, let us write $2x^2 - 3x + 4$ as a power series around $x_0 = 1$:

$$2x^2 - 3x + 4 = 3 + (x - 1) + 2(x - 1)^2.$$

In other words $a_0 = 3$, $a_1 = 1$, $a_2 = 2$, and all other $a_k = 0$. To do this, we know that $a_k = 0$ for all $k \geq 3$ as the polynomial is of degree 2. We write $a_0 + a_1(x - 1) + a_2(x - 1)^2$, we expand, and we solve for a_0, a_1, and a_2. We could have also differentiated at $x = 1$ and used the Taylor series formula (7.2).

Let us look at rational functions, that is, ratios of polynomials. An important fact is that a series for a function only defines the function on an interval even if the function is defined elsewhere. For example, for $-1 < x < 1$ we have

$$\frac{1}{1 - x} = \sum_{k=0}^{\infty} x^k = 1 + x + x^2 + \cdots.$$

This series is called the *geometric series*. The ratio test tells us that the radius of convergence is 1. The series diverges for $x \leq -1$ and $x \geq 1$, even though $\frac{1}{1-x}$ is defined for all $x \neq 1$.

We can use the geometric series together with rules for addition and multiplication of power series to expand rational functions around a point, as long as the denominator is not zero at x_0. Note that as for polynomials, we could equivalently use the Taylor series expansion (7.2).

Example 7.1.5: Expand $\frac{x}{1+2x+x^2}$ as a power series around the origin ($x_0 = 0$) and find the radius of convergence.

First, write $1 + 2x + x^2 = (1 + x)^2 = (1 - (-x))^2$. Now we compute

$$\frac{x}{1 + 2x + x^2} = x\left(\frac{1}{1 - (-x)}\right)^2$$

$$= x\left(\sum_{k=0}^{\infty}(-1)^k x^k\right)^2$$

$$= x\left(\sum_{k=0}^{\infty} c_k x^k\right)$$

$$= \sum_{k=0}^{\infty} c_k x^{k+1},$$

where using the formula for the product of series we obtain, $c_0 = 1$, $c_1 = -1 - 1 = -2$, $c_2 = 1 + 1 + 1 = 3$, etc.... Therefore

$$\frac{x}{1 + 2x + x^2} = \sum_{k=1}^{\infty}(-1)^{k+1} k x^k = x - 2x^2 + 3x^3 - 4x^4 + \cdots$$

The radius of convergence is at least 1. We use the ratio test

$$\lim_{k\to\infty}\left|\frac{a_{k+1}}{a_k}\right| = \lim_{k\to\infty}\left|\frac{(-1)^{k+2}(k+1)}{(-1)^{k+1}k}\right| = \lim_{k\to\infty}\frac{k+1}{k} = 1.$$

So the radius of convergence is actually equal to 1.

When the rational function is more complicated, it is also possible to use method of partial fractions. For example, to find the Taylor series for $\frac{x^3+x}{x^2-1}$, we write

$$\frac{x^3 + x}{x^2 - 1} = x + \frac{1}{1 + x} - \frac{1}{1 - x} = x + \sum_{k=0}^{\infty}(-1)^k x^k - \sum_{k=0}^{\infty} x^k = -x + \sum_{\substack{k=3 \\ k\text{ odd}}}^{\infty}(-2)x^k.$$

7.1.6   Exercises

Exercise 7.1.1: *Is the power series* $\sum_{k=0}^{\infty} e^k x^k$ *convergent? If so, what is the radius of convergence?*

Exercise 7.1.2: *Is the power series* $\sum_{k=0}^{\infty} k x^k$ *convergent? If so, what is the radius of convergence?*

Exercise 7.1.3: *Is the power series* $\sum_{k=0}^{\infty} k! x^k$ *convergent? If so, what is the radius of convergence?*

Exercise 7.1.4: *Is the power series* $\sum_{k=0}^{\infty} \dfrac{1}{(2k)!}(x-10)^k$ *convergent? If so, what is the radius of convergence?*

Exercise 7.1.5: *Determine the Taylor series for* $\sin x$ *around the point* $x_0 = \pi$.

Exercise 7.1.6: *Determine the Taylor series for* $\ln x$ *around the point* $x_0 = 1$, *and find the radius of convergence.*

Exercise 7.1.7: *Determine the Taylor series and its radius of convergence of* $\dfrac{1}{1+x}$ *around* $x_0 = 0$.

Exercise 7.1.8: *Determine the Taylor series and its radius of convergence of* $\dfrac{x}{4-x^2}$ *around* $x_0 = 0$. *Hint: You will not be able to use the ratio test.*

Exercise 7.1.9: *Expand* $x^5 + 5x + 1$ *as a power series around* $x_0 = 5$.

Exercise 7.1.10: *Suppose that the ratio test applies to a series* $\sum_{k=0}^{\infty} a_k x^k$. *Show, using the ratio test, that the radius of convergence of the differentiated series is the same as that of the original series.*

Exercise 7.1.11: *Suppose that* f *is an analytic function such that* $f^{(n)}(0) = n$. *Find* $f(1)$.

Exercise 7.1.101: *Is the power series* $\sum_{n=1}^{\infty} (0.1)^n x^n$ *convergent? If so, what is the radius of convergence?*

Exercise 7.1.102 (challenging)***:*** *Is the power series* $\sum_{n=1}^{\infty} \dfrac{n!}{n^n} x^n$ *convergent? If so, what is the radius of convergence?*

Exercise 7.1.103: *Using the geometric series, expand* $\frac{1}{1-x}$ *around* $x_0 = 2$. *For what* x *does the series converge?*

Exercise 7.1.104 (challenging)***:*** *Find the Taylor series for* $x^7 e^x$ *around* $x_0 = 0$.

Exercise 7.1.105 (challenging)***:*** *Imagine* f *and* g *are analytic functions such that* $f^{(k)}(0) = g^{(k)}(0)$ *for all large enough* k. *What can you say about* $f(x) - g(x)$?

7.2 Series solutions of linear second order ODEs

Note: 1 or 1.5 lecture, §8.2 in [EP], §5.2 and §5.3 in [BD]

Suppose we have a linear second order homogeneous ODE of the form

$$p(x)y'' + q(x)y' + r(x)y = 0.$$

Suppose that $p(x)$, $q(x)$, and $r(x)$ are polynomials. We will try a solution of the form

$$y = \sum_{k=0}^{\infty} a_k (x - x_0)^k$$

and solve for the a_k to try to obtain a solution defined in some interval around x_0.

The point x_0 is called an *ordinary point* if $p(x_0) \neq 0$. That is, the functions

$$\frac{q(x)}{p(x)} \quad \text{and} \quad \frac{r(x)}{p(x)}$$

are defined for x near x_0. If $p(x_0) = 0$, then we say x_0 is a *singular point*. Handling singular points is harder than ordinary points and so we now focus only on ordinary points.

Example 7.2.1: Let us start with a very simple example

$$y'' - y = 0.$$

Let us try a power series solution near $x_0 = 0$, which is an ordinary point. Every point is an ordinary point in fact, as the equation is constant coefficient. We already know we should obtain exponentials or the hyperbolic sine and cosine, but let us pretend we do not know this.

We try

$$y = \sum_{k=0}^{\infty} a_k x^k.$$

If we differentiate, the $k = 0$ term is a constant and hence disappears. We therefore get

$$y' = \sum_{k=1}^{\infty} k a_k x^{k-1}.$$

We differentiate yet again to obtain (now the $k = 1$ term disappears)

$$y'' = \sum_{k=2}^{\infty} k(k-1) a_k x^{k-2}.$$

We reindex the series (replace k with $k + 2$) to obtain

$$y'' = \sum_{k=0}^{\infty} (k+2)(k+1) a_{k+2} x^k.$$

Now we plug y and y'' into the differential equation

$$0 = y'' - y = \left(\sum_{k=0}^{\infty} (k+2)(k+1) a_{k+2} x^k \right) - \left(\sum_{k=0}^{\infty} a_k x^k \right)$$

$$= \sum_{k=0}^{\infty} \left((k+2)(k+1) a_{k+2} x^k - a_k x^k \right)$$

$$= \sum_{k=0}^{\infty} \left((k+2)(k+1) a_{k+2} - a_k \right) x^k.$$

As $y'' - y$ is supposed to be equal to 0, we know that the coefficients of the resulting series must be equal to 0. Therefore,

$$(k+2)(k+1) a_{k+2} - a_k = 0, \qquad \text{or} \qquad a_{k+2} = \frac{a_k}{(k+2)(k+1)}.$$

The above equation is called a *recurrence relation* for the coefficients of the power series. It did not matter what a_0 or a_1 was. They can be arbitrary. But once we pick a_0 and a_1, then all other coefficients are determined by the recurrence relation.

Let us see what the coefficients must be. First, a_0 and a_1 are arbitrary

$$a_2 = \frac{a_0}{2}, \quad a_3 = \frac{a_1}{(3)(2)}, \quad a_4 = \frac{a_2}{(4)(3)} = \frac{a_0}{(4)(3)(2)}, \quad a_5 = \frac{a_3}{(5)(4)} = \frac{a_1}{(5)(4)(3)(2)}, \quad \dots$$

So we note that for even k, that is $k = 2n$ we get

$$a_k = a_{2n} = \frac{a_0}{(2n)!},$$

and for odd k, that is $k = 2n + 1$ we have

$$a_k = a_{2n+1} = \frac{a_1}{(2n+1)!}.$$

Let us write down the series

$$y = \sum_{k=0}^{\infty} a_k x^k = \sum_{n=0}^{\infty} \left(\frac{a_0}{(2n)!} x^{2n} + \frac{a_1}{(2n+1)!} x^{2n+1} \right) = a_0 \sum_{n=0}^{\infty} \frac{1}{(2n)!} x^{2n} + a_1 \sum_{n=0}^{\infty} \frac{1}{(2n+1)!} x^{2n+1}.$$

We recognize the two series as the hyperbolic sine and cosine. Therefore,

$$y = a_0 \cosh x + a_1 \sinh x.$$

Of course, in general we will not be able to recognize the series that appears, since usually there will not be any elementary function that matches it. In that case we will be content with the series.

Example 7.2.2: Let us do a more complex example. Suppose we wish to solve *Airy's equation**,
that is

$$y'' - xy = 0,$$

near the point $x_0 = 0$. Note that $x_0 = 0$ is an ordinary point.

We try

$$y = \sum_{k=0}^{\infty} a_k x^k.$$

We differentiate twice (as above) to obtain

$$y'' = \sum_{k=2}^{\infty} k(k-1) a_k x^{k-2}.$$

We plug y into the equation

$$0 = y'' - xy = \left(\sum_{k=2}^{\infty} k(k-1) a_k x^{k-2} \right) - x \left(\sum_{k=0}^{\infty} a_k x^k \right)$$

$$= \left(\sum_{k=2}^{\infty} k(k-1) a_k x^{k-2} \right) - \left(\sum_{k=0}^{\infty} a_k x^{k+1} \right).$$

We reindex to make things easier to sum

$$0 = y'' - xy = \left(2a_2 + \sum_{k=1}^{\infty} (k+2)(k+1) a_{k+2} x^k \right) - \left(\sum_{k=1}^{\infty} a_{k-1} x^k \right).$$

$$= 2a_2 + \sum_{k=1}^{\infty} \left((k+2)(k+1) a_{k+2} - a_{k-1} \right) x^k.$$

Again $y'' - xy$ is supposed to be 0 so first we notice that $a_2 = 0$ and also

$$(k+2)(k+1) a_{k+2} - a_{k-1} = 0, \qquad \text{or} \qquad a_{k+2} = \frac{a_{k-1}}{(k+2)(k+1)}.$$

Now we jump in steps of three. First we notice that since $a_2 = 0$ we must have that, $a_5 = 0$, $a_8 = 0$,
$a_{11} = 0$, etc.... In general $a_{3n+2} = 0$.

The constants a_0 and a_1 are arbitrary and we obtain

$$a_3 = \frac{a_0}{(3)(2)}, \quad a_4 = \frac{a_1}{(4)(3)}, \quad a_6 = \frac{a_3}{(6)(5)} = \frac{a_0}{(6)(5)(3)(2)}, \quad a_7 = \frac{a_4}{(7)(6)} = \frac{a_1}{(7)(6)(4)(3)}, \quad \cdots$$

For a_k where k is a multiple of 3, that is $k = 3n$ we notice that

$$a_{3n} = \frac{a_0}{(2)(3)(5)(6) \cdots (3n-1)(3n)}.$$

---

*Named after the English mathematician Sir George Biddell Airy (1801–1892).

For a_k where $k = 3n + 1$, we notice

$$a_{3n+1} = \frac{a_1}{(3)(4)(6)(7)\cdots(3n)(3n+1)}.$$

In other words, if we write down the series for y we notice that it has two parts

$$y = \left(a_0 + \frac{a_0}{6}x^3 + \frac{a_0}{180}x^6 + \cdots + \frac{a_0}{(2)(3)(5)(6)\cdots(3n-1)(3n)}x^{3n} + \cdots\right)$$

$$+ \left(a_1 x + \frac{a_1}{12}x^4 + \frac{a_1}{504}x^7 + \cdots + \frac{a_1}{(3)(4)(6)(7)\cdots(3n)(3n+1)}x^{3n+1} + \cdots\right)$$

$$= a_0\left(1 + \frac{1}{6}x^3 + \frac{1}{180}x^6 + \cdots + \frac{1}{(2)(3)(5)(6)\cdots(3n-1)(3n)}x^{3n} + \cdots\right)$$

$$+ a_1\left(x + \frac{1}{12}x^4 + \frac{1}{504}x^7 + \cdots + \frac{1}{(3)(4)(6)(7)\cdots(3n)(3n+1)}x^{3n+1} + \cdots\right).$$

We define

$$y_1(x) = 1 + \frac{1}{6}x^3 + \frac{1}{180}x^6 + \cdots + \frac{1}{(2)(3)(5)(6)\cdots(3n-1)(3n)}x^{3n} + \cdots,$$

$$y_2(x) = x + \frac{1}{12}x^4 + \frac{1}{504}x^7 + \cdots + \frac{1}{(3)(4)(6)(7)\cdots(3n)(3n+1)}x^{3n+1} + \cdots,$$

and write the general solution to the equation as $y(x) = a_0 y_1(x) + a_1 y_2(x)$. Notice from the power series that $y_1(0) = 1$ and $y_2(0) = 0$. Also, $y_1'(0) = 0$ and $y_2'(0) = 1$. Therefore $y(x)$ is a solution that satisfies the initial conditions $y(0) = a_0$ and $y'(0) = a_1$.

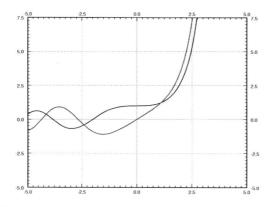

Figure 7.3: The two solutions y_1 and y_2 to Airy's equation.

The functions y_1 and y_2 cannot be written in terms of the elementary functions that you know. See Figure 7.3 for the plot of the solutions y_1 and y_2. These functions have many interesting properties. For example, they are oscillatory for negative x (like solutions to $y'' + y = 0$) and for positive x they grow without bound (like solutions to $y'' - y = 0$).

Sometimes a solution may turn out to be a polynomial.

Example 7.2.3: Let us find a solution to the so-called *Hermite's equation of order n*[*] is the equation

$$y'' - 2xy' + 2ny = 0.$$

Let us find a solution around the point $x_0 = 0$. We try

$$y = \sum_{k=0}^{\infty} a_k x^k.$$

We differentiate (as above) to obtain

$$y' = \sum_{k=1}^{\infty} k a_k x^{k-1},$$

$$y'' = \sum_{k=2}^{\infty} k(k-1) a_k x^{k-2}.$$

Now we plug into the equation

$$
\begin{aligned}
0 = y'' - 2xy' + 2ny &= \left(\sum_{k=2}^{\infty} k(k-1) a_k x^{k-2}\right) - 2x\left(\sum_{k=1}^{\infty} k a_k x^{k-1}\right) + 2n\left(\sum_{k=0}^{\infty} a_k x^k\right) \\
&= \left(\sum_{k=2}^{\infty} k(k-1) a_k x^{k-2}\right) - \left(\sum_{k=1}^{\infty} 2k a_k x^k\right) + \left(\sum_{k=0}^{\infty} 2n a_k x^k\right) \\
&= \left(2a_2 + \sum_{k=1}^{\infty} (k+2)(k+1) a_{k+2} x^k\right) - \left(\sum_{k=1}^{\infty} 2k a_k x^k\right) + \left(2na_0 + \sum_{k=1}^{\infty} 2n a_k x^k\right) \\
&= 2a_2 + 2na_0 + \sum_{k=1}^{\infty} \left((k+2)(k+1) a_{k+2} - 2k a_k + 2n a_k\right) x^k.
\end{aligned}
$$

As $y'' - 2xy' + 2ny = 0$ we have

$$(k+2)(k+1) a_{k+2} + (-2k + 2n) a_k = 0, \qquad \text{or} \qquad a_{k+2} = \frac{(2k - 2n)}{(k+2)(k+1)} a_k.$$

---

[*]Named after the French mathematician Charles Hermite (1822–1901).

This recurrence relation actually includes $a_2 = -na_0$ (which comes about from $2a_2 + 2na_0 = 0$). Again a_0 and a_1 are arbitrary.

$$a_2 = \frac{-2n}{(2)(1)}a_0, \qquad a_3 = \frac{2(1-n)}{(3)(2)}a_1,$$

$$a_4 = \frac{2(2-n)}{(4)(3)}a_2 = \frac{2^2(2-n)(-n)}{(4)(3)(2)(1)}a_0,$$

$$a_5 = \frac{2(3-n)}{(5)(4)}a_3 = \frac{2^2(3-n)(1-n)}{(5)(4)(3)(2)}a_1, \qquad \cdots$$

Let us separate the even and odd coefficients. We find that

$$a_{2m} = \frac{2^m(-n)(2-n)\cdots(2m-2-n)}{(2m)!},$$

$$a_{2m+1} = \frac{2^m(1-n)(3-n)\cdots(2m-1-n)}{(2m+1)!}.$$

Let us write down the two series, one with the even powers and one with the odd.

$$y_1(x) = 1 + \frac{2(-n)}{2!}x^2 + \frac{2^2(-n)(2-n)}{4!}x^4 + \frac{2^3(-n)(2-n)(4-n)}{6!}x^6 + \cdots,$$

$$y_2(x) = x + \frac{2(1-n)}{3!}x^3 + \frac{2^2(1-n)(3-n)}{5!}x^5 + \frac{2^3(1-n)(3-n)(5-n)}{7!}x^7 + \cdots.$$

We then write

$$y(x) = a_0 y_1(x) + a_1 y_2(x).$$

We also notice that if n is a positive even integer, then $y_1(x)$ is a polynomial as all the coefficients in the series beyond a certain degree are zero. If n is a positive odd integer, then $y_2(x)$ is a polynomial. For example, if $n = 4$, then

$$y_1(x) = 1 + \frac{2(-4)}{2!}x^2 + \frac{2^2(-4)(2-4)}{4!}x^4 = 1 - 4x^2 + \frac{4}{3}x^4.$$

7.2.1   Exercises

In the following exercises, when asked to solve an equation using power series methods, you should find the first few terms of the series, and if possible find a general formula for the kth coefficient.

Exercise 7.2.1: *Use power series methods to solve $y'' + y = 0$ at the point $x_0 = 1$.*

Exercise 7.2.2: *Use power series methods to solve $y'' + 4xy = 0$ at the point $x_0 = 0$.*

Exercise 7.2.3: *Use power series methods to solve $y'' - xy = 0$ at the point $x_0 = 1$.*

Exercise 7.2.4: *Use power series methods to solve* $y'' + x^2 y = 0$ *at the point* $x_0 = 0$.

Exercise 7.2.5: *The methods work for other orders than second order. Try the methods of this section to solve the first order system* $y' - xy = 0$ *at the point* $x_0 = 0$.

Exercise 7.2.6 (Chebyshev's equation of order p)**:** *a) Solve* $(1 - x^2)y'' - xy' + p^2 y = 0$ *using power series methods at* $x_0 = 0$. *b) For what p is there a polynomial solution?*

Exercise 7.2.7: *Find a polynomial solution to* $(x^2 + 1)y'' - 2xy' + 2y = 0$ *using power series methods.*

Exercise 7.2.8: *a) Use power series methods to solve* $(1 - x)y'' + y = 0$ *at the point* $x_0 = 0$. *b) Use the solution to part a) to find a solution for* $xy'' + y = 0$ *around the point* $x_0 = 1$.

Exercise 7.2.101: *Use power series methods to solve* $y'' + 2x^3 y = 0$ *at the point* $x_0 = 0$.

Exercise 7.2.102 (challenging)**:** *We can also use power series methods in nonhomogeneous equations. a) Use power series methods to solve* $y'' - xy = \frac{1}{1-x}$ *at the point* $x_0 = 0$. *Hint: Recall the geometric series. b) Now solve for the initial condition* $y(0) = 0$, $y'(0) = 0$.

Exercise 7.2.103: *Attempt to solve* $x^2 y'' - y = 0$ *at* $x_0 = 0$ *using the power series method of this section (* x_0 *is a singular point). Can you find at least one solution? Can you find more than one solution?*

7.3   Singular points and the method of Frobenius

Note: 1 or 1.5 lectures, §8.4 and §8.5 in [EP], §5.4–§5.7 in [BD]

While behavior of ODEs at singular points is more complicated, certain singular points are not especially difficult to solve. Let us look at some examples before giving a general method. We may be lucky and obtain a power series solution using the method of the previous section, but in general we may have to try other things.

7.3.1   Examples

Example 7.3.1: Let us first look at a simple first order equation

$$2xy' - y = 0.$$

Note that $x = 0$ is a singular point. If we only try to plug in

$$y = \sum_{k=0}^{\infty} a_k x^k,$$

we obtain

$$0 = 2xy' - y = 2x \left(\sum_{k=1}^{\infty} k a_k x^{k-1} \right) - \left(\sum_{k=0}^{\infty} a_k x^k \right)$$

$$= a_0 + \sum_{k=1}^{\infty} (2k a_k - a_k) x^k.$$

First, $a_0 = 0$. Next, the only way to solve $0 = 2k a_k - a_k = (2k - 1) a_k$ for $k = 1, 2, 3, \ldots$ is for $a_k = 0$ for all k. Therefore we only get the trivial solution $y = 0$. We need a nonzero solution to get the general solution.

Let us try $y = x^r$ for some real number r. Consequently our solution—if we can find one—may only make sense for positive x. Then $y' = rx^{r-1}$. So

$$0 = 2xy' - y = 2xrx^{r-1} - x^r = (2r - 1)x^r.$$

Therefore $r = \frac{1}{2}$, or in other words $y = x^{1/2}$. Multiplying by a constant, the general solution for positive x is

$$y = Cx^{1/2}.$$

If $C \neq 0$ then the derivative of the solution "blows up" at $x = 0$ (the singular point). There is only one solution that is differentiable at $x = 0$ and that's the trivial solution $y = 0$.

Not every problem with a singular point has a solution of the form $y = x^r$, of course. But perhaps we can combine the methods. What we will do is to try a solution of the form

$$y = x^r f(x)$$

where $f(x)$ is an analytic function.

Example 7.3.2: Suppose that we have the equation

$$4x^2 y'' - 4x^2 y' + (1 - 2x)y = 0,$$

and again note that $x = 0$ is a singular point.

Let us try

$$y = x^r \sum_{k=0}^{\infty} a_k x^k = \sum_{k=0}^{\infty} a_k x^{k+r},$$

where r is a real number, not necessarily an integer. Again if such a solution exists, it may only exist for positive x. First let us find the derivatives

$$y' = \sum_{k=0}^{\infty} (k + r) a_k x^{k+r-1},$$

$$y'' = \sum_{k=0}^{\infty} (k + r)(k + r - 1) a_k x^{k+r-2}.$$

Plugging into our equation we obtain

$$0 = 4x^2 y'' - 4x^2 y' + (1 - 2x)y$$

$$= 4x^2 \left(\sum_{k=0}^{\infty} (k + r)(k + r - 1) a_k x^{k+r-2} \right) - 4x^2 \left(\sum_{k=0}^{\infty} (k + r) a_k x^{k+r-1} \right) + (1 - 2x) \left(\sum_{k=0}^{\infty} a_k x^{k+r} \right)$$

$$= \left(\sum_{k=0}^{\infty} 4(k + r)(k + r - 1) a_k x^{k+r} \right) - \left(\sum_{k=0}^{\infty} 4(k + r) a_k x^{k+r+1} \right) + \left(\sum_{k=0}^{\infty} a_k x^{k+r} \right) - \left(\sum_{k=0}^{\infty} 2a_k x^{k+r+1} \right)$$

$$= \left(\sum_{k=0}^{\infty} 4(k + r)(k + r - 1) a_k x^{k+r} \right) - \left(\sum_{k=1}^{\infty} 4(k + r - 1) a_{k-1} x^{k+r} \right) + \left(\sum_{k=0}^{\infty} a_k x^{k+r} \right) - \left(\sum_{k=1}^{\infty} 2a_{k-1} x^{k+r} \right)$$

$$= 4r(r - 1) a_0 x^r + a_0 x^r + \sum_{k=1}^{\infty} \left(4(k + r)(k + r - 1) a_k - 4(k + r - 1) a_{k-1} + a_k - 2a_{k-1} \right) x^{k+r}$$

$$= (4r(r - 1) + 1) a_0 x^r + \sum_{k=1}^{\infty} \left((4(k + r)(k + r - 1) + 1) a_k - (4(k + r - 1) + 2) a_{k-1} \right) x^{k+r}.$$

To have a solution we must first have $(4r(r - 1) + 1) a_0 = 0$. Supposing that $a_0 \neq 0$ we obtain

$$4r(r - 1) + 1 = 0.$$

This equation is called the *indicial equation*. This particular indicial equation has a double root at $r = 1/2$.

OK, so we know what r has to be. That knowledge we obtained simply by looking at the coefficient of x^r. All other coefficients of x^{k+r} also have to be zero so

$$(4(k + r)(k + r - 1) + 1) a_k - (4(k + r - 1) + 2) a_{k-1} = 0.$$

If we plug in $r = 1/2$ and solve for a_k we get

$$a_k = \frac{4(k + 1/2 - 1) + 2}{4(k + 1/2)(k + 1/2 - 1) + 1} a_{k-1} = \frac{1}{k} a_{k-1}.$$

Let us set $a_0 = 1$. Then

$$a_1 = \frac{1}{1} a_0 = 1, \qquad a_2 = \frac{1}{2} a_1 = \frac{1}{2}, \qquad a_3 = \frac{1}{3} a_2 = \frac{1}{3 \cdot 2}, \qquad a_4 = \frac{1}{4} a_3 = \frac{1}{4 \cdot 3 \cdot 2}, \qquad \cdots$$

Extrapolating, we notice that

$$a_k = \frac{1}{k(k - 1)(k - 2) \cdots 3 \cdot 2} = \frac{1}{k!}.$$

In other words,

$$y = \sum_{k=0}^{\infty} a_k x^{k+r} = \sum_{k=0}^{\infty} \frac{1}{k!} x^{k+1/2} = x^{1/2} \sum_{k=0}^{\infty} \frac{1}{k!} x^k = x^{1/2} e^x.$$

That was lucky! In general, we will not be able to write the series in terms of elementary functions.

We have one solution, let us call it $y_1 = x^{1/2} e^x$. But what about a second solution? If we want a general solution, we need two linearly independent solutions. Picking a_0 to be a different constant only gets us a constant multiple of y_1, and we do not have any other r to try; we only have one solution to the indicial equation. Well, there are powers of x floating around and we are taking derivatives, perhaps the logarithm (the antiderivative of x^{-1}) is around as well. It turns out we want to try for another solution of the form

$$y_2 = \sum_{k=0}^{\infty} b_k x^{k+r} + (\ln x) y_1,$$

which in our case is

$$y_2 = \sum_{k=0}^{\infty} b_k x^{k+1/2} + (\ln x) x^{1/2} e^x.$$

We now differentiate this equation, substitute into the differential equation and solve for b_k. A long computation ensues and we obtain some recursion relation for b_k. The reader can (and should) try this to obtain for example the first three terms

$$b_1 = b_0 - 1, \qquad b_2 = \frac{2b_1 - 1}{4}, \qquad b_3 = \frac{6b_2 - 1}{18}, \qquad \cdots$$

We then fix b_0 and obtain a solution y_2. Then we write the general solution as $y = A y_1 + B y_2$.

7.3.2 The method of Frobenius

Before giving the general method, let us clarify when the method applies. Let

$$p(x)y'' + q(x)y' + r(x)y = 0$$

be an ODE. As before, if $p(x_0) = 0$, then x_0 is a singular point. If, furthermore, the limits

$$\lim_{x \to x_0} (x - x_0)\frac{q(x)}{p(x)} \quad \text{and} \quad \lim_{x \to x_0} (x - x_0)^2 \frac{r(x)}{p(x)}$$

both exist and are finite, then we say that x_0 is a *regular singular point.*

Example 7.3.3: Often, and for the rest of this section, $x_0 = 0$. Consider

$$x^2 y'' + x(1 + x)y' + (\pi + x^2)y = 0.$$

Write

$$\lim_{x \to 0} x\frac{q(x)}{p(x)} = \lim_{x \to 0} x\frac{x(1 + x)}{x^2} = \lim_{x \to 0} (1 + x) = 1,$$

$$\lim_{x \to 0} x^2\frac{r(x)}{p(x)} = \lim_{x \to 0} x^2\frac{(\pi + x^2)}{x^2} = \lim_{x \to 0} (\pi + x^2) = \pi.$$

So $x = 0$ is a regular singular point.

On the other hand if we make the slight change

$$x^2 y'' + (1 + x)y' + (\pi + x^2)y = 0,$$

then

$$\lim_{x \to 0} x\frac{q(x)}{p(x)} = \lim_{x \to 0} x\frac{(1 + x)}{x^2} = \lim_{x \to 0} \frac{1 + x}{x} = \text{DNE}.$$

Here DNE stands for *does not exist.* The point 0 is a singular point, but not a regular singular point.

Let us now discuss the general *Method of Frobenius*[*]. Let us only consider the method at the point $x = 0$ for simplicity. The main idea is the following theorem.

Theorem 7.3.1 (Method of Frobenius). *Suppose that*

$$p(x)y'' + q(x)y' + r(x)y = 0 \tag{7.3}$$

has a regular singular point at $x = 0$, then there exists at least one solution of the form

$$y = x^r \sum_{k=0}^{\infty} a_k x^k.$$

A solution of this form is called a Frobenius-type solution.

---

[*]Named after the German mathematician Ferdinand Georg Frobenius (1849–1917).

The method usually breaks down like this.

(i) We seek a Frobenius-type solution of the form

$$y = \sum_{k=0}^{\infty} a_k x^{k+r}.$$

We plug this y into equation (7.3). We collect terms and write everything as a single series.

(ii) The obtained series must be zero. Setting the first coefficient (usually the coefficient of x^r) in the series to zero we obtain the *indicial equation*, which is a quadratic polynomial in r.

(iii) If the indicial equation has two real roots r_1 and r_2 such that $r_1 - r_2$ is not an integer, then we have two linearly independent Frobenius-type solutions. Using the first root, we plug in

$$y_1 = x^{r_1} \sum_{k=0}^{\infty} a_k x^k,$$

and we solve for all a_k to obtain the first solution. Then using the second root, we plug in

$$y_2 = x^{r_2} \sum_{k=0}^{\infty} b_k x^k,$$

and solve for all b_k to obtain the second solution.

(iv) If the indicial equation has a doubled root r, then there we find one solution

$$y_1 = x^r \sum_{k=0}^{\infty} a_k x^k,$$

and then we obtain a new solution by plugging

$$y_2 = x^r \sum_{k=0}^{\infty} b_k x^k + (\ln x) y_1,$$

into equation (7.3) and solving for the constants b_k.

(v) If the indicial equation has two real roots such that $r_1 - r_2$ is an integer, then one solution is

$$y_1 = x^{r_1} \sum_{k=0}^{\infty} a_k x^k,$$

and the second linearly independent solution is of the form

$$y_2 = x^{r_2} \sum_{k=0}^{\infty} b_k x^k + C(\ln x) y_1,$$

where we plug y_2 into (7.3) and solve for the constants b_k and C.

(vi) Finally, if the indicial equation has complex roots, then solving for a_k in the solution

$$y = x^{r_1} \sum_{k=0}^{\infty} a_k x^k$$

results in a complex-valued function—all the a_k are complex numbers. We obtain our two linearly independent solutions[*] by taking the real and imaginary parts of y.

The main idea is to find at least one Frobenius-type solution. If we are lucky and find two, we are done. If we only get one, we either use the ideas above or even a different method such as reduction of order (Exercise 2.1.8 on page 66) to obtain a second solution.

7.3.3 Bessel functions

An important class of functions that arises commonly in physics are the *Bessel functions*[†]. For example, these functions appear when solving the wave equation in two and three dimensions. First we have *Bessel's equation* of order p:

$$x^2 y'' + x y' + \left(x^2 - p^2\right) y = 0.$$

We allow p to be any number, not just an integer, although integers and multiples of $1/2$ are most important in applications.

When we plug

$$y = \sum_{k=0}^{\infty} a_k x^{k+r}$$

into Bessel's equation of order p we obtain the indicial equation

$$r(r-1) + r - p^2 = (r-p)(r+p) = 0.$$

Therefore we obtain two roots $r_1 = p$ and $r_2 = -p$. If p is not an integer following the method of Frobenius and setting $a_0 = 1$, we obtain linearly independent solutions of the form

$$y_1 = x^p \sum_{k=0}^{\infty} \frac{(-1)^k x^{2k}}{2^{2k} k! (k+p)(k-1+p) \cdots (2+p)(1+p)},$$

$$y_2 = x^{-p} \sum_{k=0}^{\infty} \frac{(-1)^k x^{2k}}{2^{2k} k! (k-p)(k-1-p) \cdots (2-p)(1-p)}.$$

---

[*]See Joseph L. Neuringera, *The Frobenius method for complex roots of the indicial equation*, International Journal of Mathematical Education in Science and Technology, Volume 9, Issue 1, 1978, 71–77.

[†]Named after the German astronomer and mathematician Friedrich Wilhelm Bessel (1784–1846).

***Exercise* 7.3.1:** *a) Verify that the indicial equation of Bessel's equation of order p is $(r-p)(r+p) = 0$. b) Suppose that p is not an integer. Carry out the computation to obtain the solutions y_1 and y_2 above.*

Bessel functions will be convenient constant multiples of y_1 and y_2. First we must define the *gamma function*

$$\Gamma(x) = \int_0^\infty t^{x-1} e^{-t}\, dt.$$

Notice that $\Gamma(1) = 1$. The gamma function also has a wonderful property

$$\Gamma(x + 1) = x\Gamma(x).$$

From this property, one can show that $\Gamma(n) = (n - 1)!$ when n is an integer, so the gamma function is a continuous version of the factorial. We compute:

$$\Gamma(k + p + 1) = (k + p)(k - 1 + p)\cdots(2 + p)(1 + p)\Gamma(1 + p),$$
$$\Gamma(k - p + 1) = (k - p)(k - 1 - p)\cdots(2 - p)(1 - p)\Gamma(1 - p).$$

***Exercise* 7.3.2:** *Verify the above identities using $\Gamma(x + 1) = x\Gamma(x)$.*

We define the *Bessel functions of the first kind* of order p and $-p$ as

$$J_p(x) = \frac{1}{2^p\Gamma(1 + p)} y_1 = \sum_{k=0}^\infty \frac{(-1)^k}{k!\,\Gamma(k + p + 1)}\left(\frac{x}{2}\right)^{2k+p},$$

$$J_{-p}(x) = \frac{1}{2^{-p}\Gamma(1 - p)} y_2 = \sum_{k=0}^\infty \frac{(-1)^k}{k!\,\Gamma(k - p + 1)}\left(\frac{x}{2}\right)^{2k-p}.$$

As these are constant multiples of the solutions we found above, these are both solutions to Bessel's equation of order p. The constants are picked for convenience.

When p is not an integer, J_p and J_{-p} are linearly independent. When n is an integer we obtain

$$J_n(x) = \sum_{k=0}^\infty \frac{(-1)^k}{k!\,(k + n)!}\left(\frac{x}{2}\right)^{2k+n}.$$

In this case it turns out that

$$J_n(x) = (-1)^n J_{-n}(x),$$

and so we do not obtain a second linearly independent solution. The other solution is the so-called *Bessel function of second kind*. These make sense only for integer orders n and are defined as limits of linear combinations of $J_p(x)$ and $J_{-p}(x)$ as p approaches n in the following way:

$$Y_n(x) = \lim_{p \to n} \frac{\cos(p\pi)J_p(x) - J_{-p}(x)}{\sin(p\pi)}.$$

As each linear combination of $J_p(x)$ and $J_{-p}(x)$ is a solution to Bessel's equation of order p, then as we take the limit as p goes to n, $Y_n(x)$ is a solution to Bessel's equation of order n. It also turns out that $Y_n(x)$ and $J_n(x)$ are linearly independent. Therefore when n is an integer, we have the general solution to Bessel's equation of order n

$$y = AJ_n(x) + BY_n(x),$$

for arbitrary constants A and B. Note that $Y_n(x)$ goes to negative infinity at $x = 0$. Many mathematical software packages have these functions $J_n(x)$ and $Y_n(x)$ defined, so they can be used just like say $\sin(x)$ and $\cos(x)$. In fact, they have some similar properties. For example, $-J_1(x)$ is a derivative of $J_0(x)$, and in general the derivative of $J_n(x)$ can be written as a linear combination of $J_{n-1}(x)$ and $J_{n+1}(x)$. Furthermore, these functions oscillate, although they are not periodic. See Figure 7.4 for graphs of Bessel functions.

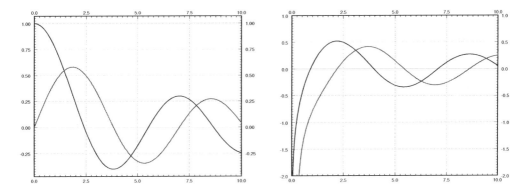

Figure 7.4: Plot of the $J_0(x)$ and $J_1(x)$ in the first graph and $Y_0(x)$ and $Y_1(x)$ in the second graph.

Example 7.3.4: Other equations can sometimes be solved in terms of the Bessel functions. For example, given a positive constant λ,

$$xy'' + y' + \lambda^2 xy = 0,$$

can be changed to $x^2 y'' + xy' + \lambda^2 x^2 y = 0$. Then changing variables $t = \lambda x$ we obtain via chain rule the equation in y and t:

$$t^2 y'' + ty' + t^2 y = 0,$$

which can be recognized as Bessel's equation of order 0. Therefore the general solution is $y(t) = AJ_0(t) + BY_0(t)$, or in terms of x:

$$y = AJ_0(\lambda x) + BY_0(\lambda x).$$

This equation comes up for example when finding fundamental modes of vibration of a circular drum, but we digress.

7.3.4   Exercises

Exercise 7.3.3: *Find a particular (Frobenius-type) solution of* $x^2y'' + xy' + (1 + x)y = 0$.

Exercise 7.3.4: *Find a particular (Frobenius-type) solution of* $xy'' - y = 0$.

Exercise 7.3.5: *Find a particular (Frobenius-type) solution of* $y'' + \frac{1}{x}y' - xy = 0$.

Exercise 7.3.6: *Find the general solution of* $2xy'' + y' - x^2y = 0$.

Exercise 7.3.7: *Find the general solution of* $x^2y'' - xy' - y = 0$.

Exercise 7.3.8: *In the following equations classify the point* $x = 0$ *as* ordinary, regular singular, *or* singular but not regular singular.

a) $x^2(1 + x^2)y'' + xy = 0$

b) $x^2y'' + y' + y = 0$

c) $xy'' + x^3y' + y = 0$

d) $xy'' + xy' - e^xy = 0$

e) $x^2y'' + x^2y' + x^2y = 0$

Exercise 7.3.101: *In the following equations classify the point* $x = 0$ *as* ordinary, regular singular, *or* singular but not regular singular.

a) $y'' + y = 0$

b) $x^3y'' + (1 + x)y = 0$

c) $xy'' + x^5y' + y = 0$

d) $\sin(x)y'' - y = 0$

e) $\cos(x)y'' - \sin(x)y = 0$

Exercise 7.3.102: *Find the general solution of* $x^2y'' - y = 0$.

Exercise 7.3.103: *Find a particular solution of* $x^2y'' + (x - 3/4)y = 0$.

Exercise 7.3.104 (tricky)***:*** *Find the general solution of* $x^2y'' - xy' + y = 0$.

Chapter 8

Nonlinear systems

8.1 Linearization, critical points, and equilibria

Note: 1 lecture, §6.1–§6.2 in [EP], §9.2–§9.3 in [BD]

Except for a few brief detours in chapter 1, we considered mostly linear equations. Linear equations suffice in many applications, but in reality most phenomena require nonlinear equations. Nonlinear equations, however, are notoriously more difficult to understand than linear ones, and many strange new phenomena appear when we allow our equations to be nonlinear.

Not to worry, we did not waste all this time studying linear equations. Nonlinear equations can often be approximated by linear ones if we only need a solution "locally," for example, only for a short period of time, or only for certain parameters. Understanding linear equations can also give us qualitative understanding about a more general nonlinear problem. The idea is similar to what you did in calculus in trying to approximate a function by a line with the right slope.

In § 2.4 we looked at the pendulum of length L. The goal was to solve for the angle $\theta(t)$ as a function of the time t. The equation for the setup is the nonlinear equation

$$\theta'' + \frac{g}{L}\sin\theta = 0.$$

Instead of solving this equation, we solved the rather easier linear equation

$$\theta'' + \frac{g}{L}\theta = 0.$$

While the solution to the linear equation is not exactly what we were looking for, it is rather close to the original, as long as the angle θ is small and the time period involved is short.

You might ask: Why don't we just solve the nonlinear problem? Well, it might be very difficult, impractical, or impossible to solve analytically, depending on the equation in question. We may not even be interested in the actual solution, we might only be interested in some qualitative idea of what the solution is doing. For example, what happens as time goes to infinity?

8.1.1   Autonomous systems and phase plane analysis

We restrict our attention to a two dimensional autonomous system

$$x' = f(x, y), \qquad y' = g(x, y),$$

where $f(x, y)$ and $g(x, y)$ are functions of two variables, and the derivatives are taken with respect to time t. Solutions are functions $x(t)$ and $y(t)$ such that

$$x'(t) = f(x(t), y(t)), \qquad y'(t) = g(x(t), y(t)).$$

The way we will analyze the system is very similar to § 1.6, where we studied a single autonomous equation. The ideas in two dimensions are the same, but the behavior can be far more complicated.

It may be best to think of the system of equations as the single vector equation

$$\begin{bmatrix} x \\ y \end{bmatrix}' = \begin{bmatrix} f(x, y) \\ g(x, y) \end{bmatrix}. \tag{8.1}$$

As in § 3.1 we draw the *phase portrait* (or *phase diagram*), where each point (x, y) corresponds to a specific state of the system. We draw the *vector field* given at each point (x, y) by the vector $\begin{bmatrix} f(x,y) \\ g(x,y) \end{bmatrix}$. And as before if we find solutions, we draw the trajectories by plotting all points $(x(t), y(t))$ for a certain range of t.

Example 8.1.1: Consider the second order equation $x'' = -x + x^2$. Write this equation as a first order nonlinear system

$$x' = y, \qquad y' = -x + x^2.$$

The phase portrait with some trajectories is drawn in Figure 8.1.

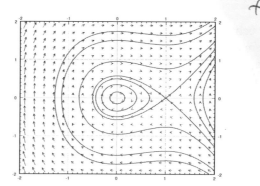

Figure 8.1: Phase portrait with some trajectories of $x' = y$, $y' = -x + x^2$.

From the phase portrait it should be clear that even this simple system has fairly complicated behavior. Some trajectories keep oscillating around the origin, and some go off towards infinity. We will return to this example often, and analyze it completely in this (and the next) section.

Let us concentrate on those points in the phase diagram above where the trajectories seem to start, end, or go around. We see two such points: $(0,0)$ and $(1,0)$. The trajectories seem to go around the point $(0,0)$, and they seem to either go in or out of the point $(1,0)$. These points are precisely those points where the derivatives of both x and y are zero. Let us define the *critical points* as the points (x, y) such that

$$\begin{bmatrix} f(x,y) \\ g(x,y) \end{bmatrix} = \vec{0}.$$

In other words, the points where both $f(x, y) = 0$ and $g(x, y) = 0$.

The critical points are where the behavior of the system is in some sense the most complicated. If $\begin{bmatrix} f(x,y) \\ g(x,y) \end{bmatrix}$ is zero, then nearby, the vector can point in any direction whatsoever. Also, the trajectories are either going towards, away from, or around these points, so if we are looking for long term behavior of the system, we should look at what happens there.

Critical points are also sometimes called *equilibria*, since we have so-called *equilibrium solutions* at critical points. If (x_0, y_0) is a critical point, then we have the solutions

$$x(t) = x_0, \quad y(t) = y_0.$$

In Example 8.1.1 on the facing page, there are two equilibrium solutions:

$$x(t) = 0, \quad y(t) = 0, \quad \text{and} \quad x(t) = 1, \quad y(t) = 0.$$

Compare this discussion on equilibria to the discussion in § 1.6. The underlying concept is exactly the same.

8.1.2 Linearization

In § 3.5 we studied the behavior of a homogeneous linear system of two equations near a critical point. For a linear system of two variables the only critical point is generally the origin $(0,0)$. Let us put the understanding we gained in that section to good use understanding what happens near critical points of nonlinear systems.

In calculus we learned to estimate a function by taking its derivative and linearizing. We work similarly with nonlinear systems of ODE. Suppose (x_0, y_0) is a critical point. First change variables to (u, v), so that $(u, v) = (0,0)$ corresponds to (x_0, y_0). That is,

$$u = x - x_0, \quad v = y - y_0.$$

Next we need to find the derivative. In multivariable calculus you may have seen that the several variables version of the derivative is the *Jacobian matrix*[*]. The Jacobian matrix of the vector-valued

---

[*]Named for the German mathematician Carl Gustav Jacob Jacobi (1804–1851).

function $\begin{bmatrix} f(x,y) \\ g(x,y) \end{bmatrix}$ at (x_0, y_0) is

$$\begin{bmatrix} \frac{\partial f}{\partial x}(x_0, y_0) & \frac{\partial f}{\partial y}(x_0, y_0) \\ \frac{\partial g}{\partial x}(x_0, y_0) & \frac{\partial g}{\partial y}(x_0, y_0) \end{bmatrix}.$$

This matrix gives the best linear approximation as u and v (and therefore x and y) vary. We define the *linearization* of the equation (8.1) as the linear system

$$\begin{bmatrix} u \\ v \end{bmatrix}' = \begin{bmatrix} \frac{\partial f}{\partial x}(x_0, y_0) & \frac{\partial f}{\partial y}(x_0, y_0) \\ \frac{\partial g}{\partial x}(x_0, y_0) & \frac{\partial g}{\partial y}(x_0, y_0) \end{bmatrix} \begin{bmatrix} u \\ v \end{bmatrix}.$$

Example 8.1.2: Let us keep with the same equations as Example 8.1.1: $x' = y$, $y' = -x + x^2$. There are two critical points, $(0,0)$ and $(1,0)$. The Jacobian matrix at any point is

$$\begin{bmatrix} \frac{\partial f}{\partial x}(x,y) & \frac{\partial f}{\partial y}(x,y) \\ \frac{\partial g}{\partial x}(x,y) & \frac{\partial g}{\partial y}(x,y) \end{bmatrix} = \begin{bmatrix} 0 & 1 \\ -1 + 2x & 0 \end{bmatrix}.$$

Therefore at $(0,0)$ the linearization is

$$\begin{bmatrix} u \\ v \end{bmatrix}' = \begin{bmatrix} 0 & 1 \\ -1 & 0 \end{bmatrix} \begin{bmatrix} u \\ v \end{bmatrix},$$

where $u = x$ and $v = y$.

At the point $(1,0)$, we have $u = x - 1$ and $v = y$, and the linearization is

$$\begin{bmatrix} u \\ v \end{bmatrix}' = \begin{bmatrix} 0 & 1 \\ 1 & 0 \end{bmatrix} \begin{bmatrix} u \\ v \end{bmatrix}.$$

The phase diagrams of the two linearizations at the point $(0,0)$ and $(1,0)$ are given in Figure 8.2 on the next page. Note that the variables are now u and v. Compare Figure 8.2 with Figure 8.1 on page 320, and look especially at the behavior near the critical points.

8.1.3 Exercises

Exercise 8.1.1: *Sketch the phase plane vector field for:*
a) $x' = x^2$, $y' = y^2$,
b) $x' = (x - y)^2$, $y' = -x$,
c) $x' = e^y$, $y' = e^x$.

Exercise 8.1.2: *Match systems*
1) $x' = x^2$, $y' = y^2$,   2) $x' = xy$, $y' = 1 + y^2$,   3) $x' = \sin(\pi y)$, $y' = x$,
to the vector fields below. Justify.

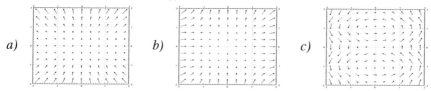

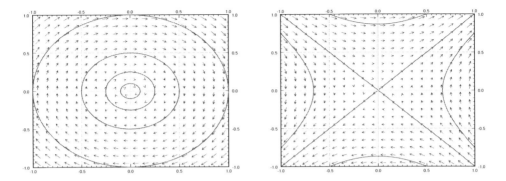

Figure 8.2: Phase diagram with some trajectories of linearizations at the critical points $(0,0)$ (left) and $(1,0)$ (right) of $x' = y$, $y' = -x + x^2$.

Exercise 8.1.3: *Find the critical points and linearizations of the following systems.*
a) $x' = x^2 - y^2$, $y' = x^2 + y^2 - 1$,
b) $x' = -y$, $y' = 3x + yx^2$,
c) $x' = x^2 + y$, $y' = y^2 + x$.

Exercise 8.1.4: *For the following systems, verify they have critical point at $(0,0)$, and find the linearization at $(0,0)$.*
a) $x' = x + 2y + x^2 - y^2$, $y' = 2y - x^2$
b) $x' = -y$, $y' = x - y^3$
c) $x' = ax + by + f(x, y)$, $y' = cx + dy + g(x, y)$, where $f(0,0) = 0$, $g(0,0) = 0$, and all first partial derivatives of f and g are also zero at $(0,0)$, that is, $\frac{\partial f}{\partial x}(0,0) = \frac{\partial f}{\partial y}(0,0) = \frac{\partial g}{\partial x}(0,0) = \frac{\partial g}{\partial y}(0,0) = 0$.

Exercise 8.1.5: *Take* $x' = (x - y)^2$, $y' = (x + y)^2$.
a) Find the set of critical points.
b) Sketch a phase diagram and describe the behavior near the critical point(s).
c) Find the linearization. Is it helpful in understanding the system?

Exercise 8.1.6: *Take* $x' = x^2$, $y' = x^3$.
a) Find the set of critical points.
b) Sketch a phase diagram and describe the behavior near the critical point(s).
c) Find the linearization. Is it helpful in understanding the system?

Exercise 8.1.101: *Find the critical points and linearizations of the following systems.*
a) $x' = \sin(\pi y) + (x - 1)^2$, $y' = y^2 - y$,
b) $x' = x + y + y^2$, $y' = x$,
c) $x' = (x - 1)^2 + y$, $y' = x^2 + y$.

Exercise **8.1.102:** *Match systems*
1) $x' = y^2$, $y' = -x^2$,   2) $x' = y$, $y' = (x - 1)(x + 1)$,   3) $x' = y + x^2$, $y' = -x$,
to the vector fields below. Justify.

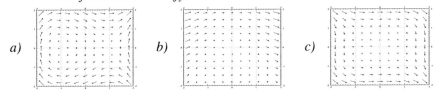

a)   b)   c)

Exercise **8.1.103:** *The idea of critical points and linearization works in higher dimensions as well. You simply make the Jacobian matrix bigger by adding more functions and more variables. For the following system of 3 equations find the critical points and their linearizations:*
$$x' = x + z^2,$$
$$y' = z^2 - y,$$
$$z' = z + x^2.$$

Exercise **8.1.104:** *Any two-dimensional non-autonomous system $x' = f(x, y, t)$, $y' = g(x, y, t)$ can be written as a three-dimensional autonomous system (three equations). Write down this autonomous system using the variables u, v, w.*

8.2 Stability and classification of isolated critical points

Note: 2 lectures, §6.1–§6.2 in [EP], §9.2–§9.3 in [BD]

8.2.1 Isolated critical points and almost linear systems

A critical point is *isolated* if it is the only critical point in some small "neighborhood" of the point. That is, if we zoom in far enough it is the only critical point we see. In the above example, the critical point was isolated. If on the other hand there would be a whole curve of critical points, then it would not be isolated.

A system is called *almost linear* (at a critical point (x_0, y_0)) if the critical point is isolated and the Jacobian at the point is invertible, or equivalently if the linearized system has an isolated critical point. In such a case, the nonlinear terms will be very small and the system will behave like its linearization, at least if we are close to the critical point.

In particular the system we have just seen in Examples 8.1.1 and 8.1.2 has two isolated critical points $(0, 0)$ and $(0, 1)$, and is almost linear at both critical points as both of the Jacobian matrices $\left[\begin{smallmatrix} 0 & 1 \\ -1 & 0 \end{smallmatrix} \right]$ and $\left[\begin{smallmatrix} 0 & 1 \\ 1 & 0 \end{smallmatrix} \right]$ are invertible.

On the other hand a system such as $x' = x^2$, $y' = y^2$ has an isolated critical point at $(0, 0)$, however the Jacobian matrix

$$\begin{bmatrix} 2x & 0 \\ 0 & 2y \end{bmatrix}$$

is zero when $(x, y) = (0, 0)$. Therefore the system is not almost linear. Even a worse example is the system $x' = x$, $y' = x^2$, which does not have an isolated critical point, as x' and y' are both zero whenever $x = 0$, that is, the entire y axis.

Fortunately, most often critical points are isolated, and the system is almost linear at the critical points. So if we learn what happens here, we have figured out the majority of situations that arise in applications.

8.2.2 Stability and classification of isolated critical points

Once we have an isolated critical point, the system is almost linear at that critical point, and we computed the associated linearized system, we can classify what happens to the solutions. We more or less use the classification for linear two-variable systems from § 3.5, with one minor caveat. Let us list the behaviors depending on the eigenvalues of the Jacobian matrix at the critical point in Table 8.1 on the next page. This table is very similar to Table 3.1 on page 131, with the exception of missing "center" points. We will discuss centers later, as they are more complicated.

In the new third column, we have marked points as *asymptotically stable* or *unstable*. Formally, a *stable critical point* (x_0, y_0) is one where given any small distance ϵ to (x_0, y_0), and any initial condition within a perhaps smaller radius around (x_0, y_0), the trajectory of the system will never go further away from (x_0, y_0) than ϵ. An *unstable critical point* is one that is not stable. Informally, a

Eigenvalues of the Jacobian matrix	Behavior	Stability
real and both positive	source / unstable node	unstable
real and both negative	sink / stable node	asymptotically stable
real and opposite signs	saddle	unstable
complex with positive real part	spiral source	unstable
complex with negative real part	spiral sink	asymptotically stable

Table 8.1: Behavior of an almost linear system near an isolated critical point.

point is stable if we start close to a critical point and follow a trajectory we will either go towards, or at least not get away from, this critical point.

A stable critical point (x_0, y_0) is called *asymptotically stable* if given any initial condition sufficiently close to (x_0, y_0) and any solution $(x(t), y(t))$ given that condition, then

$$\lim_{t \to \infty}(x(t), y(t)) = (x_0, y_0).$$

That is, the critical point is asymptotically stable if any trajectory for a sufficiently close initial condition goes towards the critical point (x_0, y_0).

Example 8.2.1: Consider $x' = -y - x^2$, $y' = -x + y^2$. See Figure 8.3 on the next page for the phase diagram. Let us find the critical points. These are the points where $-y - x^2 = 0$ and $-x + y^2 = 0$. The first equation means $y = -x^2$, and so $y^2 = x^4$. Plugging into the second equation we obtain $-x + x^4 = 0$. Factoring we obtain $x(1 - x^3) = 0$. Since we are looking only for real solutions we get either $x = 0$ or $x = 1$. Solving for the corresponding y using $y = -x^2$, we get two critical points, one being $(0, 0)$ and the other being $(1, -1)$. Clearly the critical points are isolated. Let us compute the Jacobian matrix:

$$\begin{bmatrix} -2x & -1 \\ -1 & 2y \end{bmatrix}.$$

At the point $(0, 0)$ we get the matrix $\begin{bmatrix} 0 & -1 \\ -1 & 0 \end{bmatrix}$ and so the two eigenvalues are 1 and -1. As the matrix is invertible, the system is almost linear at $(0, 0)$. As the eigenvalues are real and of opposite signs, we get a saddle point, which is an unstable equilibrium point.

At the point $(1, -1)$ we get the matrix $\begin{bmatrix} -2 & -1 \\ -1 & -2 \end{bmatrix}$ and computing the eigenvalues we get -1, -3. The matrix is invertible, and so the system is almost linear at $(1, -1)$. As we have real eigenvalues both negative, the critical point is a sink, and therefore an asymptotically stable equilibrium point. That is, if we start with any point (x_i, y_i) close to $(1, -1)$ as an initial condition and plot a trajectory, it will approach $(1, -1)$. In other words,

$$\lim_{t \to \infty}(x(t), y(t)) = (1, -1).$$

As you can see from the diagram, this behavior is true even for some initial points quite far from $(1, -1)$, but it is definitely not true for all initial points.

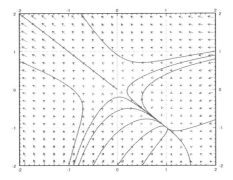

Figure 8.3: The phase portrait with few sample trajectories of $x' = -y - x^2$, $y' = -x + y^2$.

Example 8.2.2: Let us look at $x' = y + y^2 e^x$, $y' = x$. First let us find the critical points. These are the points where $y + y^2 e^x = 0$ and $x = 0$. Simplifying we get $0 = y + y^2 = y(y + 1)$. So the critical points are $(0, 0)$ and $(0, -1)$, and hence are isolated. Let us compute the Jacobian matrix:

$$\begin{bmatrix} y^2 e^x & 1 + 2ye^x \\ 1 & 0 \end{bmatrix}.$$

At the point $(0, 0)$ we get the matrix $\begin{bmatrix} 0 & 1 \\ 1 & 0 \end{bmatrix}$ and so the two eigenvalues are 1 and -1. As the matrix is invertible, the system is almost linear at $(0, 0)$. And, as the eigenvalues are real and of opposite signs, we get a saddle point, which is an unstable equilibrium point.

At the point $(0, -1)$ we get the matrix $\begin{bmatrix} 1 & -1 \\ 1 & 0 \end{bmatrix}$ whose eigenvalues are $\frac{1}{2} \pm i\frac{\sqrt{3}}{2}$. The matrix is invertible, and so the system is almost linear at $(0, -1)$. As we have complex eigenvalues with positive real part, the critical point is a spiral source, and therefore an unstable equilibrium point.

See Figure 8.4 on the following page for the phase diagram. Notice the two critical points, and the behavior of the arrows in the vector field around these points.

8.2.3 The trouble with centers

Recall, a linear system with a center meant that trajectories travelled in closed elliptical orbits in some direction around the critical point. Such a critical point we would call a *center* or a *stable center*. It would not be an asymptotically stable critical point, as the trajectories would never approach the critical point, but at least if you start sufficiently close to the critical point, you will stay close to the critical point. The simplest example of such behavior is the linear system with a center. Another example is the critical point $(0, 0)$ in Example 8.1.1 on page 320.

The trouble with a center in a nonlinear system is that whether the trajectory goes towards or away from the critical point is governed by the sign of the real part of the eigenvalues of the Jacobian.

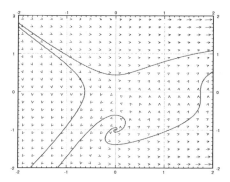

Figure 8.4: The phase portrait with few sample trajectories of $x' = y + y^2 e^x$, $y' = x$.

Since this real part is zero at the critical point itself, it can have either sign nearby, meaning the trajectory could be pulled towards or away from the critical point.

Example 8.2.3: An easy example where such a problematic behavior is exhibited is the system $x' = y, y' = -x + y^3$. The only critical point is the origin $(0,0)$. The Jacobian matrix is

$$\begin{bmatrix} 0 & 1 \\ -1 & 3y^2 \end{bmatrix}.$$

At $(0,0)$ the Jacobian matrix is $\begin{bmatrix} 0 & 1 \\ -1 & 0 \end{bmatrix}$, which has eigenvalues $\pm i$. Therefore, the linearization has a center.

Using the quadratic equation, the eigenvalues of the Jacobian matrix at any point (x, y) are

$$\lambda = \frac{3}{2}y^2 \pm i\frac{\sqrt{4 - 9y^4}}{2}.$$

At any point where $y \neq 0$ (so at most points near the origin), the eigenvalues have a positive real part (y^2 can never be negative). This positive real part will pull the trajectory away from the origin. A sample trajectory for an initial condition near the origin is given in Figure 8.5 on the facing page.

The moral of the example is that further analysis is needed when the linearization has a center. The analysis will in general be more complicated than in the above example, and is more likely to involve case-by-case consideration. Such a complication should not be surprising to you. By now in your mathematical career, you have seen many places where a simple test is inconclusive, perhaps starting with the second derivative test for maxima or minima, and requires more careful, and perhaps ad hoc analysis of the situation.

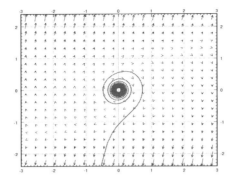

Figure 8.5: An unstable critical point (spiral source) at the origin for $x' = y, y' = -x + y^3$, even if the linearization has a center.

8.2.4   Conservative equations

An equation of the form

$$x'' + f(x) = 0$$

for an arbitrary function $f(x)$ is called a *conservative equation*. For example the pendulum equation is a conservative equation. The equations are conservative as there is no friction in the system so the energy in the system is "conserved." Let us write this equation as a system of nonlinear ODE.

$$x' = y, \qquad y' = -f(x).$$

These types of equations have the advantage that we can solve for their trajectories easily.

The trick is to first think of y as a function of x for a moment. Then use the chain rule

$$x'' = y' = y\frac{dy}{dx},$$

where the prime indicates a derivative with respect to t. We obtain $y\frac{dy}{dx} + f(x) - 0$. We integrate with respect to x to get $\int y\frac{dy}{dx}\,dx + \int f(x)\,dx = C$. In other words

$$\frac{1}{2}y^2 + \int f(x)\,dx = C.$$

We obtained an implicit equation for the trajectories, with different C giving different trajectories. The value of C is conserved on any trajectory. This expression is sometimes called the *Hamiltonian* or the energy of the system. If you look back to § 1.8, you will notice that $y\frac{dy}{dx} + f(x) = 0$ is an exact equation, and we just found a potential function.

Example 8.2.4: Let us find the trajectories for the equation $x'' + x - x^2 = 0$, which is the equation from Example 8.1.1 on page 320. The corresponding first order system is

$$x' = y, \qquad y' = -x + x^2.$$

Trajectories satisfy

$$\frac{1}{2}y^2 + \frac{1}{2}x^2 - \frac{1}{3}x^3 = C.$$

We solve for y

$$y = \pm\sqrt{-x^2 + \frac{2}{3}x^3 + 2C}.$$

Plotting these graphs we get exactly the trajectories in Figure 8.1 on page 320. In particular we notice that near the origin the trajectories are *closed curves*: they keep going around the origin, never spiraling in or out. Therefore we discovered a way to verify that the critical point at $(0,0)$ is a stable center. The critical point at $(0,1)$ is a saddle as we already noticed. This example is typical for conservative equations.

Consider an arbitrary conservative equation. The trajectories are given by

$$y = \pm\sqrt{-2\int f(x)\,dx + 2C}.$$

So all trajectories are mirrored across the x-axis. In particular, there can be no spiral sources nor sinks. All critical points occur when $y = 0$ (the x-axis), that is when $x' = 0$. The critical points are simply those points on the x-axis where $f(x) = 0$. The Jacobian matrix is

$$\begin{bmatrix} 0 & 1 \\ -f'(x) & 0 \end{bmatrix}.$$

So the critical point is almost linear if $f'(x) \neq 0$ at the critical point. Let J denote the Jacobian matrix, then the eigenvalues of J are solutions to

$$0 = \det(J - \lambda I) = \lambda^2 + f'(x).$$

Therefore $\lambda = \pm\sqrt{-f'(x)}$. In other words, either we get real eigenvalues of opposite signs, or we get purely imaginary eigenvalues. There are only two possibilities for critical points, either an unstable saddle point, or a stable center. There are never any asymptotically stable points.

8.2.5 Exercises

Exercise 8.2.1: For the systems below, find and classify the critical points, also indicate if the equilibria are stable, asymptotically stable, or unstable.
a) $x' = -x + 3x^2, y' = -y$ b) $x' = x^2 + y^2 - 1, y' = x$ c) $x' = ye^x, y' = y - x + y^2$

Exercise 8.2.2: *Find the implicit equations of the trajectories of the following conservative systems. Next find their critical points (if any) and classify them.*
a) $x'' + x + x^3 = 0$ b) $\theta'' + \sin\theta = 0$ c) $z'' + (z-1)(z+1) = 0$ d) $x'' + x^2 + 1 = 0$

Exercise 8.2.3: *Find and classify the critical point(s) of $x' = -x^2$, $y' = -y^2$.*

Exercise 8.2.4: *Suppose $x' = -xy$, $y' = x^2 - 1 - y$. a) Show there are two spiral sinks at $(-1, 0)$ and $(1, 0)$. b) For any initial point of the form $(0, y_0)$, find what is the trajectory. c) Can a trajectory starting at (x_0, y_0) where $x_0 > 0$ spiral into the critical point at $(-1, 0)$? Why or why not?*

Exercise 8.2.5: *In the example $x' = y$, $y' = y^3 - x$ show that for any trajectory, the distance from the origin is an increasing function. Conclude that the origin behaves like is a spiral source. Hint: Consider $f(t) = (x(t))^2 + (y(t))^2$ and show it has positive derivative.*

Exercise 8.2.6: *Suppose f is always positive. Find the trajectories of $x'' + f(x') = 0$. Are there any critical points?*

Exercise 8.2.7: *Suppose that $x' = f(x, y)$, $y' = g(x, y)$. Suppose that $g(x, y) > 1$ for all x and y. Are there any critical points? What can we say about the trajectories at t goes to infinity?*

Exercise 8.2.101: *For the systems below, find and classify the critical points.*
a) $x' = -x + x^2$, $y' = y$ b) $x' = y - y^2 - x$, $y' = -x$ c) $x' = xy$, $y' = x + y - 1$

Exercise 8.2.102: *Find the implicit equations of the trajectories of the following conservative systems. Next find their critical points (if any) and classify them.*
a) $x'' + x^2 = 4$ b) $x'' + e^x = 0$ c) $x'' + (x+1)e^x = 0$

Exercise 8.2.103: *The conservative system $x'' + x^3 = 0$ is not almost linear. Classify its critical point(s) nonetheless.*

Exercise 8.2.104: *Derive an analogous classification of critical points for equations in one dimension, such as $x' = f(x)$ based on the derivative. A point x_0 is critical when $f(x_0) = 0$ and almost linear if in addition $f'(x_0) \neq 0$. Figure out if the critical point is stable or unstable depending on the sign of $f'(x_0)$. Explain. Hint: see § 1.6.*

8.3 Applications of nonlinear systems

Note: 2 lectures, §6.3–§6.4 in [EP], §9.3, §9.5 in [BD]

In this section we will study two very standard examples of nonlinear systems. First, we will look at the nonlinear pendulum equation. We saw the pendulum equation's linearization before, but we noted it was only valid for small angles and short times. Now we will find out what happens for large angles. Next, we will look at the predator-prey equation, which finds various applications in modeling problems in biology, chemistry, economics and elsewhere.

8.3.1 Pendulum

The first example we will study is the pendulum equation $\theta'' + \frac{g}{L} \sin\theta = 0$. Here, θ is the angular displacement, g is the gravitational constant, and L is the length of the pendulum. In this equation we disregard friction, so we are talking about an idealized pendulum.

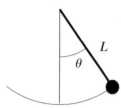

As we have mentioned before, this equation is a conservative equation, so we will be able to use our analysis of conservative equations from the previous section. Let us change the equation to a two-dimensional system in variables (θ, ω) by introducing the new variable ω:

$$\begin{bmatrix} \theta \\ \omega \end{bmatrix}' = \begin{bmatrix} \omega \\ -\frac{g}{L} \sin\theta \end{bmatrix}.$$

The critical points of this system are when $\omega = 0$ and $-\frac{g}{L} \sin\theta = 0$, or in other words if $\sin\theta = 0$. So the critical points are when $\omega = 0$ and θ is a multiple of π. That is the points are $\ldots (-2\pi, 0), (-\pi, 0), (0, 0), (\pi, 0), (2\pi, 0) \ldots$. While there are infinitely many critical points, they are all isolated. Let us compute the Jacobian matrix:

$$\begin{bmatrix} \frac{\partial}{\partial\theta}(\omega) & \frac{\partial}{\partial\omega}(\omega) \\ \frac{\partial}{\partial\theta}\left(-\frac{g}{L}\sin\theta\right) & \frac{\partial}{\partial\omega}\left(-\frac{g}{L}\sin\theta\right) \end{bmatrix} = \begin{bmatrix} 0 & 1 \\ -\frac{g}{L}\cos\theta & 0 \end{bmatrix}.$$

For conservative equations, there are two types of critical points. Either stable centers, or saddle points. The eigenvalues of the Jacobian are $\lambda = \pm\sqrt{-\frac{g}{L}\cos\theta}$.

The eigenvalues are going to be real when $\cos\theta < 0$. This happens at the odd multiples of π. The eigenvalues are going to be purely imaginary when $\cos\theta > 0$. This happens at the even multiples of π. Therefore the system has a stable center at the points $\ldots (-2\pi, 0), (0, 0), (2\pi, 0) \ldots$, and it has an unstable saddle at the points $\ldots (-3\pi, 0), (-\pi, 0), (\pi, 0), (3\pi, 0) \ldots$. Look at the phase diagram in Figure 8.6 on the next page, where for simplicity we let $\frac{g}{L} = 1$.

In the linearized equation we only had a single critical point, the center at $(0, 0)$. We can now see more clearly what we meant when we said the linearization was good for small angles. The horizontal axis is the deflection angle. The vertical axis is the angular velocity of the pendulum.

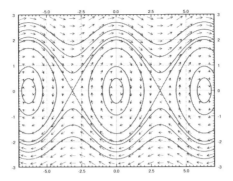

Figure 8.6: Phase plane diagram and some trajectories of the nonlinear pendulum equation.

Suppose we start at $\theta = 0$ (no deflection), and we start with a small angular velocity ω. Then the trajectory keeps going around the critical point $(0, 0)$ in an approximate circle. This corresponds to short swings of the pendulum back and forth. When θ stays small, the trajectories really look like circles and hence are very close to our linearization.

When we give the pendulum a big enough push, it will go across the top and keep spinning about its axis. This behavior corresponds to the wavy curves that do not cross the horizontal axis in the phase diagram. Let us suppose we look at the top curves, when the angular velocity ω is large and positive. Then the pendulum is going around and around its axis. The velocity is going to be large when the pendulum is near the bottom, and the velocity is the smallest when the pendulum is close to the top of its loop.

At each critical point, there is an equilibrium solution. The solution $\theta = 0$ is a stable solution. That is when the pendulum is not moving and is hanging straight down. Clearly this is a stable place for the pendulum to be, hence this is a *stable* equilibrium.

The other type of equilibrium solution is at the unstable point, for example $\theta = \pi$. Here the pendulum is upside down. Sure you can balance the pendulum this way and it will stay, but this is an *unstable* equilibrium. Even the tiniest push will make the pendulum start swinging wildly.

See Figure 8.7 on the following page for a diagram. The first picture is the stable equilibrium $\theta = 0$. The second picture corresponds to those "almost circles" in the phase diagram around $\theta = 0$ when the angular velocity is small. The next picture is the unstable equilibrium $\theta = \pi$. The last picture corresponds to the wavy lines for large angular velocities.

The quantity

$$\frac{1}{2}\omega^2 - \frac{g}{L}\cos\theta$$

is conserved by any solution. This is the energy or the Hamiltonian of the system.

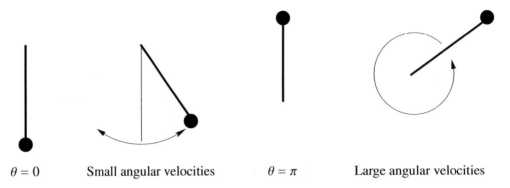

$\theta = 0$ Small angular velocities $\theta = \pi$ Large angular velocities

Figure 8.7: Various possibilities for the motion of the pendulum.

We have a conservative equation and so (exercise) the trajectories are given by

$$\omega = \pm \sqrt{\frac{2g}{L} \cos \theta + C},$$

for various values of C. Let us look at the initial condition of $(\theta_0, 0)$, that is, we take the pendulum to angle θ_0, and just let it go (initial angular velocity 0). We plug the initial conditions into the above and solve for C to obtain

$$C = -\frac{2g}{L} \cos \theta_0.$$

Thus the expression for the trajectory is

$$\omega = \pm \sqrt{\frac{2g}{L}} \sqrt{\cos \theta - \cos \theta_0}.$$

Let us figure out the period. That is, the time it takes for the pendulum to swing back and forth. We notice that the oscillation about the origin in the phase plane is symmetric about both the θ and the ω axis. That is, in terms of θ, the time it takes from θ_0 to $-\theta_0$ is the same as it takes from $-\theta_0$ back to θ_0. Furthermore, the time it takes from $-\theta_0$ to 0 is the same as to go from 0 to θ_0. Therefore, let us find how long it takes for the pendulum to go from angle 0 to angle θ_0, which is a quarter of the full oscillation and then multiply by 4.

We figure out this time by finding $\frac{dt}{d\theta}$ and integrating from 0 to ω_0. The period is four times this integral. Let us stay in the region where ω is positive. Since $\omega = \frac{d\theta}{dt}$, inverting we get

$$\frac{dt}{d\theta} = \sqrt{\frac{L}{2g}} \frac{1}{\sqrt{\cos \theta - \cos \theta_0}}.$$

Therefore the period T is given by

$$T = 4 \sqrt{\frac{L}{2g}} \int_0^{\theta_0} \frac{1}{\sqrt{\cos\theta - \cos\theta_0}} \, d\theta.$$

The integral is an improper integral, and we cannot in general evaluate it symbolically. We must resort to numerical approximation if we want to compute a particular T.

Recall from § 2.4, the linearized equation $\theta'' + \frac{g}{L}\theta = 0$ has period

$$T_{\text{linear}} = 2\pi \sqrt{\frac{L}{g}}.$$

We plot T, T_{linear}, and the relative error $\frac{T - T_{\text{linear}}}{T}$ in Figure 8.8. The relative error says how far is our approximation from the real period percentage-wise. Note that T_{linear} is simply a constant, it does not change with the initial angle θ_0. The actual period T gets larger and larger as θ_0 gets larger. Notice how the relative error is small when θ_0 is small. It is still only 15% when $\theta_0 = \frac{\pi}{2}$, that is, a 90 degree angle. The error is 3.8% when starting at $\frac{\pi}{4}$, a 45 degree angle. At a 5 degree initial angle, the error is only 0.048%.

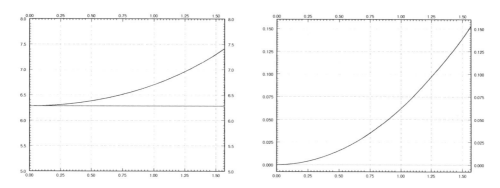

Figure 8.8: *The plot of T and T_{linear} with $\frac{g}{L} = 1$ (left), and the plot of the relative error $\frac{T - T_{linear}}{T}$ (right), for θ_0 between 0 and $\pi/2$.*

While it is not immediately obvious from the formula, it is true that

$$\lim_{\theta_0 \uparrow \pi} T = \infty.$$

That is, the period goes to infinity as the initial angle approaches the unstable equilibrium point. So if we put the pendulum almost upside down it may take a very long time before it gets down. This is consistent with the limiting behavior, where the exactly upside down pendulum never makes an oscillation, so we could think of that as infinite period.

8.3.2   Predator-prey or Lotka-Volterra systems

One of the most common simple applications of nonlinear systems are the so-called *predator-prey* or *Lotka-Volterra*[*] systems. For example, these systems arise when two species interact, one as a prey and one as a predator. It is then no surprise that the equations also see applications in economics. This simple system of equations explains the natural periodic variations of populations of different species in nature. Before the application of differential equations, these periodic variations in the population baffled biologists. Another example where the system arises is in chemical reactions.

Let us keep with the classical example of hares and foxes in a forest, as it is the easiest to understand.

$$x = \text{\# of hares (the prey)},$$
$$y = \text{\# of foxes (the predator)}.$$

When there are a lot of hares, there is plenty of food for the foxes, so the fox population grows. However, when the fox population grows, the foxes eat more hares, so when there are lots of foxes, the hare population should go down, and vice versa. The Lotka-Volterra model proposes that this behavior is described by the system of equations

$$x' = (a - by)x,$$
$$y' = (cx - d)y,$$

where a, b, c, d are some parameters that describe the interaction of the foxes and hares[†]. In this model, these are all positive numbers.

Let us analyze the idea behind this model. The model is a slightly more complicated idea based on the exponential population model. First expand,

$$x' = (a - by)x = ax - byx.$$

The hares are expected to simply grow exponentially in the absence of foxes, that is where the ax term comes in, the growth in population is proportional to the population itself. We are assuming the hares will always find enough food and have enough space to reproduce. However, there is another component $-byx$, that is, the population also is decreasing proportionally to the number of foxes. Together we can write the equation as $(a - by)x$, so it is like exponential growth or decay but the constant depends on the number of foxes.

The equation for foxes is very similar, expand again

$$y' = (cx - d)y = cxy - dy.$$

The foxes need food (hares) to reproduce: the more food, the bigger the rate of growth, hence the cxy term. On the other hand, there are natural deaths in the fox population, and hence the $-dy$ term.

---

[*]Named for the American mathematician, chemist, and statistician Alfred James Lotka (1880–1949) and the Italian mathematician and physicist Vito Volterra (1860–1940).

[†]This interaction does not end well for the hare.

Without further delay, let us start with an explicit example. Suppose the equations are

$$x' = (0.4 - 0.01y)x, \qquad y' = (0.003x - 0.3)y.$$

See Figure 8.9 for the phase portrait. In this example it makes sense to also plot x and y as graphs with respect to time. Therefore the second graph in Figure 8.9 is the graph of x and y on the vertical axis (the prey x is the thinner line with taller peaks), against time on the horizontal axis. The particular trajectory graphed was with initial conditions of 20 foxes and 50 hares.

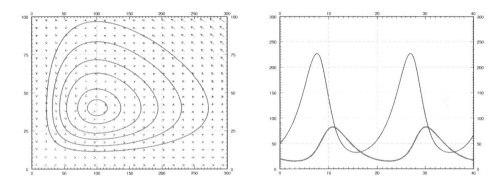

Figure 8.9: The phase portrait (left) and graphs of x and y for a sample trajectory (right).

Let us analyze what we see on the graphs. We work in the general setting rather than putting in specific numbers. We start with finding the critical points. Set $(a - by)x = 0$, and $(cx - d)y = 0$. The first equation is satisfied if either $x = 0$ or $y = a/b$. If $x = 0$, the second equation implies $y = 0$. If $y = a/b$, the second equation implies $x = d/c$. There are two equilibria: at $(0,0)$ when there are no animals at all, and at $(d/c, a/b)$.

In our specific example $x = d/c = 100$, and $y = a/b = 40$. This is the point where there are 100 hares and 40 foxes.

Let us compute the Jacobian matrix:

$$\begin{bmatrix} a - by & -bx \\ cy & cx - d \end{bmatrix}.$$

At the origin $(0,0)$ we get the matrix $\begin{bmatrix} a & 0 \\ 0 & -d \end{bmatrix}$, so the eigenvalues are a and $-d$, hence real and of opposite signs. So the critical point at the origin is a saddle. This makes sense. If you started with some foxes but no hares, then the foxes would go extinct, that is, you would approach the origin. If you started with no foxes and a few hares, then the hares would keep multiplying without check, and so you would go away from the origin.

OK, how about the other critical point at $(d/c, a/b)$. Here the Jacobian matrix becomes

$$\begin{bmatrix} 0 & -\frac{bd}{c} \\ \frac{ac}{b} & 0 \end{bmatrix}.$$

Computing the eigenvalues we get the equation $\lambda^2 + ad = 0$. In other words, $\lambda = \pm i\sqrt{ad}$. The eigenvalues being purely imaginary, we are in the case where we cannot quite decide using only linearization. We could have a stable center, spiral sink, or a spiral source. That is, the equilibrium could be asymptotically stable, stable, or unstable. Of course I gave you a picture above that seems to imply it is a stable center. But never trust a picture only. Perhaps the oscillations are getting larger and larger, but only *very* slowly. Of course this would be bad as it would imply something will go wrong with our population sooner or later. And I only graphed a very specific example with very specific trajectories.

How can we be sure we are in the stable situation? As we said before, in the case of purely imaginary eigenvalues, we have to do a bit more work. Previously we found that for conservative systems, there was a certain quantity that was conserved on the trajectories, and hence the trajectories had to go in closed loops. We can use a similar technique here. We just have to figure out what is the conserved quantity. After some trial and error we find the constant

$$C = \frac{y^a x^d}{e^{cx+by}} = y^a x^d e^{-cx-by}$$

is conserved. Such a quantity is called the *constant of motion*. Let us check C really is a constant of motion. How do we check, you say? Well, a constant is something that does not change with time, so let us compute the derivative with respect to time:

$$C' = a y^{a-1} y' x^d e^{-cx-by} + y^a d x^{d-1} x' e^{-cx-by} + y^a x^d e^{-cx-by}(-cx' - by').$$

Our equations give us what x' and y' are so let us plug those in:

$$\begin{aligned}
C' &= a y^{a-1}(cx-d)y x^d e^{-cx-by} + y^a d x^{d-1}(a-by)x e^{-cx-by} + y^a x^d e^{-cx-by}(-c(a-by)x - b(cx-d)y) \\
&= y^a x^d e^{-cx-by}\Big(a(cx-d) + d(a-by) + (-c(a-by)x - b(cx-d)y) \Big) \\
&= 0.
\end{aligned}$$

So along the trajectories C is constant. In fact, the expression $C = \frac{y^a x^d}{e^{cx+by}}$ gives us an implicit equation for the trajectories. In any case, once we have found this constant of motion, it must be true that the trajectories are simple curves, that is, the level curves of $\frac{y^a x^d}{e^{cx+by}}$. It turns out, the critical point at $(d/c, a/b)$ is a maximum for C (left as an exercise). So $(d/c, a/b)$ is a stable equilibrium point, and we do not have to worry about the foxes and hares going extinct or their populations exploding.

One blemish on this wonderful model is that the number of foxes and hares are discrete quantities and we are modeling with continuous variables. Our model has no problem with there being 0.1 fox in the forest for example, while in reality that makes no sense. The approximation is a reasonable one as long as the number of foxes and hares are large, but it does not make much sense for small numbers. One must be careful in interpreting any results from such a model.

An interesting consequence (perhaps counterintuitive) of this model is that adding animals to the forest might lead to extinction, because the variations will get too big, and one of the populations will get close to zero. For example, suppose there are 20 foxes and 50 hares as before, but now we bring in more foxes, bringing their number to 200. If we run the computation, we will find the number of hares will plummet to just slightly more than 1 hare in the whole forest. In reality that will most likely mean the hares die out, and then the foxes will die out as well as they will have nothing to eat.

Showing that a system of equations has a stable solution can be a very difficult problem. In fact, when Isaac Newton put forth his laws of planetary motions, he proved that a single planet orbiting a single sun is a stable system. But any solar system with more than 1 planet proved very difficult indeed. In fact, such a system will behave chaotically (see § 8.5), meaning small changes in initial conditions will lead to very different long term outcomes. From numerical experimentation and measurements, we know the earth will not fly out into the empty space or crash into the sun, for at least some millions of years or so. But we do not know what happens beyond that.

8.3.3   Exercises

Exercise 8.3.1: *Take the* damped nonlinear pendulum equation $\theta'' + \mu\theta' + (g/L)\sin\theta = 0$ *for some* $\mu > 0$ *(that is, there is some friction). a) Suppose $\mu = 1$ and $g/L = 1$ for simplicity, find and classify the critical points. b) Do the same for any $\mu > 0$ and any g and L, but such that the damping is small, in particular, $\mu^2 < 4(g/L)$. c) Explain what your findings mean, and if it agrees with what you expect in reality.*

Exercise 8.3.2: *Suppose the hares do not grow exponentially, but logistically. In particular consider*

$$x' = (0.4 - 0.01y)x - \gamma x^2, \qquad y' = (0.003x - 0.3)y.$$

For the following two values of γ, find and classify all the critical points in the positive quadrant, that is, for $x \geq 0$ and $y \geq 0$. Then sketch the phase diagram. Discuss the implication for the long term behavior of the population. a) $\gamma = 0.001$, b) $\gamma = 0.01$.

Exercise 8.3.3: *a) Suppose x and y are positive variables. Show $\frac{yx}{e^{x+y}}$ attains a maximum at $(1, 1)$. b) Suppose a, b, c, d are positive constants, and also suppose x and y are positive variables. Show $\frac{y^a x^d}{e^{cx+by}}$ attains a maximum at $(d/c, a/b)$.*

Exercise 8.3.4: *Suppose that for the pendulum equation we take a trajectory giving the spinning-around motion, for example $\omega = \sqrt{\frac{2g}{L}\cos\theta + \frac{2g}{L} + \omega_0^2}$. This is the trajectory where the lowest angular velocity is ω_0^2. Find an integral expression for how long it takes the pendulum to go all the way around.*

Exercise 8.3.5 (challenging)*: Take the pendulum, suppose the initial position is $\theta = 0$. a) Find the expression for ω giving the trajectory with initial condition $(0, \omega_0)$. Hint: Figure out what C should be in terms of ω_0. b) Find the crucial angular velocity ω_1, such that for any higher initial angular velocity, the*

pendulum will keep going around its axis, and for any lower initial angular velocity, the pendulum will simply swing back and forth. Hint: When the pendulum doesn't go over the top the expression for ω will be undefined for some θs.
c) What do you think happens if the initial condition is $(0, \omega_1)$, that is, the initial angle is 0, and the initial angular velocity is exactly ω_1.

Exercise 8.3.101: *Take the damped nonlinear pendulum equation $\theta'' + \mu\theta' + (g/L)\sin\theta = 0$ for some $\mu > 0$ (that is, there is friction). Suppose the friction is large, in particular $\mu^2 > 4(g/L)$. a) Find and classify the critical points. b) Explain what your findings mean, and if it agrees with what you expect in reality.*

Exercise 8.3.102: *Suppose we have the system predator-prey system where the foxes are also killed at a constant rate h (h foxes killed per unit time): $x' = (a - by)x$, $y' = (cx - d)y - h$. a) Find the critical points and the Jacobin matrices of the system. b) Put in the constants $a = 0.4$, $b = 0.01$, $c = 0.003$, $d = 0.3$, $h = 10$. Analyze the critical points. What do you think it says about the forest?*

Exercise 8.3.103 (challenging)*: Suppose the foxes never die. That is, we have the system $x' = (a - by)x$, $y' = cxy$. Find the critical points and notice they are not isolated. What will happen to the population in the forest if it starts at some positive numbers. Hint: Think of the constant of motion.*

8.4 Limit cycles

Note: 1 lecture , discussed in §6.1 and §6.4 in [EP] , §9.7 in [BD]

For nonlinear systems, trajectories do not simply need to approach or leave a single point. They may in fact approach a larger set, such as a circle or another closed curve.

Example 8.4.1: The *Van der Pol oscillator*[*] is the following equation

$$x'' - \mu(1 - x^2)x' + x = 0,$$

where μ is some positive constant. The Van der Pol oscillator comes up often in applications, for example in electrical circuits.

For simplicity, let us use $\mu = 1$. A phase diagram is given in the left hand plot in Figure 8.10. Notice how the trajectories seem to very quickly settle on a closed curve. On the right hand plot we have the plot of a single solution for $t = 0$ to $t = 30$ with initial conditions $x(0) = 0.1$ and $x'(0) = 0.1$. Notice how the solution quickly tends to a periodic solution.

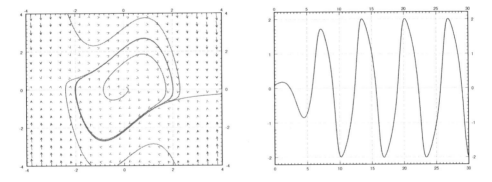

Figure 8.10: The phase portrait (left) and graphs of sample solutions of the Van der Pol oscillator.

The Van der Pol oscillator is an example of so-called *relaxation oscillation*. The word relaxation comes from the sudden jump (the very steep part of the solution). For larger μ the steep part becomes even more pronounced, for small μ the limit cycle looks more like a circle. In fact setting $\mu = 0$, we get $x'' + x = 0$, which is a linear system with a center and all trajectories become circles.

The closed curve in the phase portrait above is called a *limit cycle*. A limit cycle is a closed trajectory such that at least one other trajectory spirals into it (or spirals out of it). If all trajectories that start near the limit cycle spiral into it, the limit cycle is called *asymptotically stable*. The limit cycle in the Van der Pol oscillator is asymptotically stable.

---

[*]Named for the Dutch physicist Balthasar van der Pol (1889–1959).

Given a limit cycle on an autonomous system, any solution that starts on it is periodic. In fact, this is true for any trajectory that is a closed curve (a so-called *closed trajectory*). Such a curve is called a *periodic orbit*. More precisely, if $(x(t), y(t))$ is a solution such that for some t_0 the point $(x(t_0), y(t_0))$ lies on a periodic orbit, then both $x(t)$ and $y(t)$ are periodic functions (with the same period). That is, there is some number P such that $x(t) = x(t + P)$ and $y(t) = y(t + P)$.

Consider the system

$$x' = f(x, y), \qquad y' = g(x, y), \tag{8.2}$$

where the functions f and g have continuous derivatives.

Theorem 8.4.1 (Poincarè-Bendixson[*]). *Suppose R is a closed bounded region (a region in the plane that includes its boundary and does not have points arbitrarily far from the origin). Suppose $(x(t), y(t))$ is a solution of (8.2) in R that exists for all $t \geq t_0$. Then either the solution is a periodic function, or the solution spirals towards a periodic solution in R.*

The main point of the theorem is that if you find one solution that exists for all t large enough (that is, we can let t go to infinity) and stays within a bounded region, then you have found either a periodic orbit, or a solution that spirals towards a limit cycle. That is, in the long term, the behavior will be very close to a periodic function. We should take the theorem more as a qualitative statement rather than something to help us in computations. In practice it is hard to find solutions and therefore hard to show rigorously that they exist for all time. Another caveat to consider is that the theorem only works in two dimensions. In three dimensions and higher, there is simply too much room.

Let us next look when limit cycles (or periodic orbits) do not exist. We will assume the equation (8.2) is defined on a *simply connected region*, that is, a region with no holes we could go around. For example the entire plane is a simply connected region, and so is the inside of the unit disc. However, the entire plane minus a point is not a simply connected domain as it has a "hole" at the origin.

Theorem 8.4.2 (Bendixson-Dulac[†]). *Suppose f and g are defined in a simply connected region R. If the expression[‡]*

$$\frac{\partial f}{\partial x} + \frac{\partial g}{\partial y}$$

is either always positive or always negative on R (except perhaps a small set such as on isolated points or curves) then the system (8.2) has no closed trajectory inside R.

The theorem gives us a way of ruling out the existence of a closed trajectory, and hence a way of ruling out limit cycles. The exception about points or lines really means that we can allow the expression to be zero at a few points, or perhaps on a curve, but not on any larger set.

---

[*]Ivar Otto Bendixson (1861–1935) was a Swedish mathematician.

[†]Henri Dulac (1870–1955) was a French mathematician.

[‡]Sometimes the expression in the Poincarè-Dulac Theorem is $\frac{\partial(\varphi f)}{\partial x} + \frac{\partial(\varphi g)}{\partial y}$ for some continuously differentiable function φ. For simplicity let us just consider the case $\varphi = 1$.

Example 8.4.2: Let us look at $x' = y + y^2 e^x$, $y' = x$ in the entire plane (see Example 8.2.2 on page 327). The entire plane is simply connected and so we can apply the theorem. We compute $\frac{\partial f}{\partial x} + \frac{\partial g}{\partial y} = y^2 e^x + 0$. The function $y^2 e^x$ is always positive except on the line $y = 0$. Therefore, via the theorem, the system has no closed trajectories.

In some books (or the internet) the theorem is not stated carefully and it concludes there are no periodic solutions. That is not quite right. The above example has two critical points and hence it has constant solutions, and constant functions are periodic. The conclusion of the theorem should be that there exist no trajectories that form closed curves. Another way to state the conclusion of the theorem would be to say that there exist no nonconstant periodic solutions that stay in R.

Example 8.4.3: Let us look at a somewhat more complicated example. Take the system $x' = -y - x^2$, $y' = -x + y^2$ (see Example 8.2.1 on page 326). We compute $\frac{\partial f}{\partial x} + \frac{\partial g}{\partial y} = 2x + 2y$. This expression takes on both signs, so if we are talking about the whole plane we cannot simply apply the theorem. However, we could apply it on the set where $x + y > 0$. Via the theorem, there is no closed trajectory in that set. Similarly, there is no closed trajectory in the set $x + y < 0$. We cannot conclude (yet) that there is no closed trajectory in the entire plane. Perhaps half of it is in the set where $x + y > 0$ and the other half is in the set where $x + y < 0$.

The key is to look at the set $x + y = 0$, or $x = -y$. Let us make a substitution $x = z$ and $y = -z$ (so that $x = -y$). Both equations become $z' = z - z^2$. So any solution of $z' = z - z^2$, gives us a solution $x(t) = z(t)$, $y(t) = -z(t)$. In particular, any solution that starts out on the line $x + y = 0$, stays on the line $x + y = 0$. In other words, there cannot be a closed trajectory that starts on the set where $x + y > 0$ and goes through the set where $x + y < 0$, as it would have to pass through $x + y = 0$.

8.4.1 Exercises

Exercise 8.4.1: *Show that the following systems have no closed trajectories.*
a) $x' = x^3 + y$, $y' = y^3 + x^2$,
b) $x' = e^{x-y}$, $y' = e^{x+y}$,
c) $x' = x + 3y^2 - y^3$, $y' = y^3 + x^2$.

Exercise 8.4.2: *Formulate a condition for a 2-by-2 linear system $\vec{x}' = A\vec{x}$ to not be a center using the Bendixson-Dulac theorem. That is, the theorem says something about certain elements of A.*

Exercise 8.4.3: *Explain why the Bendixson-Dulac Theorem does not apply for any conservative system $x'' + h(x) = 0$.*

Exercise 8.4.4: *A system such as $x' = x$, $y' = y$ has solutions that exist for all time t, yet there are no closed trajectories or other limit cycles. Explain why the Poincarè-Bendixson Theorem does not apply.*

Exercise 8.4.5: *Differential equations can also be given in different coordinate systems. Suppose we have the system $r' = 1 - r^2$, $\theta' = 1$ given in polar coordinates. Find all the closed trajectories and check if they are limit cycles and if so, if they are asymptotically stable or not.*

Exercise 8.4.101: *Show that the following systems have no closed trajectories.*
a) $x' = x + y^2$, $y' = y + x^2$,   b) $x' = -x\sin^2(y)$, $y' = e^x$,   c) $x' = xy$, $y' = x + x^2$.

Exercise 8.4.102: *Suppose an autonomous system in the plane has a solution $x = \cos(t) + e^{-t}$, $y = \sin(t) + e^{-t}$. What can you say about the system (in particular about limit cycles and periodic solutions)?*

Exercise 8.4.103: *Show that the limit cycle of the Van der Pol oscillator (for $\mu > 0$) must not lie completely in the set where $-\sqrt{\frac{1+\mu}{\mu}} < x < \sqrt{\frac{1+\mu}{\mu}}$.*

Exercise 8.4.104: *Suppose we have the system $r' = \sin(r)$, $\theta' = 1$ given in polar coordinates. Find all the closed trajectories.*

8.5 Chaos

Note: 1 lecture, §6.5 in [EP], §9.8 in [BD]

You have surely heard the story about the flap of a butterfly wing in the Amazon causing hurricanes in the North Atlantic. In a prior section, we mentioned that a small change in initial conditions of the planets can lead to very different configuration of the planets in the long term. These are examples of *chaotic systems*. Mathematical chaos is not really chaos, there is precise order behind the scenes. Everything is still deterministic. However a chaotic system is extremely sensitive to initial conditions. This also means even small errors induced via numerical approximation create large errors very quickly, so it is almost impossible to numerically approximate for long times. This is large part of the trouble as chaotic systems cannot be in general solved analytically.

Take the weather for example. As a small change in the initial conditions (the temperature at every point of the atmosphere for example) produces drastically different predictions in relatively short time, we cannot accurately predict weather. This is because we do not actually know the exact initial conditions, we measure temperatures at a few points with some error and then we somehow estimate what is in between. There is no way we can accurately measure the effects of every butterfly wing. Then we will solve numerically introducing new errors. That is why you should not trust weather prediction more than a few days out.

The idea of chaotic behavior was first noticed by Edward Lorenz[*] in the 1960s when trying to model thermally induced air convection (movement). The equations Lorentz was looking at form the relatively simple looking system:

$$x' = -10x + 10y, \qquad y' = 28x - y - xz, \qquad z' = -\frac{8}{3}z + xy.$$

A small change in the initial conditions yield a very different solution after a reasonably short time.

A very simple example the reader can experiment with, which displays chaotic behavior, is a double pendulum. The equations that govern this system are somewhat complicated and their derivation is quite tedious, so we will not bother to write them down. The idea is to put a pendulum on the end of another pendulum. If you look at the movement of the bottom mass, the movement will appear chaotic. This type of system is a basis for a whole number of office novelty desk toys. It is very simple to build a version. Take a piece of a string, and tie two heavy nuts at different points of the string; one at the end, and one a bit above. Now give the bottom nut a little push, as long as the swings are not too big and the string stays tight, you have a double pendulum system.

8.5.1 Duffing equation and strange attractors

Let us study the so-called *Duffing equation*:

$$x'' + ax' + bx + cx^3 = C\cos(\omega t).$$

---

[*]Edward Norton Lorenz (1917–2008) was an American mathematician and meteorologist.

Here a, b, c, C, and ω are constants. You will recognize that except for the cx^3 term, this equation looks like a forced mass-spring system. The cx^3 term comes up when the spring does not exactly obey Hooke's law (which no real-world spring actually does obey exactly). When c is not zero, the equation does not have a nice closed form solution, so we have to resort to numerical solutions as is usual for nonlinear systems. Not all choices of constants and initial conditions will exhibit chaotic behavior. Let us study

$$x'' + 0.05x' + x^3 = 8\cos(t).$$

The equation is not autonomous, so we will not be able to draw the vector field in the phase plane. We can still draw the trajectories however.

In Figure 8.11 we plot trajectories for t going from 0 to 15, for two very close initial conditions $(2, 3)$ and $(2, 2.9)$, and also the solutions in the (x, t) space. The two trajectories are close at first, but after a while diverge significantly. This sensitivity to initial conditions is precisely what we mean by the system behaving chaotically.

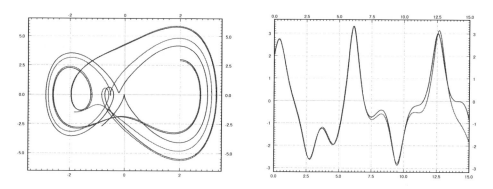

Figure 8.11: On left, two trajectories in phase space for $0 \le t \le 15$, for the Duffing equation one with initial conditions $(2, 3)$ and the other with $(2, 2.9)$. On right the two solutions in (x, t)-space.

Let us see the long term behavior. In Figure 8.12 on the next page, we plot the behavior of the system for initial conditions $(2, 3)$, but for much longer period of time. Note that for this period of time it was necessary to use a ridiculously large number of steps in the numerical algorithm used to produce the graph, as even small errors quickly propagate[*]. From the graph it is hard to see any particular pattern in the shape of the solution except that it seems to oscillate, but each oscillation appears quite unique. The oscillation is expected due to the forcing term.

In general it is very difficult to analyze chaotic systems, or to find the order behind the madness, but let us try to do something that we did for the standard mass-spring system. One way we analyzed what happens is that we figured out what was the long term behavior (not dependent on initial conditions). From the figure above it is clear that we will not get a nice description of the long term

---

[*]In fact for reference, 30,000 steps were used with the Runge-Kutta algorithm, see exercises in § 1.7.

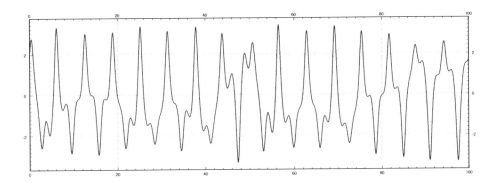

Figure 8.12: The solution to the given Duffing equation for t from 0 to 100.

behavior, but perhaps we can figure out some order to what happens on each "oscillation" and what do these oscillations have in common.

The concept we will explore is that of a *Poincarè section**. Instead of looking at t in a certain interval, we will look at where the system is at a certain sequence of points in time. Imagine flashing a strobe at a certain fixed frequency and drawing the points where the solution is during the flashes. The right strobing frequency depends on the system in question. The correct frequency to use for the forced Duffing equation (and other similar systems) is the frequency of the forcing term. For the Duffing equation above, find a solution $(x(t), y(t))$, and look at the points

$$(x(0), y(0)), \quad (x(2\pi), y(2\pi)), \quad (x(4\pi), y(4\pi)), \quad (x(6\pi), y(6\pi)), \quad \ldots$$

As we are really not interested in the transient part of the solution, that is, the part of the solution that depends on the initial condition we skip some number of steps in the beginning. For example, we might skip the first 100 such steps and start plotting points at $t = 100(2\pi)$, that is

$$(x(200\pi), y(200\pi)), \quad (x(202\pi), y(202\pi)), \quad (x(204\pi), y(204\pi)), \quad (x(206\pi), y(206\pi)), \quad \ldots$$

The plot of these points is the Poincarè section. After plotting enough points, a curious pattern emerges in Figure 8.13 on the following page (the left hand picture), a so-called *strange attractor*.

If we have a sequence of points, then an *attractor* is a set towards which the points in the sequence eventually get closer and closer to, that is, they are attracted. The Poincarè section above is not really the attractor itself, but as the points are very close to it, we can see its shape. The strange attractor in the figure is a very complicated set, and it in fact has fractal structure, that is, if you would zoom in as far as you want, you would keep seeing the same complicated structure.

The initial condition does not really make any difference. If we started with different initial condition, the points would eventually gravitate towards the attractor, and so as long as we throw away the first few points, we always get the same picture.

*Named for the French polymath Jules Henri Poincarè (1854–1912).

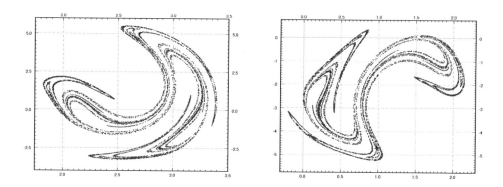

Figure 8.13: Strange attractor. The left plot is with no phase shift, the right plot has phase shift $\pi/4$.

An amazing thing is that a chaotic system such as the Duffing equation is not random at all. There is a very complicated order to it, and the strange attractor says something about this order. We cannot quite say what state the system will be in eventually, but given a fixed strobing frequency we can narrow it down to the points on the attractor.

If you would use a phase shift, for example $\pi/4$, and look at the times

$$\pi/4, \quad 2\pi + \pi/4, \quad 4\pi + \pi/4, \quad 6\pi + \pi/4, \quad \ldots$$

you would obtain a slightly different looking attractor. The picture is the right hand side of Figure 8.13. It is as if we had rotated, distorted slightly, and then moved the original. Therefore for each phase shift you can find the set of points towards which the system periodically keeps coming back to.

You should study the pictures and notice especially the scales—where are these attractors located in the phase plane. Notice the regions where the strange attractor lives and compare it to the plot of the trajectories in Figure 8.11 on page 346.

Let us compare the discussion in this section to the discussion in § 2.6 about forced oscillations. Take the equation

$$x'' + 2px' + \omega_0^2 x = \frac{F_0}{m} \cos(\omega t).$$

This is like the Duffing equation, but with no x^3 term. The steady periodic solution is of the form

$$x = C \cos(\omega t + \gamma).$$

Strobing using the frequency ω we would obtain a single point in the phase space. So the attractor in this setting is a single point—an expected result as the system is not chaotic. In fact it was the opposite of chaotic. Any difference induced by the initial conditions dies away very quickly, and we settle into always the same steady periodic motion.

8.5.2   The Lorenz system

In two dimensions to have the kind of chaotic behavior we are looking for, we have to study forced, or non-autonomous, systems such as the Duffing equation. Due to the Poincarè-Bendoxson Theorem, if an autonomous two-dimensional system has a solution that exists for all time in the future and does not go towards infinity, then we obtain a limit cycle or a closed trajectory. Hardly the chaotic behavior we are looking for.

Let us very briefly return to the Lorenz system

$$x' = -10x + 10y, \qquad y' = 28x - y - xz, \qquad z' = -\frac{8}{3}z + xy.$$

The Lorenz system is an autonomous system in three dimensions exhibiting chaotic behavior. See the Figure 8.14 for a sample trajectory.

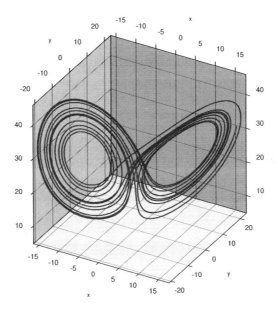

Figure 8.14: A trajectory in the Lorenz system.

The solutions will tend to an *attractor* in space, the so-called *Lorenz attractor*. In this case no strobing is necessary. Again we cannot quite see the attractor itself, but if we try to follow a solution for long enough, as in the figure, we will get a pretty good picture of what the attractor looks like.

The path is not just a repeating figure-eight. The trajectory will spin some seemingly random number of times on the left, then spin a number of times on the right, and so on. As this system arose in weather prediction, one can perhaps imagine a few days of warm weather and then a few days of cold weather, where it is not easy to predict when the weather will change, just as it is not really easy to predict far in advance when the solution will jump onto the other side. See Figure 8.15 for a plot of the x component of the solution drawn above.

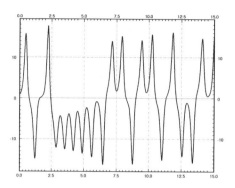

Figure 8.15: Graph of the $x(t)$ component of the solution.

8.5.3   Exercises

Exercise 8.5.1: *For the non-chaotic equation $x'' + 2px' + \omega_0^2 x = \frac{F_0}{m}\cos(\omega t)$, suppose we strobe with frequency ω as we mentioned above. Use the known steady periodic solution to find precisely the point which is the attractor for the Poincarè section.*

Exercise 8.5.2 (project): *A simple fractal attractor can be drawn via the following chaos game. Draw three points of a triangle (just the vertices) and number them, say p_1, p_2 and p_3. Start with some random point p (does not have to be one of the three points above) and draw it. Roll a die, and use it to pick of the p_1, p_2, or p_3 randomly (for example 1 and 4 mean p_1, 2 and 5 mean p_2, and 3 and 6 mean p_3). Suppose we picked p_2, then let p_{new} be the point exactly halfway between p and p_2. Draw this point and let p now refer to this new point p_{new}. Rinse, repeat. Try to be precise and draw as many iterations as possible. Your points should be attracted to the so-called Sierpinski triangle. A computer was used to run the game for 10,000 iterations to obtain the picture in Figure 8.16 on the next page.*

Exercise 8.5.3 (project): *Construct the double pendulum described in the text with a string and two nuts (or heavy beads). Play around with the position of the middle nut, and perhaps use different weight nuts. Describe what you find.*

Exercise 8.5.4 (computer project): *Use a computer software (such as Matlab, Octave, or perhaps even a spreadsheet), plot the solution of the given forced Duffing equation with Euler's method. Plotting the solution for t from 0 to 100 with several different (small) step sizes. Discuss.*

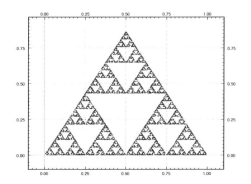

Figure 8.16: 10,000 iterations of the chaos game producing the Sierpinski triangle.

Exercise 8.5.101: *Find critical points of the Lorenz system and the associated linearizations.*

Further Reading

[BM]   Paul W. Berg and James L. McGregor, *Elementary Partial Differential Equations*, Holden-Day, San Francisco, CA, 1966.

[BD]   William E. Boyce and Richard C. DiPrima, *Elementary Differential Equations and Boundary Value Problems*, 9th edition, John Wiley & Sons Inc., New York, NY, 2008.

[EP]   C.H. Edwards and D.E. Penney, *Differential Equations and Boundary Value Problems: Computing and Modeling*, 4th edition, Prentice Hall, 2008.

[F]   Stanley J. Farlow, *An Introduction to Differential Equations and Their Applications*, McGraw-Hill, Inc., Princeton, NJ, 1994. (Published also by Dover Publications, 2006.)

[I]   E.L. Ince, *Ordinary Differential Equations*, Dover Publications, Inc., New York, NY, 1956.

[T]   William F. Trench, *Elementary Differential Equations with Boundary Value Problems*. Books and Monographs. Book 9. 2013. `http://digitalcommons.trinity.edu/mono/9`

Solutions to Selected Exercises

0.2.101: Compute $x' = -2e^{-2t}$ and $x'' = 4e^{-2t}$. Then $(4e^{-2t}) + 4(-2e^{-2t}) + 4(e^{-2t}) = 0$.

0.2.102: Yes.

0.2.103: $y = x^r$ is a solution for $r = 0$ and $r = 2$.

0.2.104: $C_1 = 100$, $C_2 = -90$

0.2.105: $\varphi = -9e^{8s}$

0.3.101: a) PDE, equation, second order, linear, nonhomogeneous, constant coefficient.
b) ODE, equation, first order, linear, nonhomogeneous, not constant coefficient, not autonomous.
c) ODE, equation, seventh order, linear, homogeneous, constant coefficient, autonomous.
d) ODE, equation, second order, linear, nonhomogeneous, constant coefficient, autonomous.
e) ODE, system, second order, nonlinear.
f) PDE, equation, second order, nonlinear.

0.3.102: equation: $a(x)y = b(x)$, solution: $y = \frac{b(x)}{a(x)}$.

1.1.101: $y = e^x + \frac{x^2}{2} + 9$

1.1.102: $x = (3t - 2)^{1/3}$

1.1.103: $x = \sin^{-1}(t + 1)$

1.1.104: 170

1.1.105: If $n \neq 1$, then $y = ((1 - n)x + 1)^{1/(1-n)}$. If $n = 1$, then $y = e^x$.

1.1.106: The equation is $r' = -C$ for some constant C. The snowball will be completely melted in 25 minutes from time $t = 0$.

1.1.107: $y = Ax^3 + Bx^2 + Cx + D$, so 4 constants.

1.2.101:

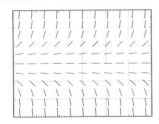

$y = 0$ is a solution such that $y(0) = 0$.

1.2.102: Yes a solution exists. $y' = f(x, y)$ where $f(x, y) = xy$. The function $f(x, y)$ is continuous and $\frac{\partial f}{\partial y} = x$, which is also continuous near $(0, 0)$. So a solution exists and is unique. (In fact $y = 0$ is the solution).

1.2.103: No, the equation is not defined at $(x, y) = (1, 0)$.

1.2.104: a) $y' = \cos y$, b) $y' = y \cos(x)$, c) $y' = \sin x$. Justification left to reader.

1.2.105: Picard does not apply as f is not continuous at $y = 0$. The equation does not have a continuously solution. If it did notice that $y'(0) = 1$, by first derivative test, $y(x) > 0$ for small positive x, but then for those x we would have $y'(x) = 0$, so clearly the derivative cannot be continuous.

1.3.101: $y = Ce^{x^2}$

1.3.102: $x = e^{t^3} + 1$

1.3.103: $x^3 + x = t + 2$

1.3.104: $y = \frac{1}{1 - \ln x}$

1.3.105: $\sin(y) = -\cos(x) + C$

1.3.106: The range is approximately 7.45 to 12.15 minutes.

1.3.107: a) $x = \frac{1000e^t}{e^t + 24}$. b) 102 rabbits after one month, 861 after 5 months, 999 after 10 months, 1000 after 15 months.

1.4.101: $y = Ce^{-x^3} + \frac{1}{3}$

1.4.102: $y = 2e^{\cos(2x)+1} + 1$

1.4.103: 250 grams

1.4.104: $P(5) = 1000e^{2 \times 5 - 0.05 \times 5^2} = 1000e^{8.75} \approx 6.31 \times 10^6$

1.4.105: $Ah' = I - kh$, where k is a constant with units $m^2 s$.

1.5.101: $y = \frac{2}{3x - 2}$

1.5.102: $y = \frac{3 - x^2}{2x}$

1.5.103: $y = (7e^{3x} + 3x + 1)^{1/3}$

1.5.104: $y = \sqrt{x^2 - \ln(C - x)}$

1.6.101:
a) 0, 1, 2 are critical points.
b) $x = 0$ is unstable (semistable), $x = 1$ is stable, and $x = 2$ is unstable.
c) 1

1.6.102: a) There are no critical points.   b) ∞

1.6.103: a) $\frac{dx}{dt} = kx(M - x) + A$ b) $\frac{kM + \sqrt{(kM)^2 + 4Ak}}{2k}$

1.6.104: a) α is a stable critical point, β is an unstable one.   b) α,   c) α,   d) ∞ or DNE.

1.7.101: Approximately: 1.0000, 1.2397, 1.3829

1.7.102:
a) 0, 8, 12
b) $x(4) = 16$, so errors are: 16, 8, 4.
c) Factors are 0.5, 0.5, 0.5.

1.7.103: a) 0, 0, 0   b) $x = 0$ is a solution so errors are: 0, 0, 0.

1.7.104: a) Improved Euler: $y(1) \approx 3.3897$ for $h = 1/4$, $y(1) \approx 3.4237$ for $h = 1/8$,   b) Standard Euler: $y(1) \approx 2.8828$ for $h = 1/4$, $y(1) \approx 3.1316$ for $h = 1/8$,   c) $y = 2e^x - x - 1$, so $y(2)$ is approximately 3.4366.   d) Approximate errors for improved Euler: 0.046852 for $h = 1/4$, and 0.012881 for $h = 1/8$.   For standard Euler: 0.55375 for $h = 1/4$, and 0.30499 for $h = 1/8$.   Factor is approximately 0.27 for improved Euler, and 0.55 for standard Euler.

1.8.101: a) $e^{xy} + \sin(x) = C$,   b) $x^2 + xy - 2y^2 = C$,   c) $e^x + e^y = C$,   d) $x^3 + 3xy + y^3 = C$.

1.8.102: a) Integrating factor is y, the equation becomes $dx + 3y^2\, dx = 0$.   b) Integrating factor is e^x, the equation becomes $e^x\, dx - e^{-y}\, dx = 0$.   c) Integrating factor is y^2, the equation becomes $\cos(x) + y\, dx + x\, dy = 0$.   d) Integrating factor is x, the equation becomes $(2xy + y^2)\, dx + (x^2 + 2xy)\, dx = 0$.

1.8.103: a) The equation is $-f(x)\, dx + \frac{1}{g(y)}\, dy$, and this is exact because $M = -f(x)$, $N = \frac{1}{g(y)}$, so $M_y = 0 = N_x$.   b) $-x\, dx + \frac{1}{y}\, dy = 0$, leads to potential function $F(x, y) = -\frac{x^2}{2} + \ln|y|$, solving $F(x, y) = C$ leads to the same solution as the example.

2.1.101: Yes. To justify try to find a constant A such that $\sin(x) = Ae^x$ for all x.

2.1.102: No. $e^{x+2} = e^2 e^x$.

2.1.103: $y = 5$

2.1.104: $y = C_1 \ln(x) + C_2$

2.1.105: $y'' - 3y' + 2y = 0$

2.2.101: $y = C_1 e^{(-2+\sqrt{2})x} + C_2 e^{(-2-\sqrt{2})x}$

2.2.102: $y = C_1 e^{3x} + C_2 x e^{3x}$

2.2.103: $y = e^{-x/4} \cos((\sqrt{7}/4)x) - \sqrt{7} e^{-x/4} \sin((\sqrt{7}/4)x)$

2.2.104: $y = \frac{2(a-b)}{5} e^{-3x/2} + \frac{3a+2b}{5} e^x$

2.2.105: $z(t) = 2e^{-t} \cos(t)$

2.2.106: $y = \frac{a\beta - b}{\beta - \alpha} e^{\alpha x} + \frac{b - a\alpha}{\beta - \alpha} e^{\beta x}$

2.2.107: $y'' - y' - 6y = 0$

2.3.101: $y = C_1 e^x + C_2 x^3 + C_3 x^2 + C_4 x + C_5$

2.3.102: a) $r^3 - 3r^2 + 4r - 12 = 0$, b) $y''' - 3y'' + 4y' - 12y = 0$, c) $y = C_1 e^{3x} + C_2 \sin(2x) + C_3 \cos(2x)$

2.3.103: $y = 0$

2.3.104: No. $e^1 e^x - e^{x+1} = 0$.

2.3.105: Yes. (Hint: First note that $\sin(x)$ is bounded. Then note that x and $x \sin(x)$ cannot be multiples of each other.)

2.3.106: $y''' - y'' + y' - y = 0$

2.4.101: $k = 8/9$ (and larger)

2.4.102:
a) $0.05I'' + 0.1I' + (1/5)I = 0$
b) $x(t) = Ce^{-t} \cos(\sqrt{3}\,t - \gamma)$
c) $x(t) = 10e^{-t} \cos(\sqrt{3}\,t) + \frac{10}{\sqrt{3}} e^{-t} \sin(\sqrt{3}\,t)$

2.4.103: a) $k = 500000$, b) $\frac{1}{5\sqrt{2}} \approx 0.141$, c) $45000\,\text{kg}$, d) $11250\,\text{kg}$

2.4.104: $m_0 = \frac{1}{3}$. If $m < m_0$, then the system is overdamped and will not oscillate.

2.5.101: $y = \frac{-16\sin(3x) + 6\cos(3x)}{73}$

2.5.102: a) $y = \frac{2e^x + 3x^3 - 9x}{6}$, b) $y = C_1 \cos(\sqrt{2}x) + C_2 \sin(\sqrt{2}x) + \frac{2e^x + 3x^3 - 9x}{6}$

2.5.103: $y(x) = x^2 - 4x + 6 + e^{-x}(x - 5)$

2.5.104: $y = \frac{2xe^x - (e^x + e^{-x})\log(e^{2x} + 1)}{4}$

2.5.105: $y = \frac{-\sin(x+c)}{3} + C_1 e^{\sqrt{2}x} + C_2 e^{-\sqrt{2}x}$

2.6.101: $\omega = \frac{\sqrt{31}}{4\sqrt{2}} \approx 0.984 \quad C(\omega) = \frac{16}{3\sqrt{7}} \approx 2.016$

2.6.102: $x_{sp} = \frac{(\omega_0^2 - \omega^2)F_0}{m(2\omega p)^2 + m(\omega_0^2 - \omega^2)^2} \cos(\omega t) + \frac{2\omega p F_0}{m(2\omega p)^2 + m(\omega_0^2 - \omega^2)^2} \sin(\omega t) + \frac{A}{k}$, where $p = \frac{c}{2m}$ and $\omega_0 = \sqrt{\frac{k}{m}}$.

2.6.103: a) $\omega = 2$, b) 25

3.1.101: $y_1 = C_1 e^{3x}$, $y_2 = y(x) = C_2 e^x + \frac{C_1}{2} e^{3x}$, $y_3 = y(x) = C_3 e^x + \frac{C_1}{2} e^{3x}$

3.1.102: $x = \frac{5}{3}e^{2t} - \frac{2}{3}e^{-t}$, $y = \frac{5}{3}e^{2t} + \frac{4}{3}e^{-t}$

3.1.103: $x_1' = x_2$, $x_2' = x_3$, $x_3' = x_1 + t$

3.1.104: $y_3' + y_1 + y_2 = t$, $y_4' + y_1 - y_2 = t^2$, $y_1' = y_3$, $y_2' = y_4$

3.1.105: $x_1 = x_2 = at$. Explanation of the intuition is left to reader.

3.2.101: -15

3.2.102: -2

3.2.103: $\vec{x} = \begin{bmatrix} 15 \\ -5 \end{bmatrix}$

3.2.104: a) $\begin{bmatrix} 1/a & 0 \\ 0 & 1/b \end{bmatrix}$　b) $\begin{bmatrix} 1/a & 0 & 0 \\ 0 & 1/b & 0 \\ 0 & 0 & 1/c \end{bmatrix}$

3.3.101: Yes.

3.3.102: No. $2 \begin{bmatrix} \cosh(t) \\ 1 \end{bmatrix} - \begin{bmatrix} e^t \\ 1 \end{bmatrix} - \begin{bmatrix} e^{-t} \\ 1 \end{bmatrix} = \vec{0}$

3.3.103: $\begin{bmatrix} x \\ y \end{bmatrix}' = \begin{bmatrix} 3 & -1 \\ 1 & 0 \end{bmatrix} \begin{bmatrix} x \\ y \end{bmatrix} + \begin{bmatrix} e^t \\ 0 \end{bmatrix}$

3.3.104: a) $\vec{x}' = \begin{bmatrix} 0 & 2t \\ 0 & 2t \end{bmatrix} \vec{x}$　b) $\vec{x} = \begin{bmatrix} C_2 e^{t^2} + C_1 \\ C_2 e^{t^2} \end{bmatrix}$

3.4.101:

a) Eigenvalues: $4, 0, -1$ Eigenvectors: $\begin{bmatrix} 1 \\ 0 \\ 1 \end{bmatrix}, \begin{bmatrix} 0 \\ 1 \\ 0 \end{bmatrix}, \begin{bmatrix} 3 \\ 5 \\ -2 \end{bmatrix}$

b) $\vec{x} = C_1 \begin{bmatrix} 1 \\ 0 \\ 1 \end{bmatrix} e^{4t} + C_2 \begin{bmatrix} 0 \\ 1 \\ 0 \end{bmatrix} + C_3 \begin{bmatrix} 3 \\ 5 \\ -2 \end{bmatrix} e^{-t}$

3.4.102:

a) Eigenvalues: $\frac{1+\sqrt{3}i}{2}, \frac{1-\sqrt{3}i}{2},$ Eigenvectors: $\begin{bmatrix} -2 \\ 1-\sqrt{3}i \end{bmatrix}, \begin{bmatrix} -2 \\ 1+\sqrt{3}i \end{bmatrix}$

b) $\vec{x} = C_1 e^{t/2} \begin{bmatrix} -2\cos(\frac{\sqrt{3}t}{2}) \\ \cos(\frac{\sqrt{3}t}{2}) + \sqrt{3}\sin(\frac{\sqrt{3}t}{2}) \end{bmatrix} + C_2 e^{t/2} \begin{bmatrix} -2\sin(\frac{\sqrt{3}t}{2}) \\ \sin(\frac{\sqrt{3}t}{2}) - \sqrt{3}\cos(\frac{\sqrt{3}t}{2}) \end{bmatrix}$

3.4.103: $\vec{x} = C_1 \begin{bmatrix} 1 \\ 1 \end{bmatrix} e^t + C_2 \begin{bmatrix} 1 \\ -1 \end{bmatrix} e^{-t}$

3.4.104: $\vec{x} = C_1 \begin{bmatrix} \cos(t) \\ -\sin(t) \end{bmatrix} + C_2 \begin{bmatrix} \sin(t) \\ \cos(t) \end{bmatrix}$

3.5.101:   a) Two eigenvalues: $\pm\sqrt{2}$ so the behavior is a saddle.   b) Two eigenvalues: 1 and 2, so the behavior is a source.   c) Two eigenvalues: $\pm 2i$, so the behavior is a center (ellipses).   d) Two eigenvalues: -1 and -2, so the behavior is a sink.   e) Two eigenvalues: 5 and -3, so the behavior is a saddle.

3.5.102:   Spiral source.

3.5.103:

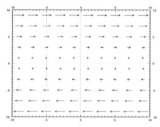

The solution will not move anywhere if $y = 0$. When y is positive, then the solution moves (with constant speed) in the positive x direction. When y is negative, then the solution moves (with constant speed) in the negative x direction. It is not one of the behaviors we have seen.

Note that the matrix has a double eigenvalue 0 and the general solution is $x = C_1 t + C_2$ and $y = C_1$, which agrees with the above description.

3.6.101: $\vec{x} = \begin{bmatrix} 1 \\ -1 \\ 1 \end{bmatrix} (a_1 \cos(\sqrt{3}\,t) + b_1 \sin(\sqrt{3}\,t)) + \begin{bmatrix} 0 \\ 1 \\ -2 \end{bmatrix} (a_2 \cos(\sqrt{2}\,t) + b_2 \sin(\sqrt{2}\,t)) + \begin{bmatrix} 0 \\ 0 \\ 1 \end{bmatrix} (a_3 \cos(t) +$

$b_3 \sin(t)) + \begin{bmatrix} -1 \\ 1/2 \\ 2/3 \end{bmatrix} \cos(2t)$

3.6.102: $\begin{bmatrix} m & 0 & 0 \\ 0 & m & 0 \\ 0 & 0 & m \end{bmatrix} \vec{x}'' = \begin{bmatrix} -k & k & 0 \\ k & -2k & k \\ 0 & k & -k \end{bmatrix} \vec{x}.$ Solution: $\vec{x} = \begin{bmatrix} 1 \\ -2 \\ 1 \end{bmatrix} (a_1 \cos(\sqrt{3k/m}\,t) + b_1 \sin(\sqrt{3k/m}\,t)) +$

$\begin{bmatrix} 1 \\ 0 \\ -1 \end{bmatrix} (a_2 \cos(\sqrt{k/m}\,t) + b_2 \sin(\sqrt{k/m}\,t)) + \begin{bmatrix} 1 \\ 1 \\ 1 \end{bmatrix} (a_3 t + b_3).$

3.6.103: $x_2 = (2/5)\cos(\sqrt{1/6}\,t) - (2/5)\cos(t)$

3.7.101:   a) $3, 0, 0$ b) No defects.   c) $\vec{x} = C_1 \begin{bmatrix} 1 \\ 1 \\ 1 \end{bmatrix} e^{3t} + C_2 \begin{bmatrix} 1 \\ 0 \\ -1 \end{bmatrix} + C_3 \begin{bmatrix} 0 \\ 1 \\ -1 \end{bmatrix}$

3.7.102:

a) $1, 1, 2$

b) Eigenvalue 1 has a defect of 1

c) $\vec{x} = C_1 \begin{bmatrix} 0 \\ 1 \\ -1 \end{bmatrix} e^t + C_2 \left(\begin{bmatrix} 1 \\ 0 \\ 0 \end{bmatrix} + t \begin{bmatrix} 0 \\ 1 \\ -1 \end{bmatrix} \right) e^t + C_3 \begin{bmatrix} 3 \\ 3 \\ -2 \end{bmatrix} e^{2t}$

3.7.103:

a) $2, 2, 2$

b) Eigenvalue 2 has a defect of 2

c) $\vec{x} = C_1 \begin{bmatrix} 0 \\ 3 \\ 1 \end{bmatrix} e^{2t} + C_2 \left(\begin{bmatrix} 0 \\ -1 \\ 0 \end{bmatrix} + t \begin{bmatrix} 0 \\ 3 \\ 1 \end{bmatrix} \right) e^{2t} + C_3 \left(\begin{bmatrix} 1 \\ 0 \\ 0 \end{bmatrix} + t \begin{bmatrix} 0 \\ -1 \\ 0 \end{bmatrix} + \frac{t^2}{2} \begin{bmatrix} 0 \\ 3 \\ 1 \end{bmatrix} \right) e^{2t}$

3.7.104: $A = \begin{bmatrix} 5 & 5 \\ 0 & 5 \end{bmatrix}$

3.8.101: $e^{tA} = \begin{bmatrix} \frac{e^{3t}+e^{-t}}{2} & \frac{e^{-t}-e^{3t}}{2} \\ \frac{e^{-t}-e^{3t}}{2} & \frac{e^{3t}+e^{-t}}{2} \end{bmatrix}$

3.8.102: $e^{tA} = \begin{bmatrix} 2e^{3t}-4e^{2t}+3e^t & \frac{3e^t}{2}-\frac{3e^{3t}}{2} & -e^{3t}+4e^{2t}-3e^t \\ 2e^t-2e^{2t} & e^t & 2e^{2t}-2e^t \\ 2e^{3t}-5e^{2t}+3e^t & \frac{3e^t}{2}-\frac{3e^{3t}}{2} & -e^{3t}+5e^{2t}-3e^t \end{bmatrix}$

3.8.103: a) $e^{tA} = \begin{bmatrix} (t+1)e^{2t} & -te^{2t} \\ te^{2t} & (1-t)e^{2t} \end{bmatrix}$ b) $\vec{x} = \begin{bmatrix} (1-t)e^{2t} \\ (2-t)e^{2t} \end{bmatrix}$

3.8.104: $\begin{bmatrix} 1+2t+5t^2 & 3t+6t^2 \\ 2t+4t^2 & 1+2t+5t^2 \end{bmatrix}$ $\quad e^{0.1A} \approx \begin{bmatrix} 1.25 & 0.36 \\ 0.24 & 1.25 \end{bmatrix}$

3.9.101: The general solution is (particular solutions should agree with one of these):

$x(t) = \frac{1}{5}C_1(e^{9t}+e^{4t}) + \frac{4}{5}C_2(e^{9t}-e^{4t}) - \frac{18t+5}{54}$ $\quad y(t) = \frac{1}{5}C_1(e^{9t}-e^{4t}) + \frac{1}{5}C_2(4e^{5t}+e^{4t}) + \frac{t}{6} + \frac{7}{216}$

3.9.102: The general solution is (particular solutions should agree with one of these):

$x(t) = \frac{1}{2}C_1(e^t+e^{-t}) + \frac{1}{2}C_2(e^t-e^{-t}) + te^t$ $\quad y(t) = \frac{1}{2}C_1(e^t-e^{-t}) + \frac{1}{2}C_2(e^t+e^{-t}) + te^t$

3.9.103: $\vec{x} = \begin{bmatrix} 1 \\ 1 \end{bmatrix}\left(\frac{5}{2}e^t - t - 1\right) + \begin{bmatrix} 1 \\ -1 \end{bmatrix}\frac{-1}{2}e^{-t}$

3.9.104: $\vec{x} = \begin{bmatrix} 1 \\ 9 \end{bmatrix}\left(\left(\frac{1}{140} + \frac{1}{120\sqrt{6}}\right)e^{\sqrt{6}t} + \left(\frac{1}{140} + \frac{1}{120\sqrt{6}}\right)e^{-\sqrt{6}t} - \frac{t}{60} - \frac{\cos(t)}{70}\right)$

$+ \begin{bmatrix} 1 \\ -1 \end{bmatrix}\left(\frac{-9}{80}\sin(2t) + \frac{1}{30}\cos(2t) + \frac{9t}{40} - \frac{\cos(t)}{30}\right)$

4.1.101: $\omega = \pi\sqrt{\frac{15}{2}}$

4.1.102: $\lambda_k = 4k^2\pi^2$ for $k = 1, 2, 3, \ldots$ $\quad x_k = \cos(2k\pi t) + B\sin(2k\pi t)$ (for any B)

4.1.103: $x(t) = -\sin(t)$

4.1.104: General solution is $x = Ce^{-\lambda t}$. Since $x(0) = 0$ then $C = 0$, and so $x(t) = 0$. Therefore, the solution is always identically zero. One condition is always enough to guarantee a unique solution for a first order equation.

4.1.105: $\frac{\sqrt{3}}{3}e^{\frac{-3}{2}\sqrt[3]{\lambda}} - \frac{\sqrt{3}}{3}\cos(\frac{\sqrt{3}\sqrt[3]{\lambda}}{2}) + \sin(\frac{\sqrt{3}\sqrt[3]{\lambda}}{2}) = 0$

4.2.101: $\sin(t)$

4.2.102: $\sum_{n=1}^{\infty} \frac{(\pi-n)\sin(\pi n+\pi^2)+(\pi+n)\sin(\pi n-\pi^2)}{\pi n^2-\pi^3} \sin(nt)$

4.2.103: $\frac{1}{2} - \frac{1}{2}\cos(2t)$

4.2.104: $\frac{\pi^4}{5} + \sum_{n=1}^{\infty} \frac{(-1)^n(8\pi^2 n^2 - 48)}{n^4} \cos(nt)$

4.3.101: a) $\frac{8}{6} + \sum_{n=1}^{\infty} \frac{16(-1)^n}{\pi^2 n^2} \cos(\frac{n\pi}{2}t)$ b) $\frac{8}{6} - \frac{16}{\pi^2}\cos(\frac{\pi}{2}t) + \frac{4}{\pi^2}\cos(\pi t) - \frac{16}{9\pi^2}\cos(\frac{3\pi}{2}t) + \cdots$

4.3.102: a) $\sum_{n=1}^{\infty} \frac{(-1)^{n+1}2\lambda}{n\pi} \sin(\frac{n\pi}{\lambda}t)$ b) $\frac{2\lambda}{\pi}\sin(\frac{\pi}{\lambda}t) - \frac{\lambda}{\pi}\sin(\frac{2\pi}{\lambda}t) + \frac{2\lambda}{3\pi}\sin(\frac{3\pi}{\lambda}t) - \cdots$

4.3.103: $f'(t) = \sum_{n=1}^{\infty} \frac{\pi}{n^2+1} \cos(n\pi t)$

4.3.104: a) $F(t) = \frac{t}{2} + C + \sum_{n=1}^{\infty} \frac{1}{n^4}\sin(nt)$ b) no.

4.3.105: a) $\sum_{n=1}^{\infty} \frac{(-1)^{n+1}}{n}\sin(nt)$ b) f is continuous at $t = \pi/2$ so the Fourier series converges to $f(\pi/2) = \pi/4$. Obtain $\pi/4 = \sum_{n=1}^{\infty} \frac{(-1)^{n+1}}{2n-1} = 1 - 1/3 + 1/5 - 1/7 + \cdots$.   c) Using the first 4 terms get $76/105 \approx 0.72$ (quite a bad approximation, you would have to take about 50 terms to start to get to within 0.01 of $\pi/4$).

4.4.101: a) $1/2 + \sum_{\substack{n=1 \\ n \text{ odd}}}^{\infty} \frac{-4}{\pi^2 n^2}\cos(\frac{n\pi}{3}t)$ b) $\sum_{n=1}^{\infty} \frac{2(-1)^{n+1}}{\pi n}\sin(\frac{n\pi}{3}t)$

4.4.102: a) $\cos(2t)$ b) $\sum_{\substack{n=1 \\ n \text{ odd}}}^{\infty} \frac{-4n}{\pi n^2 - 4\pi}\sin(nt)$

4.4.103: a) $f(t)$ b) 0

4.4.104: $\sum_{n=1}^{\infty} \frac{-1}{n^2(1+n^2)}\sin(nt)$

4.4.105: $\frac{t}{\pi} + \sum_{n=1}^{\infty} \frac{1}{2^n(\pi - n^2)}\sin(nt)$

4.5.101: $x = \frac{1}{\sqrt{2}-4\pi^2}\sin(2\pi t) + \frac{0.1}{\sqrt{2}-100\pi^2}\cos(10\pi t)$

4.5.102: $x = \sum_{n=1}^{\infty} \frac{e^{-n}}{3-2n}\cos(2nt)$

4.5.103: $x = \frac{1}{2\sqrt{3}} + \sum_{\substack{n=1 \\ n \text{ odd}}}^{\infty} \frac{-4}{n^2\pi^2(\sqrt{3}-n^2\pi^2)}\cos(n\pi t)$

4.5.104: $x = \frac{1}{2\sqrt{3}} - \frac{2}{\pi^3}t\sin(\pi t) + \sum_{\substack{n=3 \\ n \text{ odd}}}^{\infty} \frac{-4}{n^2\pi^4(1-n^2)}\cos(n\pi t)$

4.6.101: $u(x,t) = 5\sin(x)e^{-3t} + 2\sin(5x)e^{-75t}$

4.6.102: $u(x,t) = 1 + 2\cos(x)e^{-0.1t}$

4.6.103: $u(x,t) = e^{\lambda t} e^{\lambda x}$ for some λ

4.6.104: $u(x,t) = Ae^x + Be^t$

4.7.101: $y(x,t) = \sin(x)\left(\sin(t) + \cos(t)\right)$

4.7.102: $y(x,t) = \frac{1}{5\pi}\sin(\pi x)\sin(5\pi t) + \frac{1}{100\pi}\sin(2\pi x)\sin(10\pi t)$

4.7.103: $y(x,t) = \sum\limits_{n=1}^{\infty} \frac{2(-1)^{n+1}}{n}\sin(nx)\cos(n\sqrt{2}\,t)$

4.7.104: $y(x,t) = \sin(2x) + t\sin(x)$

4.8.101: $y(x,t) = \frac{\sin(2\pi(x-3t)) + \sin(2\pi(3t+x))}{2} + \frac{\cos(3\pi(x-3t)) - \cos(3\pi(3t+x))}{18\pi}$

4.8.102:   a) $y(x,0.1) = \begin{cases} x - x^2 - 0.04 & \text{if } 0.2 \le x \le 0.8 \\ 0.6x & \text{if } x \le 0.2 \\ 0.6 - 0.6x & \text{if } x \ge 0.8 \end{cases}$

b) $y(x, 1/2) = -x + x^2$ c) $y(x, 1) = x - x^2$

4.8.103:   a) $y(1,1) = -1/2$ b) $y(4,3) = 0$ c) $y(3,9) = 1/2$

4.9.101: $u(x,y) = \sum\limits_{n=1}^{\infty} \frac{1}{n^2}\sin(n\pi x)\left(\frac{\sinh(n\pi(1-y))}{\sinh(n\pi)}\right)$

4.9.102: $u(x,y) = 0.1\sin(\pi x)\left(\frac{\sinh(\pi(2-y))}{\sinh(2\pi)}\right)$

4.10.101: $u = 1 + \sum\limits_{n=1}^{\infty} \frac{1}{n^2} r^n \sin(n\theta)$

4.10.102: $u = 1 - x$

4.10.103:   a) $u = \frac{-1}{4}r^2 + \frac{1}{4}$ b) $u = \frac{-1}{4}r^2 + \frac{1}{4} + r^2\sin(2\theta)$

4.10.104: $u(r,\theta) = \frac{1}{2\pi}\int_{-\pi}^{\pi} \frac{\rho^2 - r^2}{\rho - 2r\rho\cos(\theta - \alpha) + r^2} g(\alpha)\, d\alpha$

5.1.101: $\lambda_n = \frac{(2n-1)\pi}{2}$, $n = 1, 2, 3, \ldots$, $y_n = \cos\left(\frac{(2n-1)\pi}{2}x\right)$

5.1.102:   a) $p(x) = 1$, $q(x) = 0$, $r(x) = \frac{1}{x}$, $\alpha_1 = 1$, $\alpha_2 = 0$, $\beta_1 = 1$, $\beta_2 = 0$. The problem is not regular. b) $p(x) = 1 + x^2$, $q(x) = x^2$, $r(x) = 1$, $\alpha_1 = 1$, $\alpha_2 = 0$, $\beta_1 = 1$, $\beta_2 = 1$. The problem is regular.

5.2.101: $y(x,t) = \sin(\pi x)\cos(4\pi^2 t)$

5.2.102: $9y_{xxxx} + y_{tt} = 0$ $(0 < x < 10, t > 0)$, $y(0,t) = y_x(0,t) = 0$, $y(10,t) = y_x(10,t) = 0$, $y(x,0) = \sin(\pi x)$, $y_t(x,0) = x(10-x)$.

5.3.101: $y_p(x,t) = \sum\limits_{\substack{n=1 \\ n\text{ odd}}}^{\infty} \frac{-4}{n^4\pi^4}\left(\cos(n\pi x) - \frac{\cos(n\pi)-1}{\sin(n\pi)}\sin(n\pi x) - 1\right)\cos(n\pi t)$.

5.3.102:   Approximately 1991 centimeters

6.1.101: $\frac{8}{s^3} + \frac{8}{s^2} + \frac{4}{s}$

6.1.102: $2t^2 - 2t + 1 - e^{-2t}$

6.1.103: $\frac{1}{(s+1)^2}$

6.1.104: $\frac{1}{s^2+2s+2}$

6.2.101: $f(t) = (t-1)(u(t-1) - u(t-2)) + u(t-2)$

6.2.102: $x(t) = (2e^{t-1} - t^2 - 1)u(t-1) - \frac{e^{-t}}{2} + \frac{3e^t}{2}$

6.2.103: $H(s) = \frac{1}{s+1}$

6.3.101: $\frac{1}{2}(\cos t + \sin t - e^{-t})$

6.3.102: $5t - 5\sin t$

6.3.103: $\frac{1}{2}(\sin t - t\cos t)$

6.3.104: $\int_0^t f(\tau)(1 - \cos(t - \tau))\, d\tau$

6.4.101: $x(t) = t$

6.4.102: $x(t) = e^{-at}$

6.4.103: $x(t) = (\cos * \sin)(t) = \frac{1}{2}t\sin(t)$

6.4.104: $\delta(t) - \sin(t)$

6.4.105: $3\delta(t-1) + 2t$

7.1.101: Yes. Radius of convergence is 10.

7.1.102: Yes. Radius of convergence is e.

7.1.103: $\frac{1}{1-x} = -\frac{1}{1-(2-x)}$ so $\frac{1}{1-x} = \sum_{n=0}^{\infty} (-1)^{n+1}(x-2)^n$, which converges for $1 < x < 3$.

7.1.104: $\sum_{n=7}^{\infty} \frac{1}{(n-7)!} x^n$

7.1.105: $f(x) - g(x)$ is a polynomial. Hint: Use Taylor series.

7.2.101: $a_2 = 0$, $a_3 = 0$, $a_4 = 0$, recurrence relation (for $k \geq 5$): $a_k = -2a_{k-5}$, so:
$y(x) = a_0 + a_1 x - 2a_0 x^5 - 2a_1 x^6 + 4a_0 x^{10} + 4a_1 x^{11} - 8a_0 x^{15} - 8a_1 x^{16} + \cdots$

7.2.102: a) $a_2 = \frac{1}{2}$, and for $k \geq 1$ we have $a_k = \frac{a_{k-3}+1}{k(k-1)}$, so
$y(x) = a_0 + a_1 x + \frac{1}{2}x^2 + \frac{a_0+1}{6}x^3 + \frac{a_1+1}{12}x^4 + \frac{3}{40}x^5 + \frac{a_0+2}{30}x^6 + \frac{a_1+2}{42}x^7 + \frac{5}{112}x^8 + \frac{a_0+3}{72}x^9 + \frac{a_1+3}{90}x^{10} + \cdots$
b) $y(x) = \frac{1}{2}x^2 + \frac{1}{6}x^3 + \frac{1}{12}x^4 + \frac{3}{40}x^5 + \frac{1}{15}x^6 + \frac{1}{21}x^7 + \frac{5}{112}x^8 + \frac{1}{24}x^9 + \frac{1}{30}x^{10} + \cdots$

7.2.103: Applying the method of this section directly we obtain $a_k = 0$ for all k and so $y(x) = 0$ is the only solution we find.

7.3.101: a) ordinary, b) singular but not regular singular, c) regular singular, d) regular singular, e) ordinary.

7.3.102: $y = Ax^{\frac{1+\sqrt{5}}{2}} + Bx^{\frac{1-\sqrt{5}}{2}}$

7.3.103: $y = x^{3/2} \sum_{k=0}^{\infty} \frac{(-1)^{-1}}{k!(k+2)!} x^k$ (Note that for convenience we did not pick $a_0 = 1$)

7.3.104: $y = Ax + Bx\ln(x)$

8.1.101: a) Critical points $(0,0)$ and $(0,1)$. At $(0,0)$ using $u = x$, $v = y$ the linearization is $u' = -2u - (1/\pi)v$, $v' = -v$. At $(0,1)$ using $u = x$, $v = y - 1$ the linearization is $u' = -2u + (1/\pi)v$, $v' = v$.
b) Critical point $(0,0)$. Using $u = x$, $v = y$ the linearization is $u' = u + v$, $v' = u$.
c) Critical point $(1/2, -1/4)$. Using $u = x - 1/2$, $v = y + 1/4$ the linearization is $u' = -u + v$, $v' = u + v$.

8.1.102: 1) is c), 2) is a), 3) is b)

8.1.103: Critical points are $(0,0,0)$, and $(-1,1,-1)$. The linearization at the origin using variables $u = x$, $v = y$, $w = z$ is $u' = u$, $v' = -v$, $z' = w$. The linearization at the point $(-1,1,-1)$ using variables $u = x + 1$, $v = y - 1$, $w = z + 1$ is $u' = u - 2w$, $v' = -v - 2w$, $w' = w - 2u$.

8.1.104: $u' = f(u,v,w)$, $v' = g(u,v,w)$, $w' = 1$.

8.2.101: a) $(0,0)$: saddle (unstable), $(1,0)$: source (unstable),   b) $(0,0)$: spiral sink (asymptotically stable), $(0,1)$: saddle (unstable),   c) $(1,0)$: saddle (unstable), $(0,1)$: saddle (unstable)

8.2.102: a) $\frac{1}{2}y^2 + \frac{1}{3}x^3 - 4x = C$, critical points $(-2,0)$: an unstable saddle, and $(2,0)$: a stable center. b) $\frac{1}{2}y^2 + e^x = C$, no critical points. c) $\frac{1}{2}y^2 + xe^x = C$, critical point at $(-1,0)$ is a stable center.

8.2.103: Critical point at $(0,0)$. Trajectories are $y = \pm\sqrt{2C + (1/2)x^4}$, for $C > 0$, these give closed curves around the origin, so the critical point is a stable center.

8.2.104: A critical point x_0 is stable if $f'(x_0) < 0$ and unstable when $f'(x_0) > 0$.

8.3.101: a) Critical points are $\omega = 0$, $\theta = k\pi$ for any integer k. When k is odd, we have a saddle point. When k is even we get a sink. b) The findings mean the pendulum will simply go to one of the sinks, for example $(0,0)$ and it will not swing back and forth. The friction is too high for it to oscillate, just like an overdamped mass-spring system.

8.3.102: a) Solving for the critical points we get $(0, -h/d)$ and $(\frac{bh+ad}{ac}, \frac{a}{b})$. The Jacobian at $(0, -h/d)$ is $\begin{bmatrix} a+bh/d & 0 \\ -ch/d & -d \end{bmatrix}$ whose eigenvalues are $a + bh/d$ and $-d$. So the eigenvalues are always real of opposite signs and we get a saddle (In the application however we are only looking at the positive quadrant so this critical point is not relevant). At $(\frac{bh+ad}{ac}, \frac{a}{b})$ we get Jacobian matrix $\begin{bmatrix} 0 & -\frac{b(bh+ad)}{ac} \\ \frac{ac}{b} & \frac{bh+ad}{a}-d \end{bmatrix}$. b) For the specific numbers given, the second critical point is $(\frac{550}{3}, 40)$ the matrix is $\begin{bmatrix} 0 & -11/6 \\ 3/25 & 1/4 \end{bmatrix}$, which has eigenvalues $\frac{5\pm i\sqrt{327}}{40}$. Therefore there is a spiral source. This means the solution will spiral outwards. The solution will eventually hit one of the axis $x = 0$ or $y = 0$ so something will die out in the forest.

8.3.103: The critical points are on the line $x = 0$. In the positive quadrant the y' is always positive and so the fox population always grows. The constant of motion is $C = y^a e^{-cx-by}$, for any C this curve must hit the y axis (why?), so the trajectory will simply approach a point on the y axis somewhere and the number of hares will go to zero.

8.4.101: Use Bendixson-Dulac Theorem. a) $f_x + g_y = 1 + 1 > 0$, so no closed trajectories. b) $f_x + g_y = -\sin^2(y) + 0 < 0$ for all x,y except the lines given by $y = k\pi$ (where we get zero), so no closed trajectories. c) $f_x + g_y = y + 0 > 0$ for all x,y except the line given by $y = 0$ (where we get zero), so no closed trajectories.

8.4.102:   Using Poincarè-Bendixson Theorem, the system has a limit cycle, which is the unit circle centered at the origin as $x = \cos(t) + e^{-t}$, $y = \sin(t) + e^{-t}$ gets closer and closer to the unit circle. Thus we also have that $x = \cos(t)$, $y = \sin(t)$ is the periodic solution.

8.4.103: $f(x, y) = y$, $g(x, y) = \mu(1 - x^2)y - x$. So $f_x + g_y = 1 + \mu(1 - x^2) = 1 + \mu - \mu x^2$. The Bendixson-Dulac Theorem says there is no closed trajectory lying in the set $\frac{1+\mu}{\mu} < x^2$.

8.4.104:   The closed trajectories are those where $\sin(r) = 0$, therefore, all the circles with radius a multiple of π are closed trajectories.

8.5.101:   Critical points: $(0, 0, 0)$, $(3\sqrt{8}, 3\sqrt{8}, 27)$, $(-3\sqrt{8}, -3\sqrt{8}, 27)$. Linearization at $(0, 0, 0)$ using $u = x$, $v = y$, $w = z$ is $u' = -10u + 10v$, $v' = 28u - v$, $w' = -(8/3)w$. Linearization at $(3\sqrt{8}, 3\sqrt{8}, 27)$ using $u = x - 3\sqrt{8}$, $v = y - 3\sqrt{8}$, $w = z - 27$ is $u' = -10u + 10v$, $v' = u - v - 3\sqrt{8}w$, $w' = 3\sqrt{8}u + 3\sqrt{8}v - (8/3)w$. Linearization at $(-3\sqrt{8}, -3\sqrt{8}, 27)$ using $u = x + 3\sqrt{8}$, $v = y + 3\sqrt{8}$, $w = z - 27$ is $u' = -10u + 10v$, $v' = u - v + 3\sqrt{8}w$, $w' = -3\sqrt{8}u - 3\sqrt{8}v - (8/3)w$.

Index

54127110R00208

Made in the USA
San Bernardino, CA
08 October 2017